T0349046

DEVELOPMENTS IN
EARTH SURFACE PROCESSES, 14

VOLUME FOURTEEN

THE WESTERN ALPS, FROM RIFT TO PASSIVE MARGIN TO OROGENIC BELT

AN INTEGRATED GEOSCIENCE OVERVIEW

DEVELOPMENTS IN EARTH SURFACE PROCESSES, 14

DEVELOPMENTS IN
EARTH SURFACE PROCESSES

VOLUME FOURTEEN

THE WESTERN ALPS, FROM RIFT TO PASSIVE MARGIN TO OROGENIC BELT

AN INTEGRATED GEOSCIENCE OVERVIEW

PIERRE-CHARLES DE GRACIANSKY
Ecole Nationale Supérieure des Mines de Paris (France)
DAVID G. ROBERTS
University of London, Egham, Surrey (United Kingdom)
PIERRE TRICART
University of Grenoble (France)

Amsterdam • Boston • Heidelberg • London • New York • Oxford
ELSEVIER Paris • San Diego • San Francisco • Singapore • Sydney • Tokyo

Elsevier
Radarweg 29, PO Box 211, 1000 AE Amsterdam, The Netherlands
Linacre House, Jordan Hill, Oxford OX2 8DP, UK

First edition 2011

Copyright © 2011 Elsevier B.V. All rights reserved

No part of this publication may be reproduced, stored in a retrieval system
or transmitted in any form or by any means electronic, mechanical, photocopying,
recording or otherwise without the prior written permission of the publisher

Permissions may be sought directly from Elsevier's Science & Technology Rights
Department in Oxford, UK: phone (+44) (0) 1865 843830; fax (+44) (0) 1865 853333;
email: permissions@elsevier.com. Alternatively you can submit your request online by
visiting the Elsevier web site at http://elsevier.com/locate/permissions, and selecting
Obtaining permission to use Elsevier material

Notice
No responsibility is assumed by the publisher for any injury and/or damage to persons
or property as a matter of products liability, negligence or otherwise, or from any use
or operation of any methods, products, instructions or ideas contained in the material
herein.

British Library Cataloguing in Publication Data
A catalogue record for this book is available from the British Library

Library of Congress Cataloging-in-Publication Data
A catalog record for this book is available from the Library of Congress

ISBN: 978-0-444-53724-9
ISSN: 0928-2025

For information on all Elsevier publications
visit our website at books.elsevier.com

Printed and bound by CPI Group (UK) Ltd, Croydon, CR0 4YY
Transferred to Digital Print 2012

Working together to grow
libraries in developing countries

www.elsevier.com | www.bookaid.org | www.sabre.org

ELSEVIER   BOOK AID
             International   Sabre Foundation

In memoriam of Marcel Lemoine (1924–2009)

CONTENTS

FOREWORD

The first version of this book was written under the impetus of Marcel Lemoine and published in 2000 by Gordon and Breach, under the title **De l'Ocean à la Chaine de Montagnes: Tectonique des Plaques dans les Alpes**. The book was intended for the scientifically curious, high school teachers, students and university teachers, petroleum and marine geologists and, more generally, all interested in reconstructing the history of folded belts.

From the first, the success of the book was shown by large numbers of sales despite its rather specialized nature compared with more general and popular topics. Its success seemed to reflect the curiosity of the scientific public as well as satisfying the educational needs of official secondary and tertiary teaching programmes in France.

In the first version, we were limited in page numbers and to a relatively modest format. However, readers of the first and final drafts suggested that many sections could be usefully expanded and that fuller explanation of certain results would also be desirable. In addition, colleagues suggested an English version would be appropriate.

These suggestions have led us to prepare a second and expanded English version for geoscientists interested in the evolution of mountain belts and the Alps in particular.

The first French version included a recapitulation of the Principles of Plate Tectonics necessary for comprehension of the whole book. For this second, English version, we have assumed that these basic principles are well known and they have therefore not been repeated in view of the numerous, quality English text books on Plate Tectonics. Apart from this, we have maintained the original three part organisation of the work;

- **Part one**, dealing with the major Alpine structural trends and oceanic spreading.
- **Part two**, the heart of the book, on the structure and evolution of the oceanic basins and the adjacent continental margins at whose expense the Alpine fold belt was formed.
- **Part three**, on the structural inversion of the Tethyan passive margin, which emerged from the ocean to form the characteristic topography of the Alpine mountain belt that we see today.

We have developed or added in this second version many points for which we could not find space in the French edition. Chapters expanded are thus:

the origin and development of the Tethys now partially disappeared due to Alpine orogenesis (Chapter 3); the role of Hercynian structural inheritance in the evolution of the continental margins and later Alpine structures (Chapter 4); the characteristics of the Valais (oceanic?) basin, a branch of the Tethys formed during the Early Cretaceous (or earlier? – Chapter. 7); the characteristics of Alpine ophiolites in comparison with oceanic crust generated by present day slow-spreading ridge (Chapters 11 and 12); structural inversion during the initial orogenic phase with emphasis on the role of detachment surfaces (Chapter 13); relations between the structural evolution of the future Alpine domains and the stratigraphic evolution of W. European cratonic areas during the Mesozoic (Chapter 16).

On the other hand, the Neotectonics chapter has been completely recast to include recent work (Chapter 14). We have also benefited from new studies of the geology of the Bourg d'Oisans area presented in the 2002 thesis of François Chevalier, which has extensively revised previous work on the structural evolution of the continental margin during rifting (Chapter 6).

The reader will see that the first and second versions of this book do not include detailed geological descriptions of the Western and Central Alps. In addition, there are no chapters in either version specific to tectonics, petrography or Alpine stratigraphy. Similarly, no chapter is especially dedicated to one or other of the classical disciplines of geology. Our objective has been to show how logical analysis and integration of results provided by each of the classical sub-disciplines of geoscience permits innovative reconstruction and understanding of the pre-oceanic, oceanic and orogenic phases of the fold belt while utilizing modern precepts of plate tectonics.

Throughout the book we have used the anglicised version of French place names for the convenience of English-speaking readers.

NOTE TO READERS

If we had to provide a more or less complete bibliography of the very broad subject of Alpine geology, several thick additional volumes would be necessary. We have, therefore, chosen to provide a selected bibliography with sufficient key or recent references to allow further enquiry by the interested geoscientist. However, source references for all the figures reproduced or redrafted are provided in the accompanying caption for the reader's benefit.

ACKNOWLEDGEMENTS

We wish to thank our many friends and colleagues for their generous assistance in helping us write this book and the earlier French version by Lemoine, de Graciansky and Tricart (2000).

For this new work we would like to especially acknowledge:

- The late Marcel Lemoine (1924–2009) who encouraged us to write a much revised updated and expanded version of the 2000 book published in French by Gordon and Breach which enjoyed much success in bookshops from the year of publication.
- Colleagues who have collaborated by participating in discussions during the preparation of the book and by providing us with copies of both published and unpublished work:

Albert Autran (BRGM, Orléans, France), Matthias Bernet (Grenoble, France), Romain Beucher (Grenoble, France), Anne-Marie Boulier (Grenoble, France), Julien Carcaillet (Grenoble), Jean-Daniel Champagnac (Hanover, Germany), Jean-Claude Chermette (Total, Paris, France), Francis Chevalier (Total, France), Vittorio De Zanche (Padua, Italy), Carlo Doglioni (Roma, Italy), Thierry Dumont (Grenoble, France), Maurizio Gaetani (Milan, Italy), Thierry Jacquin (Geolink, Grenoble, France), Etienne Jaillard (Grenoble, France), Claude Kerckhove (Grenoble, France), Yves Lagabrielle (Montpellier, France), Jean Marcoux (Paris, France), Henri Masson (Lausanne, Switzerland), Daniel Mercier (Paris, France), Jerôme Nomade (Grenoble, France), Anne Paul (Grenoble, France), Arnaud Pêcher (Grenoble, France), Jürgen von Raumer (Fribourg, Switzerland), François Roure (IFP, Rueil-Malmaison, France), Stéphane Schwartz (Grenoble, France), Stephan Schmid (Basel), Jean-Claude Sibuet (IFRE-MER, Brest, France), Gérard Stampfli (Lausanne, Switzerland), Christian Sue (Brest, France), Brian Tucholke (Woods Hole, USA), Peter Van der Beek (Grenoble, France).

... and all those who we may have omitted involuntarily.

- Especially our reviewers: Prof A.G. Smith (Cambridge), Dr Rodney Graham (Hess).
- We also thank Albert Autran (BRGM, Orléans, France), Vittorio De Zanche (Padua, Italy), Stéphane Guillot (Grenoble), Jean Marcoux (Paris, France) and Jean-Claude Sibuet (IFREMER, Brest, France) for reading and commenting on specific chapters.

– All those publishers and authors solicited for permission to allow us to reproduce and adapt previously published figures which have aided their research, in particular Gerardo Bautista (Gordon and Breach and Editions des Archives Contemporaines, Paris, France) and Françoise Rangin (Publications de la Société Géologique de France, Paris).

The content of the two successive versions of this book owes a great deal to the activity and enthusiasm of our young and not so young colleagues, students of Faculties or the Ecole des Mines de Paris, often during the preparation of their diplomas or thesis, and their co-workers, and, not least, the field trips and field schools of Grenoble University.

Throughout nearly thirty years of friendship, we have had the pleasure of developing, exchanging and testing ideas, principally in the course of long weeks of field work with students, Ph.D students and colleagues who have participated in field work in alphabetical order: Philippe Baron, Thierry Bas, Christophe Basile, Maurice Bourbon, Elisabeth Carrio (deceased), Jean-Daniel Champagnac, Pierre-Yves Chénet, Marie-Elisabeth Claudel, Gérard Dardeau (deceased), Marc Delorme, Thierry Dumont, Jean-Luc Faure, Thierry Grand, Henriette Lapierre (deceased), Jean-Marc Lardeaux, Joëlle Lazarre, Georges Mascle, Joséphine Mégard-Galli, Daniel Mercier, Martine Richer, Yann Rolland, Jean-Luc Rudkiewicz, Pierre Samec and Michel Trift.

– Colleagues who have participated in many discussions and who have provided copies of their published and unpublished work as well as photographs:

Jean-Claude Barféty (BRGM, Grenoble, France), Daniel Bernoulli (Basel, Switzerland), Raymond Cirio (Briançon, France), Arthur Escher (Lausanne, Switzerland), Niko Froitzeim (Bonn, Germany), Alain Gauthier (Ajaccio, France), Maurice Gidon (Grenoble, France), Yves Lagabrielle (Montpellier, France), Philippe Lesur (Paris, France), Henri Masson (Lausanne, Switzerland), Catherine Mével (Paris, France), Jean-Louis Olivet (IFREMER, Brest, France), Marc Tardy (Chambéry, France), and Rudolph Trümpy (Zurich, Switzerland; deceased).

 **IN MEMORIAM**

Marcel Lemoine 1924–2009

On the publication of this book on the geology of the Western Alps, a unique and particularly warm tribute must be rendered to Marcel Lemoine who initiated this work.

We wish to dedicate this work to his memory.

Marcel Lemoine began his career as assistant to Paul Fallot, a renowned professor at the Collège de France (1948–1952). Afterwards, he taught geology and paleontology at the Ecole Nationale Supérieure des Mines de Paris (1951–1976) where he had previously been an undergraduate student of Civil Engineering. He then became a Director of Research of the French Centre National de la Recherche Scientifique and Director of the Laboratory of Alpine geology at the University of Grenoble (1977–1985). In the final part of his career (1986–1992), he joined the Marine Geology Laboratory of the University of Paris in Villefranche. He was the author of more than 200 scientific papers published in France and elsewhere. He was also a corresponding member of several European geological societies.

Marcel Lemoine was a teacher in his heart and soul. A major sense of his purpose in his life was to communicate knowledge and understanding. Thus, he did not cease to pursue other forms of teaching as well as those usual in the University. His engagement in supporting the training of high school teachers is one example. In terms of high-level, public popularization of Alpine geology, he contributed equally, and in many ways, in the form of conferences, field geology courses exhibitions and other works amongst which the best known and popular are an initiation into the geology of the French Alps on the high footpaths of Queyras (1988) and the Briançonnais (1994). In addition, Marcel Lemoine participated in more than twenty refresher courses for geologists and geophysicists from the national and international petroleum exploration industry built around his understanding of modern passive margins and that of European passive margin of the Tethys.

Above all a field geologist, Marcel Lemoine was one of the principal workers who advanced knowledge of the structure and history of the Western Alps. Among his most important studies are:

(1) Establishment of the stratigraphic succession comprising the Piemontais *Schistes lustrés,* an assemblage of metamorphosed oceanic sediments whose age was almost unknown before his

work. This quest drove him to Corsica and Hungary and occupied him for the rest of his career and during his retirement.

(2) The reconstruction of the structural and stratigraphic evolution of the Jurassic rifted passive margin over much of the European margin of the Tethys.

(3) Reconstruction of the history of the Alpine Tethys Ocean through his comparison of Alpine ophiolites with modern analogues on ultra slow spreading mid-ocean ridges.

Marcel Lemoine had experienced the importance of integration of the various disciplines in the analysis of particular problems in geoscience. In particular, he was one of the key workers in France and Europe who was able to fruitfully reconcile Tethyan stratigraphy and structure in the Alps with results from passive margins and the ocean basins to the benefit of the worldwide geological community.

Marcel Lemoine knew how to communicate his passion for geoscience to students and colleagues in academia and industry. He offered a rigorous scientific approach, made on the basis of observations without concession to the prevailing fashion or models in vogue. World class in his breadth and depth, he challenged models made attractive by their simplicity. He always deferred with the utmost modesty to observation. For this reason, as in his aptitude to multidisciplinary problem analysis, Marcel Lemoine founded a well-known school of thought. His singular way of "making geology work" by way of understanding and explanation survives him and will survive for a long time yet.

LIST OF FIGURES

CHAPTER ONE

CHAPTER TWO

CHAPTER THREE

CHAPTER FOUR

 CHAPTER FIVE

CHAPTER SIX

CHAPTER SEVEN

CHAPTER EIGHT

CHAPTER NINE

CHAPTER TEN

 CHAPTER ELEVEN

CHAPTER TWELVE

CHAPTER THIRTEEN

CHAPTER FOURTEEN

CHAPTER FIFTEEN

CHAPTER SIXTEEN

PART ONE

Introduction

Beginning in the 18th century, the Alps were the first mountain belt in the history of European science to attract the interest of Naturalists. In consequence, many key geological principles were developed as a result of the progress of geological exploration in the Alps and, simultaneously, in the oceans.

Within mountain belts, deep sea or pelagic sediments were first discovered before being clearly recognized as such. Their origin remained enigmatic for a long period before comparison with sediments dredged from the deep sea floor by HMS Challenger which provided, at the end of the 19th century, the necessary insight into their origin. Their presence at the highest elevations of mountain belts allowed recognition of the ephemeral nature of the oceans at geological time scales though cause and effect lay beyond conception at the end of the 19th century.

It should be noted that towards the end of the 19th century, the possibility of large scale, 100 km, horizontal displacements of rock units was demonstrated for the first time. This concept was a key generality that directly resulted from the discovery of nappes in 1884.

The theory of geosynclines, interpreted as the birthplace of folded belts, emerged in 1859 from the celebrated work of James Hall in the Appalachians. It was a key driver for exploration in folded belts until 1960. By then, however, a steady flow of geophysical results from the oceans had seriously weakened the theory. The advent of palaeomagnetism had also by then confirmed the mobility of the continents thus vindicating Wegener's visionary hypothesis of continental drift. Quantification of continental drift through the Plate Tectonics hypothesis was achieved in the mid–1960s from marine geological and geophysical studies and, especially the dating of oceanic magnetic anomalies, latterly seismic reflection profiles that provided key insights into passive margin structure. These observations provide the geodynamic context and framework for the present geological understanding of Alpine evolution.

Chapter 1 of Part I is dedicated to this saga.

To understand present-day Alpine structure and the reasons for the nomenclature of the structural units summarised in Chapter 2, it is necessary to recognize the existence of a now missing Ocean (Chapter 3), the Tethys, which evolved into the Alpine fold belt from the end of the Mesozoic and through the Tertiary era to the Present.

During the Palaeozoic, the continents comprised a single supercontinent known as Pangaea. During the course of the Mesozoic, Pangaea began to fragment leading to the formation of the Tethys Ocean (Figs. 3.1–3.4). This ocean was bordered to the north by Laurussia, which then included Laurentia (North America and Eurasia). To the south, it was bordered by the Gondwana supercontinent which included the future continents of South America, Africa, Madagascar, India, Australia and Antarctica. The Alpine fold belts derived from the destruction of the Tethys are collectively oriented in a general east–west sense and extend from Gibraltar to Indonesia via the Himalayas. In this array of folded belts, the Peri-Mediterranean fold belts extend from Gibraltar to Asia Minor. The Alps comprise only a short segment of this complex system (Fig. A).

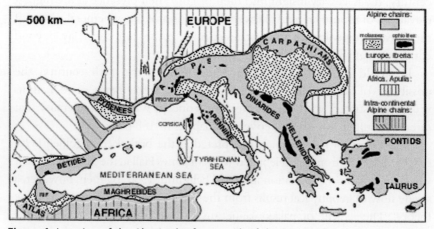

Figure A Location of the Alps in the framework of the Peri-Mediterranean mountain belts (*adapted from M. Lemoine (2000), Fig. 5.1, p. 61*).

Geosynclines, Passive Margins, Foreland Basins and Folded Belts: An Introduction

Contents

1. PROLOGUE

Although geological sampling of surficial sediments in shallow and deep water dates back to the Challenger expedition in the 19th century, detailed studies of continental margins did not begin until after World War II. While early gravity expeditions led by Meinesz (1941) among others had yielded some insights into the deeper structure of the margins of the Pacific and Atlantic Oceans, the use of sonar and other techniques developed during the war allowed for rapid mapping of the sea floor and also investigation of the deeper structure beneath the shelf continental slope and abyssal plains.

It had long been recognized that pelagic sediments identified by the Challenger expedition (Murray and Renard 1891) had their equivalents in folded belts. However, the range of possible interpretations remained large especially in view of the then established view of the permanence of the ocean basins. Simplicity versus complexity were the two bywords that differentiated the community of marine geoscientists from those concerned with terrestrial geology and geoscience.

Suess in his seminal global geology summary (1885) noted that '*the possibility was recognised of deducing from the uniform strike of the folds of a mountain chain, a mean general direction or trend line: such trend lines were seldom seen to be straight but consisted of arcs or curves, often violently bent curves of accommodation; the trend lines of central Europe were observed to possess a certain regular arrangement and to be traceable in part as far as Asia. It was further recognised that the ocean from the mouth of the Ganges to Alaska and to Cape Horn is bordered by folded mountain chains while in the other hemisphere this is not the case so that **Pacific and Atlantic types** may be recognised.*' Suess thus recognized, over a hundred years ago, the fundamental differences between the active (Pacific) and passive (Atlantic) continental margins. He noted the continuity of the circum Pacific and Alpine–Himalayan fold belts whose association with calc alkaline volcanism and deep earthquakes is now very well known and understood. Suess was also well aware of the problems of major marine transgressions especially that of the Late Cretaceous. However, Suess thought that the ocean crust was similar to that of the continents and that the oceans owed their origins to 'subsidence and collapse'.

However, the technological hurdles that had to be overcome to determine the geology of continental margins were matched by the problems imposed by the complex and intense deformation of folded belts exposed on land.

2. OROGENESIS, ROCK DEFORMATION AND DEVELOPMENT OF THE THRUST CONCEPT

H.B. de Saussure (1740–1799), the Swiss naturalist from Geneva, was one of the first to express the idea that the torsion of beds observed on the flanks of Alpine valleys might be caused by 'forcing back' of rock material. His classic interpretation of the fold of the Arpenaz waterfall (Fig. 1.1) which dominates the Arve valley near the small town of Sallanches (between Chamonix and Geneva) was made, however, without any of the present knowledge of the rheological properties of rocks.

The permanence of continents and oceans was considered a basic truth by authors of the first three quarters of the 19th century. In consequence, they found difficulty in conceiving that the rocks forming the mountains had been subjected to horizontal displacements greater than those observed

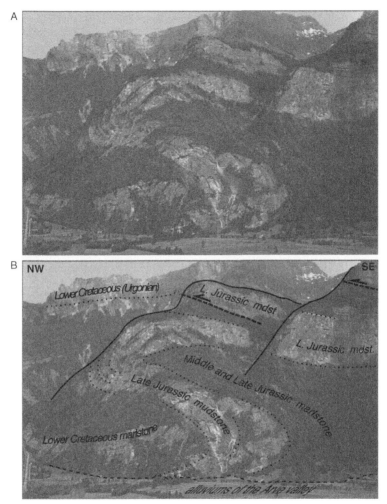

Figure 1.1 *Fold of the Arpenaz waterfall, Haute Savoie (France)*. H.B. de Saussure (1740–1799), probably the first or among the first Alpine geologists, described the fold of the Arpenaz waterfall between Geneva and Chamonix in the Arve valley. To explain the observed deformation, he proposed that the Jurassic and Cretaceous limestones were soft muds at the time of deformation. Today, one of the interesting aspects of this fold is that it shows the divergence of structure towards the external part of the fold belt, here to the NW. *Horace Benedict de Saussure (1790)*.

in associated folded beds. Not until the 1880s would the existence of thrusts be demonstrated clearly.

A comparison of concurrent structural interpretations of the Glarus area (Switzerland) classically exemplifies the debate at the end of the 19th century on this issue (Fig. 1.2).

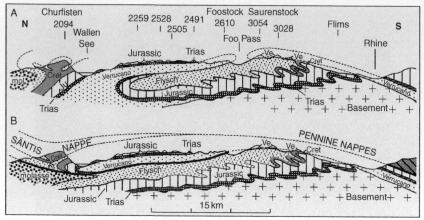

Figure 1.2 *Two structural interpretations of the Glarus Alps.* **(A)** Albert Heim sketched two large facing recumbent folds. A fan fold is required to explain the Churfisten structure on the left of section. The interpretation resolves the problem of abnormal superposition with respect to the normal way up succession at the same time minimizing the magnitude of the sub-horizontal tectonic transport of thrust material. The illustration provided by Heim in his 1878 paper is superb. It adopts and recaptures the earlier conclusions of Arnold Escher (1866). M: Molasse; f: Flysch; c: Cretaceous; J: Jurassic.; t Trias; p: Permian. *From A. Heim 1878.* **(B)** Marcel Bertrand introduced the notion of tectonic 'covering' originating the concept of the thrust nappe. He had not visited the field from the time of 1884 publication on the subject. He used structural results related to the coal basin of northern France to reinterpret the conclusions of A. Heim without contesting the validity of the observations. His sketch illustrates the concept according to which thrust nappes are derived by rupture of the inverse flank of a recumbent fold. *From Marcel Bertrand (1883–1884), reproduced after Bailey (1935), Fig. 10, p. 40.*

The interpretation of Heim (1891) which portrays two facing folded beds adroitly resolved the problem of abnormal superposition with respect to the normal bed succession while reconciling the intellectual requirement to minimize the amplitude of sub-horizontal displacements of rocks (Fig. 1.2A).

The 1884 interpretation of Bertrand in the same area subsequently gained the support of the community from the end of the 19th century. This described one of the first Alpine thrusts using modern extant ideas (Fig. 1.2B). From then on, classic thrust interpretations were made by great authors such as Schardt (1893) for the Prealps (Fig. 2.8), Lugeon (1902) for the Helvetic nappes, Termier (1903) for the Western Alps and for the Eastern Alps (Fig. 2.6A).

3. MOUNTAIN BELTS AND THE GEOSYNCLINAL THEORY (1859–1965)

The enormous thicknesses of sediments documented in fold belts and their adjacent basins caused major difficulties reconciled in the 'geosynclinal' theory (Fig. 1.3A) of Hall (1859) and Dana (1873). These thicknesses far exceeded the depths of the modern oceans and the sediments typically consisted of shallow marine deposits. Obviously subsidence had to have taken place to allow the accumulation of such thicknesses. Dana used the term 'geosynclinal' with reference to a subsiding and infilling basin resulting from his concept of crustal contraction due to a cooling earth.

The Western Alpine structural zones soon came to be interpreted in terms of the Dana's Geosynclinal model. In this way, from 1900, Haug added extra detail by invoking an elongate narrow trough between the continents whose erosion supplied the sediment. Two belts of sedimentary rocks were thought to accumulate in troughs separated by an intervening ridge called a geanticline. He designated the Dauphine geosyncline, the Brianconnais geanticline and the Piemontais geosyncline (Fig. 1.3B). The deepwater trough with abundant volcanic rocks and deep water sediments was termed a eugeosyncline while the trough with mainly shallow water sediments was called a miogeosyncline. The driving mechanism was thought to be compression between two colliding continents.

A little later, Steinmann (1927) considered that the Alpine ultrabasic and basic igneous rock suites called ophiolites were emplaced by injection and differentiation of basic and ultrabasic magmas under marine sediments well before dissection by later thrusts (Fig. 1.3B). This model reflects mistaken use of 'analogues' drawn from major igneous complexes of which the Bushveld complex of South Africa is a type example and well known at that time. Today, the ophiolites are known to be fragments of oceanic lithosphere entrained in thrust sheets.

Since the prevailing model implied that the Alpine geosynclinal was subjected to compression since its origin, the 'cordillera' or 'geanticline' of the Brianconnais–Grand Saint Bernard (Fig. 1.3B) was considered to exactly or embryonically prefigure that of the present complex of thrust sheets (Argand 1934); this notion was embodied as late as 1950 by Gignoux or even Debelmas (1955).

Application of the geosynclinal model has been mostly valuable in that it has provided guidelines for geological exploration and thinking in different fold belts worldwide and particularly those surrounding the

Mediterranean. The geosyncline model was classically taught worldwide and remained a mainstay of geology. Notwithstanding the recognition of oceanic and continental crust and the first primitive studies of the deep structure of continental margins in the 1950s, strenuous efforts were made to apply geosynclinal theory to these new observations (see for example Marshall Kay (1951, 1967) and Fig. 1.3A, this volume; Aubouin 1965; Drake et al. 1959).

The 19th and early 20th centuries saw the flowering of worldwide geological exploration and the development of the many sub-disciplines that are embodied in geoscience today. The picture of the earth seemed comprehensive where mapped in detail but was less than convincing in

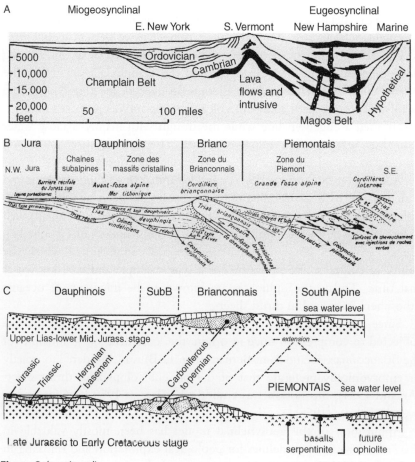

Figure 3 *(continued)*

explaining the relationship between fold belts, volcanoes and seismicity as well as the new data from passive margins in several respects.

4. GEOPHYSICAL AND GEOLOGICAL EXPLORATION IN THE OCEAN: FIRST STEPS AND RESULTS

At the turn of the 19th century, inferences from gravity observations were used to compare the gross crustal structure of the continents and oceans. Airy (1855) proposed that the weaker gravity anomalies associated with mountain belts were caused by a low density root, a hypothesis later known as isostasy. In 1909, Helmert showed that the gravity anomaly across continental margins exhibited a characteristic edge effect marked by a high gravity on the outer shelf and a low on the continental rise: away from the rise, gravity values returned to the worldwide norm. Wegener (1924) recognized the significance of this and concluded that continental crust was absent in the ocean

Figure 1.3 *(A) Interpretation of the Appalachian system in Middle Ordovician time.* This reconstruction derives from the Hall and Dana's geosynclinal model. It has been the starting point for the interpretation of mountains for about a century (1858–1965). *From Drake et al. (1959), Fig. 7, p. 7* *(B) Interpretation of the Alpine Geosynclinal, after Gignoux (1950).* The Alpine *geosynclinal* model was taught from the 1930s to the end of the 1950s. It described a *'geanticline'* for the Brianconnais cordillera characterized by littoral sediments (figured here by dotted lines) separating two troughs with thick sedimentation. This cordillera was identified as an 'embryonic nappe'. Ophiolites were supposed to have been injected at the base of the internal cordillera, which formed another 'embryonic nappe'. The idea of 'embryotectonic' interpretation comes from the Swiss author Argand (1916). *In Lemoine (1988).* *(C) Tentative palinspastic reconstructions across the palaeogeographic domains of the Western Alps, after Lemoine (1975).* These sections are taken from the foundation paper that is one of the first up-to-date interpretations of the Western Alps in Mesozoic time. With the geosynclinal model, a single section sufficed to summarize the structural fabric across the fold belt from the duration of the Mesozoic to the start of the Tertiary. With the model of derivation of the fold belt from the ocean basin and its margins, several sections are necessary. The first section illustrates the extension phase, which brought about the disintegration of the post-Hercynian platform beginning from the end of the Palaeozoic. The second section shows the opening of the ocean and contemporaneous subsidence of the newly formed margins. Other sections allow illustration of the results of inversion of the deformation field from the distension that prevailed until the middle of the Cretaceous to the subsequent shortening (see Figs. 13.1, 13.6 and 13.13 for example). The ophiolites are the remains of the oceanic lithosphere, thrust above the adjacent continental margin, the whole having been subjected to compression generally accompanied by metamorphism. *Reproduced after M. Lemoine (1975), Fig. 2, p. 211.*

basins and that the underlying crust must be very thin. He also inferred a large pressure differential between continents and oceans would be a consequence of this variation speculating that this might cause the step faulting observed on the coastal segments of margins such as South Africa and eastern South America.

Laborious measurements of gravity in submarines by Meinesz among others resulted in the conclusion that the cause of the gravity variation was abrupt thinning of the sialic continental crust across the continental margin.

The first systematic seismic refraction measurements at sea were made in the Atlantic, Pacific and Indian Oceans by Gaskell, Hill and Swallow (1959) during the post-war HMS Challenger expedition complemented by groups at Scripps Institute of Oceanography and Lamont Geological Observatory. Collectively the results showed that continental crust was completely absent in all the ocean basins and that the ocean basins were characterized by thin sediments overlying a simple tripartite layered ocean crust quite different to that observed beneath the continents. The transition from continental to oceanic crust was thought to occur beneath the continental rise.

The problem posed by the presence of only thin sediments on oceanic crust was recognized by Kuenen in 1950. He queried the absence of thick sediment accumulations that must have been eroded from folded belts and the continents during previous geological time assuming continental margins were Tertiary and Mesozoic in age. In the western Alps for example, the post-Hercynian Mesozoic succession siliciclastic sediments that would have been eroded from Hercynian terranes are almost absent and the sedimentary record is dominated by carbonates and marls. Kuenen's question thus impacts fundamentally on the processes that formed the oceanic crust.

The first results from seismic refraction studies and their value in defining the nature of the oceanic crust have been mentioned above. Allied to these studies was an explosion in mapping of the sea floor of the world's oceans using modern echo-sounding equipment from 1945 onwards. The general division of passive margin bathymetry into continent shelf, slope, rise and adjoining abyssal plain was soon recognized as was the presence and world-wide extent of the mid-ocean ridge system though its presence had been inferred earlier by oceanographers. Heezen et al. (1959) recognized that the mid-ocean ridges were characterized by an axial median valley which was the loci of shallow earthquakes. They were able to demonstrate the continuity of the mid-ocean ridge in the Indian Ocean with the East African rift system in Ethiopia. However, they felt unable to offer an explanation of their findings at that time though Heezen later proposed expansion of the earth. In a similar way, mapping of Pacific active margins had identified a shelf, slope and adjacent trench associated with shallow, intermediate and deep earthquakes adjoined onshore by either island arcs or Andean fold belts also with volcanoes.

At the same time, refraction-based sections of the passive margin off North America by Drake et al. (1959) prompted comparison with geosynclines and especially Kay's reconstruction (1951) of the miogeosyncline and eugeosyncline of the Appalachian system in middle Ordovician time (Fig. 1.3A). The comparison showed as many differences as similarities. While sediment thicknesses were broadly comparable, there were also differences in structural style and overall basin shape. In short, it was difficult to construct a section across a passive margin that resembled a classic geosyncline.

5. CONTINENTAL DRIFT AND PLATE TECTONICS: PRINCIPLES

Since the advent of modern cartography, the similarity in shape of the coastlines of the Atlantic has resulted in speculation that they might once have been joined together. The idea of continental drift first suggested by Taylor (1910) and then Wegener (1924) suggested a means of accounting for the major differences between Atlantic and Pacific type margins. Later seminal work by Du Toit (1937), which compared Atlantic margins with rifts and noted the continuity of the Samfrau 'geosyncline' in the now separated continents of the southern hemisphere, was ignored along with other supporting evidence such as Holmes' (1928) avowal of continental drift. At the time the idea of continental drift was correctly rejected on the grounds that Wegener's mechanism, which proposed continents ploughing through the ocean floors, was physically untenable. Indeed from the viewpoint of geologists in central Eurasia or on the west, Pacific, coast of the USA, continental drift seemed intrinsically unreasonable. As a result vertical and compressional tectonics held sway despite being under increasing challenge from marine geophysical studies.

A major paradigm shift resulted from palaeomagnetic studies notably those of Blackett et al. (1960) and Runcorn (1956), which showed that the continents had moved relative to each other. In addition the celebrated, successful reconstruction of the North and South Atlantic Oceans by Bullard, Everett and Smith (1965) using rotation about Euler poles confirmed the former conjugation of Africa and South America, Africa and North America and North America with Eurasia.

Although continental drift was from then on regarded as proven, the exact mechanism remained uncertain. For example, Carey (1958) in a paper that anticipated plate reconstructions by 10 years asserted that the fits of the continents could be readily explained by an expanding earth. However, Dietz (1961, 1963) and Hess (1962) proposed the 'sea floor spreading' hypothesis as a basis for understanding the tectonics of the sea floor and in consequence the margins of the ocean basins. This simple and elegant hypothesis states that new oceanic crust is formed at the axis of mid-ocean ridges by upwelling and injection of magma as dykes. Conservation of the earth's surface area is achieved by destruction of oceanic crust in the subduction zones bordering the margins of the Pacific for the most part. Passive margins were therefore considered to have formed by crustal extension that led to complete rupture of the continental crust and the formation of ocean crust by spreading as is shown by the continuity of the mid-ocean ridge in the Gulf of Aden with the actively extending northern Ethiopian rift.

In 1962, Vine and Matthews developed a critical corollary to the sea floor-spreading hypothesis (Vine and Matthews, 1963). They proposed that the by then well-mapped pattern of magnetic lineations in the ocean basins could be explained by formation of new oceanic crust during periods of normal and reversed magnetic polarity. The magnetic lineations thus preserved both a record of the history of reversals and also the rate of sea floor spreading. Their hypothesis initially used the precise chronology of magnetic reversals in the late Neogene to calibrate spreading rates. The systematic increase in age of the ocean floor away from the mid-ocean ridge axis was later confirmed by one of the first voyages of the Deep Sea Drilling ship Glomar Challenger in 1968. Le Pichon (1968) showed from a summation of global rates of plate divergence and convergence that the earth was neither contracting nor expanding.

Another key step was made by McKenzie and Parker (1967) who were able to describe the movement of the North Pacific in terms of tectonics on a sphere thus founding plate tectonics. From earthquake mechanism studies and the recognition of transform faults, three types of plate boundary were defined (Fig. 1.4):

Extensional corresponding to active rifts, mid-ocean ridges and their fossil trace represented by *passive margins* and failed rifts such as the North Sea.

Convergent represented by the subduction zones of the Pacific and the collisional fold belts of the Andes, Himalayas and Alps. Convergence first results in the formation of an accretionary prism and associated trench (e.g. Timor wedge, Figs. 1.4 and 1.5). Shortening progressively leads to emergence and the construction of fold belts by collision.

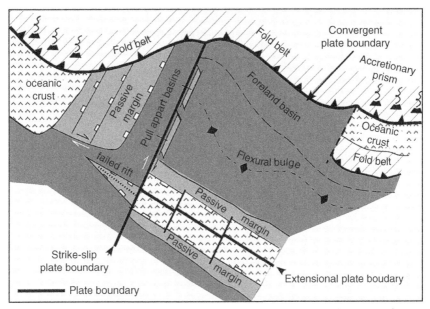

Figure 1.4 *Three types of plate boundary and associated sedimentary basins.* Theoretical. No scale. No orientation. Refer to *McKenzie and Parker (1967).*

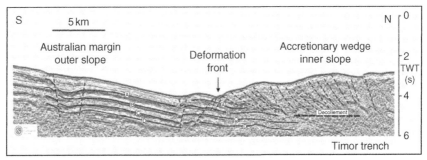

Figure 1.5 *Section across the Timor trench and accretionary wedge.* The island of Timor to the left is being obducted onto the Australian plate to the right. An accretionary prism borders the underfilled trench in front of the Timor block. Note the fracturing of the Australian plate by flexural loading. *From Hughes et al. (1996), Fig. 5, p. 80.*

Strike slip represented by oceanic transform faults, which describe plate trajectories and intracontinental transform faults such as the Dead Sea transform and the San Andreas Fault: transfer faults observed in rift systems are commonly precursors to oceanic fracture zones.

All past and present sedimentary basins can be described in terms of formation at one or other of these plate boundary types.

5.1 Basin-forming mechanisms

The plate tectonics hypothesis also integrated a number of other observations on heat flow and the nature of the lithosphere and asthenosphere. The lithosphere forms a number of rigid plates that move across the earth in relative motion to each other.

All sedimentary basins represent regions of prolonged subsidence. The driving mechanisms of subsidence are processes within the lithosphere controlled by deformation concentrated on plate boundaries. The lithosphere is relatively cool and rigid and its base is defined by the 1330°C isotherm. The strength profile with depth of the continental lithosphere exhibits a weak ductile zone in the lower crust separating the seismogenic upper crust from the upper mantle. In contrast, the strength of the oceanic lithosphere increases with depth to the brittle–ductile transition in the upper mantle (Fig. 1.6).

Classification of sedimentary basins such as that of Bally and Snelson (1981) consider the underlying lithosphere type-position relative to a plate boundary and its type, extensional or divergent, convergent and transform (Figs. 1.4 and 1.6). There are three types of driving mechanism that control the formation of sedimentary basins. It should be emphasized that sedimentary basins are composite in nature and consist of a stacked series of tectonostratigraphic megasequences that have discrete and characteristic geometries that reflect the nature of the basin forming process. In extensional basins, there is typically a syn-rift megasequence succeeded by a passive post-rift megasequence.

The three main mechanisms discussed in more detail below are:
- Lithospheric and crustal stretching caused by extension
- Thermal subsidence due to cooling and passive subsidence of the oceanic lithosphere as it moves away from the spreading axis or at the end of rifting and onset of spreading
- Loading of the lithosphere which causes flexure and subsidence as in foreland basins.

Generically sedimentary basins thus can therefore be grouped as follows (Figs. 1.4 and 1.6):
- Basins formed by stretching of the lithosphere
- Basins formed by loading and flexure of continental and or oceanic lithosphere
- Transtensional basins formed by strike slip movement along major transforms.

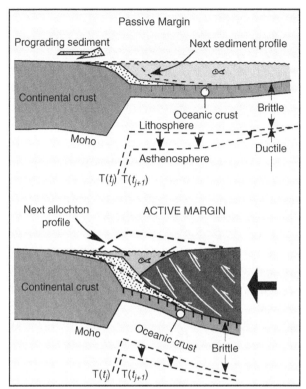

Figure 1.6 *Model construction of passive (rifted) and active (convergent) margins.* Major vertical exaggeration. $T(t_j)$ and $T(t_{j+1})$ correspond to single isotherms at two times, t_j and t_{j+1}. The model shows steps in addition of sediment on passive margin up to a specific bathymetric profile alternated with thermal time steps during which model moves towards thermal equilibration. Tectonic switch from passive (upper diagram) to active margin (lower diagram) is modelled by emplacing overthrust loads sequentially onto passive margin. Flexural response of lithosphere to increasing load through time changes because thermally controlled effective elastic thickness of lithosphere also changes. *Redrawn from Stockmal et al. (1986), Fig. 1, p. 182.*

5.2 Rifts and passive margins

The shallow Moho under rifts and passive margins clearly shows stretching of the crust and lithosphere. McKenzie (1978), stimulated by the clear seismic evidence for extension in rifted basins, developed a quantitative, uniform mechanical stretching that assumed symmetrical stretching and pure shear (Fig. 1.7A). The extension was considered to be uniform and instantaneous and the initial surface of the lithosphere was assumed to be at sea level. His model demonstrated that the total subsidence in extensional

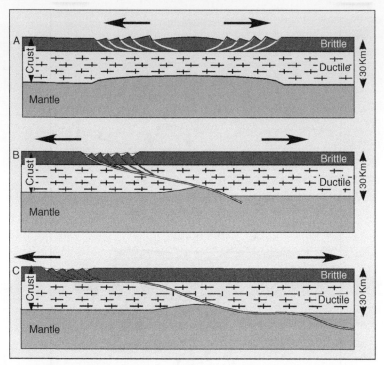

Figure 1.7 *Rifts and future passive margins.* **(A)** Model of pure shear applied to the whole continental crust. *Simplified after McKenzie (1978).* **(B)** Model of simple shear applied to the whole continental crust and mantle generalized from observations in the Basin and Range province (Western USA). *Simplified after B. Wernicke (1985).* **(C)** A more recent model also proposes simple shear. Extension is largely done by delamination along the surfaces of discontinuity between ductile and brittle layers. This model is probably more realistic in terms of rheology. *Simplified after Lister, Etheridge and Simon (1991)* Scheme A has allowed modelling of the subsidence and heat flow in rifts and newly formed margins. Model B has been criticized because the detachment fault responsible for asymmetric stretching is shown in the same way in both ductile and brittle milieu. Models B and C are probably more realistic than A in the sense that rifts or conjugate passive margins are rarely symmetric (e.g. Fig. 12.3). The three models do not take into account the existence of a ductile layer in the upper mantle (Fig. 12.2). *From Fraser et al. (2007), Fig. 3, p.101*

basins comprises two components. Firstly, an initial fault controlled or syn-rift subsidence dependent on the initial thickness of the crust and the amount of stretching and, secondly, post-rift thermal subsidence caused by cooling and contraction of the lithosphere; the magnitude of the post-rift subsidence is dependent on the amount of stretching. The rate of post-rift thermal subsidence decreases exponentially with time (50 Ma) due to a decrease in heat flow caused by cooling of the lithosphere. In later variation, Jarvis and McKenzie (1980) attempted to include the effects of finite rifting events. Later models (Fig. 1.7B and 1.7C) addressed the effects of depth dependent stretching, asymmetric extension and normal simple shear for the whole lithosphere based on the tectonics of the Basin and Range province (Wernicke 1985). Depth dependent stretching is supported by seismic studies (e.g. Avedik et al. 1982) although the exact mechanism is not established as yet. However, it is clear from deep sea drilling west of Iberia and the Tasna ocean continent transition in the Alps that the terminal stages of extreme extension resulted in exhumation of sub-continental mantle and boudinage of the upper crust above a low angle detachment fault (Figs. 10.2, 10.3, 10.8 and 12.3; Manatschal et al. 2006).

It should also be noted that sediment loading in the post-rift phase will increase the available accommodation space by flexural subsidence and results in the characteristic steer's head geometry that typifies the post-rift succession in failed rifts and passive margins (where only half is preserved). The combination of thermal subsidence and flexural loading has resulted in the accumulation of as much as 12 km of sediment on some passive margins (Figs. 1.6 and 1.9).

5.3 Foreland basins

Foreland basins such as the Alberta basin and Alpine and Apennines foredeeps (Fig. 1.10: Oman mountains; Fig. 2.4: Alpine molasse and Po basins) are associated with continental collision and formed by flexure of the lithosphere. They are elongate basins that thicken toward the adjoining contemporaneous fold or orogenic belt. The deflection of the lithosphere, which forms the accommodation space infilled by foreland basin sediments, is determined by the flexural rigidity of the lithosphere and the magnitude of the load imposed by the orogenic belt or wedge (Figs. 1.5 and 1.6). These factors also determine the location and height of the associated flexural bulge. Sediments derived from the orogenic belt infill the foreland basin further contributing to the load and thus accommodation space. Depending on the rate of convergence or shortening in the fold belt, successive foreland basins may develop as earlier foreland basins are cannibalized and incorporated in the fold belt.

5.3.1 Seismic imaging of sedimentary basins

Until the late 1950s, onshore oil exploration mainly used surface geology aided by gravity and single point seismic reflection techniques to interpret the sub-surface. There was very little offshore exploration. The advent of the Common Depth Point (CDP) technique of reflection seismology in the early 1960s (Mayne 1962) coupled with increasing power and speed of computers transformed oil exploration. Rapid acquisition of seismic profiles that clearly imaged both stratigraphy and structure of entire onshore basins became possible. Among the classic papers that ensued was that by Bally, Gordon and Stewart (1965) on balancing sections in folded belts which defined a technique still used today. The stimulus to explore offshore was led by the need to prospect for oil and gas adjacent to major discoveries made onshore in the Netherlands and Gulf of Mexico. As a result, marine seismic prospecting using the CDP technique using a variety of sound sources became available initially to industry and later in the decade 1965–1975 to research institutes mainly in Europe. The technique resulted in the first exploration across active and passive continental margins by Shell reported later by Beck and Lehner in 1974. The first deep seismic studies of the Bay of Biscay and its margins were made by the IFP in the late 1960s (Fig. 1.8). These seismic studies were made in parallel with the development of plate tectonics. In contrast, the single channel air gun seismic systems used by the academic community were of much less value on passive margins because of the low power of the sound source and an inability to remove multiples. Nonetheless, as a rapid means of mapping sediment thickness in deep water away from the margins the technique yielded results of great value in, for example, describing the increase in sediment thickness away from mid-ocean ridges as predicted by the sea floor-spreading hypothesis.

5.3.2 Structure of passive margins

The new seismic images of passive margins proved to be of fundamental importance in understanding their structural history in the context of plate tectonics and to defining drill sites where the stratigraphic record could be examined in cores. Figure 1.8 shows a typical profile across the starved passive margin of North Biscay. On this section, low angle listric faults define tilted fault blocks containing conformable pre-rift sediments overlain by a syn-rift section followed by a thin, passively deposited, post-rift cover. The major unconformity separating the pre- and syn-rift sections is known as the *rift onset unconformity* while that between the syn- and post-rift is known as the *breakup unconformity*. Recognition of the existence of extension and arrays of tilted fault blocks on present passive margins led to a reappraisal of the possible role of rifting in the Ligurian Tethys. In a seminal paper (Fig. 1.3C), Lemoine et al. (1975) recognized the importance of

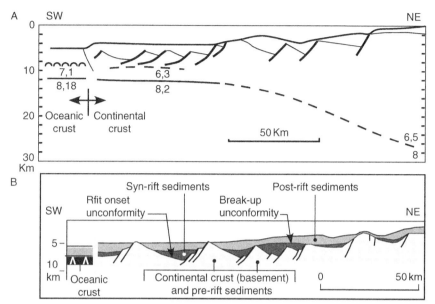

Figure 1.8 *(A) Interpreted seismic section of the Eastern Atlantic margin: Western Approaches. Adapted from Montadert, Roberts et al. (1979); Montadert in Boillot et al. (1984), Fig. II-13, p. 104. (B) Schematic crustal section through the north Biscay continental margin. Adapted from Lemoine et al. (2000), Fig. 2.2, p. 29; from Avedik et al. (1982).*

Jurassic extensional faults, previously thought to be Alpine in age, and were able to demonstrate that the margins of the Ligurian Tethys formed by rifting in Early and early Middle Jurassic times succeeded by spreading on and after the late Middle Jurassic (Lemoine 1988).

More thickly sedimented margins are characterized by thick deltaic sediments which prograded across the original fabric of tilted blocks (Figs. 1.9A and 1.9B). Others dominated by deposition of carbonates are characterized by thick carbonate platforms and relatively starved conditions on the slope and rise outside the platform (Fig. 1.9C). Intermediate cases, such as the Goban Spur (Fig. 1.9D), show partial burial of the tilted blocks which still have topographic expression today. Major palaeoslope unconformities are commonly present extending across much of the slope.

5.3.3 Palaeogeographic domains and passive margins

The palaeogeographic evolution of passive margins takes place in two phases: firstly in the syn–rift phase and secondly during the post–rift phase (Fig. 1.8), which largely inherits the palaeogeography developed at the end of rifting.

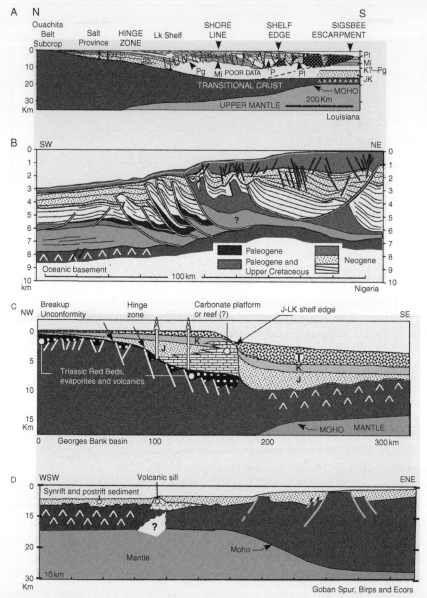

Figure 1.9 *Atlantic passive margin types*. (A): Louisiana, *redrawn from Worrall and Snelson (1989), Fig. 20, p. 114.* Very thick Palaeogene and Neogene deltaic sediments have prograded across Cretaceous and Jurassic pelagic sediments resting on salt. The sedimentary loading has mobilized the salt into diapirs and salt nappes, which form the Sigsbee escarpment. **(B):** Nigeria, *redrawn from Bally and Tari (2004), AAPG Winter Education Conference, Fig. 40.* Thick Neogene deltaic system derived from the Niger delta have prograded across the passive margin onto oceanic crust. The sediment of the delta is collapsing by up-dip extension and down-dip contraction, which forms a deep water fold belt. **(C)** Southwestern end of Georges Bank basin, *redrawn from Schlee and Klitgord (1988), Fig. 17, p. 251.* **(D):** Goban Spur, off Ireland. *Redrawn from Peddy et al. (1989), Fig. 7. p. 432.*

In the syn-rift phase, the overall palaeogeography is determined by the amount of extension and associated subsidence (Fig. 12.3). In the least extended parts of the rift, sediments on the crest of tilted fault blocks are deposited in shallow water while those in the intervening half graben are laid down in relatively deeper water. With increased extension, sediments on both the crest and the half graben will be deposited in greater water depth that progressively increase toward supposedly 2500 m near the continent ocean transition. These sediments may overly exhumed lower continental mantle.

The post-rift of passive margins can be divided into shelf (or platform), slope, rise and abyssal plain geological provinces each characterized by their own distinctive facies in the post-rift succession above continental crust and also overlying the oceanic crust.

Stacking of the syn-rift and post-rift palaeogeographies results in a series of discrete domains that are closely akin to those observed historically in the Alps (Fig. 13.2). In addition, the nature of the pre-rift sediments and their sub-crop to the rift onset unconformity as well as large intervening horst blocks (e.g. the Brianconnais platform; Figs. 6.12, 6.13 and 6.17) can add an extra 'palaeogeographic' dimensions to further characterize these domains.

5.3.4 Seismic studies of foreland basins

Seismic studies of foreland basins aided by exploration drilling results have shown that these basins differ in several key respects to passive margins (Figs. 1.10, 1.11, 14.2 and 14.3). Such basins can be divided into two tectonostratigraphic megasequences separated by the regional unconformity marking the onset of foreland basin development. The pre-foreland basin megasequence comprises the post-, syn-, and pre-rift major sequences of the passive margin including the successive Hercynian major sequences in the case of the Western Alps. The pre-foreland basin may consist of a post-rift or remnant passive margin megasequence sometimes underlain by an older syn-rift megasequence. The polarity of the foreland basin is opposite to that of the underlying pre-foreland basin as shown in figs. 1.10, 14.3 and 14.6. Foreland basins shallow upward and are also cannibalized by the adjacent fold belt (Fig. 1.10). They are also commonly associated with a peripheral or flexural bulge (Figs. 1.10 and 14.3) located several hundred kilometres away from the mountain front. There may be several foreland basin megasequences reflecting changes in convergence direction, position of the flexural load, dynamic

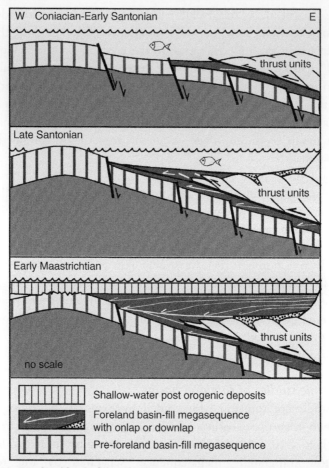

Figure 1.10 *Foreland basin fill and seismic sequences illustrated in the Sultanate of Oman mountains. Coniacian–Early Santonian:* The first sediments onlap the flexed foreland. The thrusting rate exceeds sedimentation rate. *Late Santonian:* Sedimentation rate exceeds thrusting rate. Sediments are deposited on both the flexured foreland and the frontal allochtons. Part of them are canibalized during the last thrust movements. Peripheral bulge is emerged. *Early Maastrichtian:* Post-thrusting basin fill. Bulge submergence and resumption of shallow-water deposition. The distal part of the foreland basin is at the west and is opposite polarity to that of the underlying pre-foreland basin whose distal part lay to the east on this example. *Redrawn from Warburton et al. (1990), Fig. 6, p. 424.*

uplift, and so on. In that sense, the composite Central and Western Alpine foreland basin megasequence comprises (1) the Palaeogene (=Nummulitic) marine megasequence and (2) the Miocene to Pliocene marine and continental molasses.

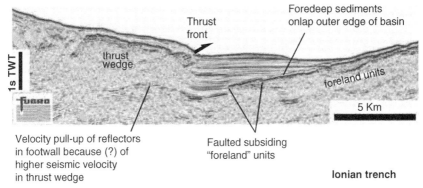

Thrust front

Foredeep sediments onlap outer edge of basin

thrust wedge

foreland units

5 Km

Velocity pull-up of reflectors in footwall because (?) of higher seismic velocity in thrust wedge

Faulted subsiding "foreland" units

Ionian trench

Figure 1.11 *The Ionian trench off Calabria and the South Apennine foreland basin.* A detail of the toes of the Calabrian subduction–accretion complex (offshore Italy) and its relationship to the 'foredeep'. Note the onlap of foredeep sediments onto dipping foreland sediments. The sedimentary infill of the basin is roughly horizontal, without any internal fanning geometry. This is probably the result of a very high rate of sedimentation with respect to the thrust advance. The foredeep is cut by normal faults presumably related to flexural loading by the adjacent thrust belt. *Section courtesy of Fugro Robertson in R.H. Butler (2008).*

6. SEDIMENTATION IN OCEANIC BASINS AND PROBLEMS IN PALAEODEPTH RECONSTRUCTION

The bioclastic or siliciclastic sediments of the margin pass laterally into thin pelagic sediments that overlie the continent–ocean transition and the oceanic crust beyond. Early deep sea drilling results obtained in the Central Atlantic Ocean prompted a comparison of the cored, pelagic Mesozoic sediments with those known at outcrop in Alpine Mediterranean fold belts (Bernoulli and Jenkyns 1974; Bosellini and Winterer 1975). The similarity in facies was compelling. It prompted reappraisal in a plate tectonic context of the assumption discovered during the first Challenger expedition that carbonate dissolution closely controls the nature and distribution of deep sea pelagic sediments. The carbonate compensation depth also varies according to the different carbonate minerals. Berger and Winterer (1974) showed that as an oceanic plate moves away and downward from the mid-ocean ridge axis the sediments deposited on the subsiding plate will change from calcareous to non-calcareous as the calcite compensation depth (CCD) is crossed. However, many local factors combine to modify the physico-chemical and biological characteristics of the sea to strongly influence the CCD and the other associated depths of dissolution both at the margins and on mid-ocean ridges.

Chenet and Francheteau (1980) have attempted a reconstruction of the variation in depth of the seafloor and the CCD with time, calculated from fifteen DSDP sites in the Central Atlantic then in continuity with the Ligurian Tethys. The underlying hypothesis was the thermal subsidence law for the oceanic crust according to Sclater (1977). The hope was that the nature of the sediments might perhaps allow deduction of the order of magnitude of their depth of deposition from knowledge of the variations in the CCD with time. These authors have shown that the response to the question cannot be simple because of uncertainties in the ages of the sediments and in the estimation of palaeodepths. Moreover the CCD level may be dependent on the structural setting. The reconstructed CCD is shallower close to the mid-ocean ridge and to the continental margin but deeper above the abyssal plains.

For example, during the period Aptian–Cenomanian, the CCD was situated at 2200 ± 100 m in a 400-km wide zone superposed on the mid-ocean ridge ('CCD axis'). In the deep ocean it was at 3150 ± 100 m ('off-axis CCD') and at 2300 ± 100 m at the continental margin (Figs. 1.12 and 1.13).

Other points, poorly understood in detail, concern the influence on the CCD of changes in climate, deep bottom water circulation, the effects of upwelling along margins, eustatic changes and the evolution of biota. For example, at a given time, a local increase in the productivity of calcareous organisms can lower the CCD and, conversely, strong productivity of benthic organisms can elevate the CCD due to carbon dioxide production resulting from degradation of organic matter. In the Late Jurassic, the CCD fell progressively in the global ocean due to the bloom of coccoliths in the Late Jurassic and the bloom of planktonic foraminifera from the middle Cretaceous had a similar effect (Fig. 1.14). Nonetheless, although there can be great temporal variations in the CCD (Fig. 1.14), it seems that pelagic sedimentary units can be correlated over large distances, an observation that pertains to the *Schistes Lustrés* discussed in Chapters 11 and 16.

7. THE WILSON CYCLE: MOUNTAIN BELTS, PASSIVE MARGINS AND FORELAND BASIN-FOLDED BELTS

The notion of ocean opening and closure was formulated by Wilson (1966). In this scenario, the cycle was proposed to consist of the following stages (Fig. 1.15):

– Rifting of continental lithosphere

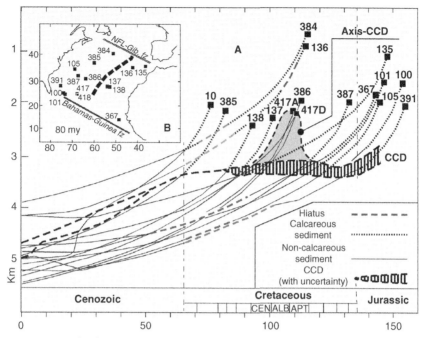

Figure 1.12 *Subsidence curves of Central Atlantic DSDP sites and evolution of CCD with time for the Late Jurassic and Cretaceous.* The curve proposed for the variation in the CCD with time shows: (1) The fall in the CCD in the Late Jurassic is perhaps a consequence of the bloom of calcareous microplankton (Coccolithophorids). (2) The modest rise in the middle Cretaceous is perhaps linked to the deposition of organic-rich black shales. (3) Again the fall to present depths (4500 or more) from Late Cretaceous times was very probably caused by the bloom of planktonic foraminifera and calcareous microplankton. *Redrawn from Chenet and Francheteau (1980), Fig. 9, p. 512.*

- Formation of an oceanic spreading axis and adjoining passive margins
- Widening of the ocean by spreading
- Subduction and contraction of the original ocean basin
- Obduction of the colliding continental crust and foreland basin development
- Continental collision and mountain building
- Post-orogenic collapse by extension and the onset of a new phase of rifting

Wilson based his scenario in part on the parallelism of the Appalachian fold belt and the margins of the central Atlantic. Here it is clear that the Triassic rifts of eastern North America have nucleated on old Appalachian thrusts

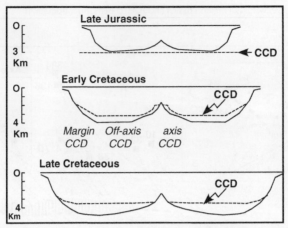

Figure 1.13 *Model of CCD evolution for Late Jurassic, Early and Late Cretaceous times*.
As the Central Atlantic and the Tethys were part of the same Ocean during Late Jurassic
and Early Cretaceous times, estimation of the CCD from the Atlantic can be tentatively
extended to the Tethys only for oceanic sites but not to the continental margins. Vertical
exaggeration is 150. *Redrawn from Chenet and Francheteau (1980), Fig. 10, p. 513.*

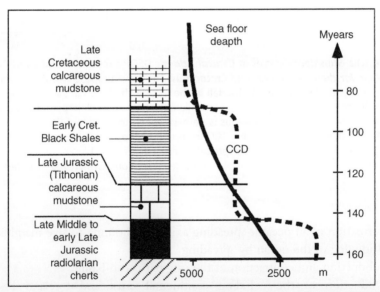

Figure 1.14 *Mesozoic facies succession of the Brianconnais as a function of age and
reconstructed oscillations in the CCD*. The diagram allows interpretation of the facies
succession observed between the end of the Jurassic and the end of the Cretaceous in
the Brianconnais. Depths, shown in metres are speculative. The calcite compensation
depth (CCD) appears to have increased in two stages since the start of the Mesozoic:
firstly towards the end of the Jurassic following the bloom of calcareous nannofossils
or coccoliths and then secondly due to the multiplication of planktonic foraminifera
from the middle Cretaceous. The Jurassic event is of major significance in global Earth
history: it was the first time that calcareous sedimentation, hitherto confined to the
shelves, occurred in the pelagic and ultimately the deep, marine domain.

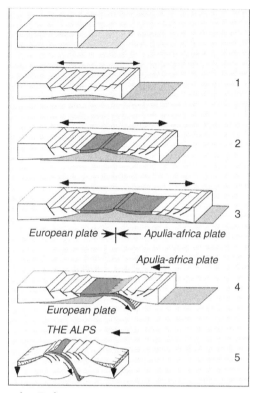

Figure 1.15 *From the Tethyan passive margin to the Alps. Main phases of the Wilson cycle.* (1) Rifting of the continental lithosphere. A simplified McKenzie model is shown here. (2) Formation of oceanic spreading axis and adjoining passive margins. (3) Widening of the ocean by spreading. (4) Subduction and contraction of the original ocean basin. Deposition of the earliest flysch in the subduction trough. (5) Continental collision and mountain building. Development of foreland and back-arc molasse basins. *Completed from Lemoine et al. (2000), Fig. 6.4, p. 80.*

and that this extensional episode later resulted in spreading in the Middle Jurassic. Although the Central Atlantic has yet to evolve into a subduction zone, the present day margin between North Australia and Timor offers a very good analogue (Figs. 1.5 and 8.8). In this area, active subduction still continues south of Java and Lombok. Further to the east, however, the Australian plate has collided with the Indonesian accretionary prism so that Timor is now being obducted onto the leading edge of the Australian plate leading to the formation of an underfilled foreland basin. Ultimately this foreland basin will be completely filled with sediment to form the foreland basin pre-foreland basin couplet. Continent–continent collision or hypercollision marks the penultimate stage of the Wilson cycle. Post-orogenic collapse (see Chapter 15) may evolve into rifting and spreading as in the Balearic basin or peneplanation may follow to be succeeded later by a new rift phase that inherits the older structure of the foreland basin and fold belt.

The Wilson cycle from mountain belts to rifts, passive margins, foreland basin and fold belts is a continuum. The Western Alps provides an ideal natural laboratory to examine this continuum from the Hercynian deformation through the formation of the Liguro-piemontais Ocean, to the formation of the Alps and adjacent foreland basins to present day post-orogenic collapse. Subsequent chapters discuss in more detail both structure and stratigraphy in terms of each of these tectonic phases.

FURTHER READING

Trümpy (2001); Bernoulli and Jenkyns (2009) Bernoulli and Lemoine (1980).

CHAPTER TWO

The Alps: Present Day Structure

Contents

Summary

The nappe pile of the Alps consists of three stacked groups of units. From the base to the top, these groups are (1) of European origin, (2) of oceanic origin, the Liguro-piemontais ophiolites and (3) of Apulo-African origin comprising both Austro-alpine nappes and south Alpine units.

To the east, the Eastern Alps are separated from the Southern Alps by the Peri-adriatic fault array. Both constitute about half the majority of the fold belt in map view. Thus 90% of the outcrop in the Eastern Alps consists of Apulo-African units underlain by so-called Penninic units of both oceanic and European origin, which appear in a few tectonic windows. To the west, the Central Alps form a rectilinear segment and the Western Alps a curved segment.

The Western Alps, from Rift to Passive Margin to Orogenic Belt, Volume 14
ISSN 0928-2025, DOI 10.1016/S0928-2025(11)14002-X
© 2011 Elsevier B.V.
All rights reserved.

Outcrops in these segments consist mainly of external Helvetic or Dauphine units overlain by Penninic nappes. The Helvetic and Dauphine units are derived from the European craton. Along the transition from the Central to the Western Alps, these European units are covered locally by a small composite slice of Apulo-African origin, known as the Dent Blanche nappe. Further again to the southwest, the Western Alps consists only of Dauphine and Penninic units.

The major thrusts separating the principal structural zones have different roles and importance. For example, the term 'Penninic Frontal Thrust' has been historically used to describe the frontal thrust of nappes of internal origin comprised of Helminthoid flysch beds and associated slices of Embrun–Ubaye units which were emplaced at the surface. This sense of emplacement has been lost or forgotten.

Today, the term Penninic Frontal Thrust is used for the deep crustal and lithospheric scale thrust along which nappes of internal origin, except the Flysch nappes mentioned above, were thrust over the external Dauphine–Helvetic zones.

In the Central and the northern part of the Western Alps, the Penninic Frontal Thrust coincides with the suture of the small Valais subsiding basin. In the southern part of the Western Alps where the Valais zone is absent, it corresponds to the front of the Brianconnais and Subbrianconnais nappes.

In the Eastern Alps, an ophiolitic suture cannot be easily located due to early subduction of part of the ophiolitic nappes and, also, the superposed folding of the nappe pile. By contrast, the frontal thrust of units of Austro-alpine origin is a major clearly defined structural boundary.

 1. THE ALPS: MAIN SUBDIVISIONS

1.1 Geographic definition

Geographically, the 200–400 km wide Alpine fold belt extends from the Swiss and Bavarian plateaus in the north to the Venetian and Po plains in the south (Figs. 2.1 and 2.2). Along strike, it extends from the Mediterranean Sea in the west as far as the longitude of Vienna and continues eastward into the Carpathians. In plan view, a semi-rectilinear part comprises the Eastern, Central and Southern Alps between Vienna and Lake Geneva. The arcuate segment corresponds to the Western Alps between Geneva in the north, the lower valley of the Rhone in the west and Provence, and the Mediterranean Sea to the south.

1.2 Geological overview of the Alps: main structural units

Geologically, the Alpine chain consists of an assemblage of units thrust over each other from south to north (Fig. 2.3) between Vienna and Geneva and, approximately to the west, in the Western Alps (Fig. 2.2).

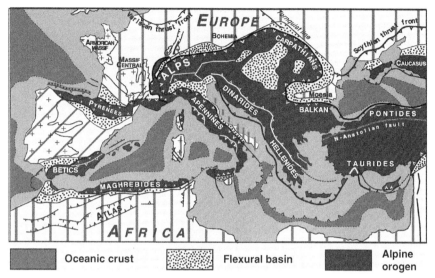

Figure 2.1 *The setting of the Alps in Circum-Mediterranean fold belts.* *Adapted from Ziegler and Roure 1996, Fig. 1, p. 18.*

The lower and external part of the nappe stack comprises units of continental European origin to the north and west. Units of oceanic origin (Chapter 3 and 11) surmount these. Above this assemblage rest Austroalpine continental units thrust from the south and derived from the microcontinent known as Apulia or Adria (Chapter 10). This continental block, which now corresponds to the Italian Peninsula and the basement of the Adriatic Sea, was a more or less independent northern promontory of Africa. For simplicity, the name Apulo-African block or microcontinent will be used in this book.

1.3 Western and Central Alps

To the north of the Central Alps, Alpine structures are thrust onto the southern edge of the Molasse basin that is infilled by thousands of metres of clastic sediments ranging in age from Oligocene to Miocene (32–10 Ma). To the south, these structures are buried under the Molasse basin of the Po plain by as much as 10 km of sediments of Oligocene to Quaternary age (32 Ma to Recent). At outcrop, European and oceanic units dominate to the west of the longitude of Lake Constance and the Swiss canton of Grisons (Graubünden), i.e. in the Central and Western Alps (Fig. 2.4).

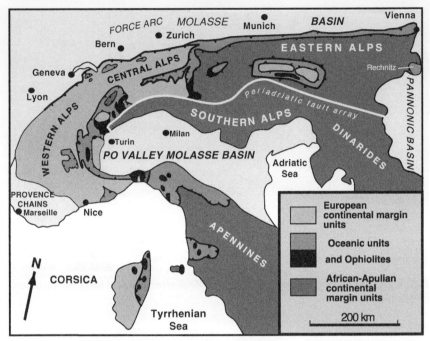

Figure 2.2 *Simplified structure of the Alps, Corsica and northern Apennines.* This sketch map shows the distribution of units of European, oceanic (Liguro-piemontais) and Apulo-African origin. The Apennine arc is composed of Apulo-African tectonic units overthrust by nappes originating from the Liguro-piemontais Ocean. By contrast, the Alps are composed, from base to top, of European units, Liguro-piemontais oceanic units and Apulo-African continental units. At outcrop, nearly half the Alps are of Apulo-African origin. *Adapted from Lemoine et al. 2000, Fig. 5.2, p. 64.*

1.4 Eastern and Southern Alps

The Eastern Alps, to the east of the Grisons, are essentially outcrops of Austro-alpine nappes of Apulo-African origin. Units of European or oceanic origin outcrop in a narrow zone at the northern front of the fold belt which includes several slivers and klippen of European and oceanic origin. Other outcrops of units of European origin are located in tectonic windows created by erosion through Austro-alpine nappes folded into large-scale anticlines. The main windows are the Lower Engadine, Hohe Tauern and Rechnitz on the Austro-Hungarian frontier.

The Peri-adriatic fault array bounds the Eastern and Central Alps to the south. The Southern Alps lie to the south of this line and are entirely composed of almost unmetamorphosed, weakly folded Apulo-African units. They are affected by major thrusts towards the south, which can be

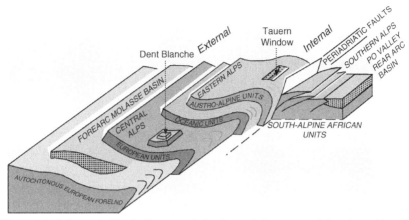

Figure 2.3 *Schematic block diagram of the Central, Eastern and Southern Alps.* The diagram illustrates the geometric dispositions of units whose origin is European, oceanic, Liguro-piemontais (= South-Penninic), Austro-alpine and South-alpine. (See colour plate 1) *Adapted from Agard and Lemoine 2003.*

considered as backthrusts given the general direction of Alpine thrusting from south to north (Fig. 2.4).

The sketch map of the Alps (Fig. 2.2) shows that half of the outcrop of the Alpine fold belt, including the major part of the Eastern Alps and the whole of the Southern Alps, is of Apulo-African origin.

For a long period, it was believed that the Central and Western Alps, as a whole, were initially completely covered by Austro-alpine nappes. This was especially because the Austro-alpine Dent Blanche nappe (Fig. 2.4, DB), a sliver of gneissic, continental basement of South-alpine origin, rests on both oceanic and European nappes (see transect Fig. 2.6C located on map, Fig. 2.4). However, the Dent Blanche sliver, together with the Ivrea zone (Figs. 2.6C and 2.12B located on Fig. 2.4), are the only remnant left by erosion of a perhaps discontinuous assemblage that is clearly different from the Austro-alpine of the Eastern Alps. Moreover, the distribution of cobbles in the molasse shows that the north–south oriented erosional edge, which bounds the Austro-alpine nappes to the west around the Grisons canton (Fig. 2.4), was probably not far from their original western limit.

The focus of this book is on the Western and Central Alps from the lower Rhone valley to the Grisons in the east. In this area, numerous, clear results from a variety of studies allow an overview of the evolution of the Liguro-piemontais Ocean and its European and Apulo-African conjugate margins.

2. THE CENTRAL AND WESTERN ALPS: MAJOR STRUCTURAL TRENDS

Overview of the Central Alps

The Central Alps are the orographic, western continuation of the Eastern Alps from the Grisons as far west as Lake Geneva.

The overall structure is that of a complex collisional prism composed of from north to south and base to top (Figs. 2.4, 2.5, 2.15):

1. An external part composed of basement units (the Gastern–Aar–Gotthard crystalline massifs), units of sediments (Helvetic nappes) and an assemblage of exotic nappes outside the prism. The exotic assemblage comprises both the Pre-alpine and Flysch nappes, which were derived respectively from the Penninic and (partly) Austro-alpine zones. These lie on the internal edge of the Molasse basin on one side (NW) and on the top of the Helvetic nappes on the other (SE).

2. A central and internal part consisting mainly of continental basement corresponding to the Penninic nappes (Fig. 2.5).

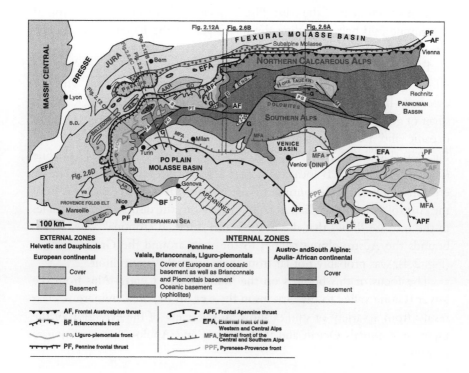

The ophiolitic nappes are incorporated within the prism. The Dent Blanche nappe covers the whole.

2.1 Penninic nappes

The *Pennine Alps* are a mountain range between the Valais part of the Rhone valley and Val d'Aoste. The word *Penninic* corresponds to a stack of tectonic units which outcrop in part of the Pennine Alps. The Penninic nappes consist of three superposed assemblages which are, respectively, from bottom to top: (1) North Penninic (or lower Penninic) derived from the Valais subsiding basin and its margins, (2) Middle Penninic corresponding to

---

Figure 2.4 *Simplified structural map of the Alps.* The Alps are situated between the stable European craton to the north, unaffected by Alpine orogenesis and to the south, the Apulo-Africa domain. The latter is represented by part of the basement under the Po basin, but further south, it was affected by Apennine folds and thrusts. **The external fronts of the Western and Central Alps** (EFA) coincide more or less with the foot of the mountains. These are the external fronts that verge northward or northwest to west; i.e. the front of the Jura and Sub-alpine chains, the front of the assemblage of Pre-alpine – Helvetic nappes and Sub-alpine molasse. In the southern Western Alps, the Sub-alpine front interferes with the Pyreno-Provencal front (PPF). **The internal front of the Central and Southern Alps** (MFA) does not coincide with the foot of the Alps at the boundary of the Po Basin. South-verging blind thrusts are known beneath Milan well south of the southern topographic edge of the Alps; these concealed thrusts are buried below molasse and alluvial sediments. **The Penninic Frontal Thrust** (PF) is a composite line which marks *sensu stricto,* the thrust front of the internal zones on the external zone. **The frontal Brianconnais thrust** (BF), which can be followed in the Western and part of the Central Alps, marks a major suture and palaeogeographic boundary with Subbrianconnais to the south and Valais to the north and east. In the Western Alps, this major structure marks a jump in the degree of deformation and intensity of metamorphism; this difference becomes blurred and disappears eastward in the Central Alps. The ophiolite nappe front is here called the Liguro-piemontais front (LFO). **The Frontal Austro-alpine thrust** (AF) appears towards the boundary between the Western and Central Alps and is best developed eastward in the Central and Eastern Alps. Underlying this thrust is the major **Liguro-piemontais ophiolitic suture** inherited from the former active Apulia continental margin. Molasse in pale yellow: small molasse basins of the Western Alps; BD: Bas Dauphine; va: Valensole-Digne; External Crystalline Massifs, AR: Argentera; Go: Gotthard; MB: Mont Blanc; P: Pelvoux; M-EST: Maures-Esterel in Provence; Internal Crystalline Massifs, DM: Dora-Maira; APF: Apennine Front; ENG: Engadine window; DINF: front of the Dinarides; G: Late Alpine granites (Oligocene); Peri-adriatic faults, PC: Canavese; PG: Gail; PT: Tonale; S, E, M: Simplon, Engadine and Molltal fault arrays. (See colour plate 2) *Adapted and completed from Lemoine et al. 2000, Fig. 5.3, Table III.*

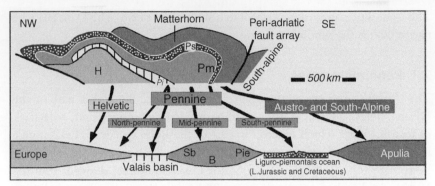

Figure 2.5 Schematic section of the major Alpine subdivisions. Upper section: schematic of the present-day superposition of the main tectonic units. Lower section: schematic Jurassic reconstruction. Note the small Valais trough is not present on all sections. H: External Helvetic and Dauphine zone. P: Penninic nappes. North Penninic or Lower Penninic (Valais basin and its borders). Pm: Mid-Penninic of the Subbrianconnais. Brianconnais and Piemontais (SBR Block). South Penninic or Upper Penninic (Liguro-piemontais Ocean). *Adapted from Lemoine et al. 2000, Fig. 5.4, p. 67.*

the Subbrianconnais, Brianconnais and Piemontais domains and (3) Upper Penninic (or south Penninic) derived from the Liguro–piemontais Ocean, not far from the Austro–alpine domain (Figs. 2.5, 2.7; Chapter 4, Table 1).

2.1.1 The North Penninic or Lower Penninic assemblage

The northern (=Lower) Penninic units derive from the Valais basin (Chapter 7). Lower Penninic nappe units include Valais ophiolites as well as the Simplon–Ticino basement nappes and their metamorphosed sedimentary cover which originated from the European (or northwest) side of the Valais basin. Units corresponding to the southern margin of the Valais basin are much less well preserved than their northern equivalents. Sediments are represented by the *Schistes Lustrés* or *Bundnerschiefer* Formation of the Valais and Grisons. The *Schistes Lustrés* represent an assemblage of metasediments of Jurassic and Cretaceous age mainly composed of calcareous calcshists. In terms of present-day structure, the remnants of the Valais basin define a North Alpine suture situated between the south European margin and the Saint-Bernard–Monte Rosa–Brianconnais continental block in the Central Alps and the northern Western Alps.

2.1.2 The Middle Penninic assemblage

The Middle Penninic assemblage comprises the two main basement nappes, the Grand-Saint-Bernard and Monte Rosa. Also included is the sedimentary cover minus the detached part of the section that now forms the Pre-alpine nappes beyond the main front of the fold belt named as the Penninic Frontal Thrust.

The Grand-Saint-Bernard nappe complex includes four sub-units which are from base to top: the *Zone Houillere* (Coal Measures Zone) and the Pontis nappe, the Siviez-Mischabel and Mont-Fort nappes (Fig. 2.7). The Monte Rosa and Saint-Bernard nappe complex both originate from the continental 'SBR' block (= **S**aint-**B**ernard Monte **R**osa), which separated the Valais (north Penninic) and Liguro-piemontais (south Penninic) oceans. On the Gotthard transect, the Schams, Tambo, Suretta and Margna nappes also originate from the 'SBR' block (see below, Fig. 12.9).

2.1.3 The South Penninic or Upper Penninic assemblage

The South Penninic assemblage includes the Liguro-piemontais ophiolites and the sedimentary cover of *Schistes Lustrés*. The latter correspond to a complex of metasediments of Jurassic and Cretaceous age consisting mainly of schists and calcschists whose stratigraphic correlation is now better understood (See Chapter 11, below).

2.1.4 Linkage between the Western and Central Alps

The southwestward continuation of the Central Alpine coal-bearing zone (Zone Houillere), Pontis and Siviez-Mischabel nappes is represented by the Brianconnais in the Western Alps (Fig. 2.7 and Table 4.1, below). The southwestward prolongation of the Monte Rosa and Mont-Fort nappes is represented by the Piemontais or distal part of the European margin of the Liguro-piemontais Ocean. Equivalents of the upper Penninic ophiolites such as those of Antrona and Zermatt–Saas Fee are the Liguro-piemontais ophiolites of the Western Alps (e.g. Chenaillet, Queyras).

2.2 Pre-alpine nappes

The Pre-alps (Figs. 2.4, 2.6, 2.7 and 2.8) are composed of sedimentary cover nappes of Ultrahelvetic, Penninic and Austro-alpine origin. The two main Pre-alpine massifs are the Chablais to the south of Lake Geneva, the Romande (French Switzerland), the Fribourg and Bernoise Pre-alps between Lake Geneva and Lake Thun. In addition, there are a series of small slivers or klippen in Savoy, to the south of the Chablais, and in central Switzerland to the east of the Romande Pre-alps.

In the two main massifs, the Pre-alpine nappes lie to the north on the edge of the foreland basin and, to the south, on the northern dip slope of the

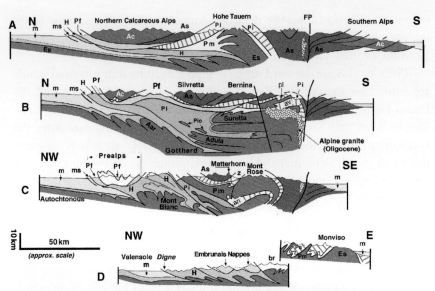

Figure 2.6 *Four extremely simplified cross sections of the Alps.* (A) Eastern and Southern Alps, from Salzburg to Venice via the Austrian and Dolomites. (B) Via the canton of Grisons (Switzerland). (C) Central Alps, from Bern or Fribourg to Ivrea to near Turin via the Matterhorn. (D) Western Alps, from Digne to Viso. Es: European basement (Autochthon, Helvetic, Brianconnais, Piemontais). As: Apulo-African (Austro- and South-alpine) basement and pre-Triassic sediments. Ac: Austro-alpine and South-alpine cover. H: Helvetic-Dauphine cover. Pi: Lower Penninic (Valais). Pio: Valais ophiolites. Pm: Middle Penninic. Ps: Upper Penninic ophiolites (Liguro-piemontais). Pf: Penninic flysch. m: Oligo-miocene molasse. ms: Sub-alpine molasse. FP: Peri-adriatic faults. pl: Platta. av: Avers. z: Zermatt–Saas Fee. an: Antrona. br: Brianconnais. (See colour plate 3) *Adapted from Lemoine et al. 2000, Fig. 5.5, Table IV.*

Helvetic nappes. The superimposed Pre-alpine units, clearly shown in the two massifs, comprise from base to top (Fig. 2.8):

Ultrahelvetic nappes in the lower part of the pile

Major Pre-alpine nappes, which include the Plastic Median Pre-alps (Subbrianconnais), Rigid Median Pre-alps (Brianconnais), Breche (Breccia) nappe (Piemontais).

These nappes are composed of unmetamorphosed sediments ranging in age from Triassic to middle Eocene. The Zone Houillere and Pontis nappe was the original basement of the Plastic Median Pre-alps. The Siviez-Mischabel nappe (Fig. 2.7) and the Mont-Fort and Monte Rosa nappes were the basement for the Rigid Median Pre-alps and the Breche nappes, respectively (Fig. 2.7 and, below, Table 4.1).

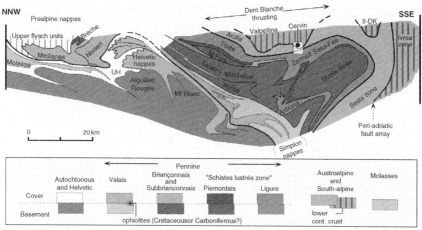

Figure 2.7 *Cross-section of the Alps near the Central-Western Alps boundary and close to the Matterhorn.* This is the same section as that of Fig. 2.6. C, but with more detail. The Grand-Saint-Bernard nappe is split into the Pontis, Siviez-Mischabel and Mont Fort nappes. UH: Ultrahelvetic. The interpretation adopted here, published in 1993 by the Geology School of Lausanne, appears to be the most compatible with classic work and the most verifiable though it will certainly be perfected in the future. Various authors from time to time have re-evaluated the deep seismic profiles resulting in different and contradictory interpretations, so demonstrating the difficulty and risks in this exercise. For example, an interpretation made in 2000 adds a supplementary 'slab' comprising the deep continental crust, which belongs to the European continental crust, on the profile across western Switzerland and to Apulia crust on the eastern Swiss profile. The same interpretation shows the Antrona ophiolites in continuity with the Valais ophiolites. This view contradicts accepted palaeogeographic reconstructions. In effect, the Antrona and Zermatt nappes which surround the Monte Rosa nappe are classically considered as originating from the Liguro-piemontais Ocean, which was separated from the Valais subsiding basin by the 'SBR' continental block to which the Mont Rosa belonged precisely. (See colour plate 4) *Simplified after Escher et al. 1987. Adapted from Lemoine et al. 2000, Fig. 5.6, Table VI.*

These nappes were detached along Triassic evaporites and thrust north-ward on the external Helvetics as far as the southern border of the foreland basin before the underlying basement was affected by Tertiary Alpine metamorphism

Nappes composed of early flysch of Late Cretaceous age: The flysch is mainly dated as late Cretaceous but might also have formed during the Palaeocene or as late as the middle Eocene. These are the nappes termed Gurnigel-Sarine, whose northern, Gurnigel element is overthrust by the Plastic Median Pre-alps, the Dranses (or Helminthoid flysch), Gets and Simme.

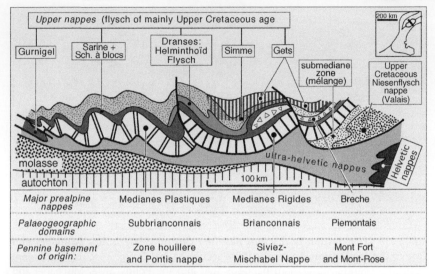

Figure 2.8 *Schematic structure of the Pre-alps between Lake Geneva and Lake Thune.* Caron 1972; Caron in Trümpy 1980. Adapted from Lemoine et al. 2000, Fig. 5.7, p. 68.

In general, the Gurnigel-Sarine and Dranses flysch nappes were derived from trenches related to subduction of the Liguro-piemontais oceanic lithosphere under the active Apulo-African continental margin from which the Austro-alpine (= South-alpine) units originate. The Simme nappe has been postulated to originate from within the Austro-alpine domain or more precisely from the present location of the Canavese zone that separates the basement of the Sesia and Ivrea zones (Figs. 2.7 and 2.12B).

The **Niesen flysch nappe** is of Valais origin and is situated on the internal border of the Romande Pre-alps where it rests on the internal Ultrahelvetic.

2.3 Ophiolitic nappes

Within the nappe stack of the Central and Western Alps, the ophiolitic nappes are derived from the Liguro-piemontais (= south Penninic) Ocean.

Whether of northern or southern Penninic origin, these nappes are characterized by a discontinuous ophiolitic sole that is generally thin (from a few tens of centimetres to 2 or 3 km maximum; Chapter 11, Fig. 11.1) and covered by metamorphosed oceanic sediments that are the classical *Schistes Lustrés* (Fig. 11.22 and 11.23). These metasediments, initially only a few hundred of metres thick, are now intensely folded and metamorphosed. Locally they contain intercalations of ophiolitic blocks and lenses.

The three main south Penninic ophiolitic nappes in the Central Alps along the Cervin (Matterhorn) transect (Fig. 2.7).

The Tsate nappe is located in the north between the basement of the Grand-Saint-Bernard and the Dent Blanche. It is very intensely sheared. In its northern part at least, it never experienced subduction and has only undergone greenschist metamorphism. It may have originated from an oceanic accretionary prism. To the south, two ophiolitic nappes were subjected to high-pressure–low-temperature (HP–LT) metamorphism during subduction. These are the Zermatt–Saas Fee and Antrona nappes, which lie above and below the Monte Rosa nappe, respectively. The Zermatt–Saas Fee nappe is considered to form the normal flank of the tectonic cover of the Monte Rosa nappe with the Antrona nappe being located on the inverted flank of the Monte Rosa nappe (Fig. 2.7). Another hypothesis, rejected here, considers the Monte Rosa to be small microcontinent separating two narrow oceanic troughs called Antrona and Zermatt–Saas Fee.

2.4 Austro-alpine and South-alpine

In the Central Alps, as noted earlier, the Austro-alpine is represented only by the Lower Austro-alpine Dent Blanche nappe (Fig. 2.7). It is derived from units located to the south of the Monte Rosa and Grand Paradis massifs of European basement and from the ophiolitic, south Penninic suture and 'Zone of Lakes'. These units are, from external to internal, the Sesia zone, Ivrea zone and Zone of Lakes. The Sesia zone is strictly not South-alpine as it is more external than the Peri-adriatic lineament, here represented by the Canavese line. The Sesia basement is composed of upper continental crust which was partly subducted (HP–LT metamorphism) at the end of the Late Cretaceous. The South Alpine (s.s.) in the Central Alps corresponds to the Ivrea zone constituted by lower crust including 'kinzigites', i.e. granulite-grade gneisses, associated with diorites, gabbros and peridotites. The Ivrea zone is thrust onto the Sesia zone as shown by the thrust sliver of the so-called II DK or 'second dioritic-kinzigite zone', the first DK being that of the Ivrea. Further to the north, the Dent Blanche nappe complex has been thrust en masse onto Penninic units. It comprises basement units of Arolla, i.e. upper crust (cf. Sesia and Valpelline) (lower crust cf. Ivrea and II DK: Figs. 2.7, and 16.9, Cervin). The Dent Blanche

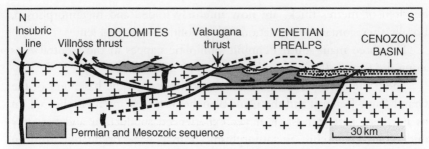

Figure 2.9 ***Schematic cross-section of the Southern Alps across the Central Dolomites.*** The Central Dolomites form the innermost part of the Neogene thrust belt of the Southern Alps. The central part of the Southern Alps is divided into two structural complex units by the east–west trending Valsugana thrust. This surface separates the basement and Permian to Mesozoic cover of the Dolomites to the north from the Venetian Pre-alps to the southeast. The southward displacement along the Valsugana thrust is estimated at 8–10 km, and the resulting uplift of the Dolomites is considerable. In this north–south section, the Central Dolomites form a vast piggy-thrust synclinorium structure which rests on the south-verging Valsugana backthrust to the south and on the associated north-verging Villnoss overthrust to the north. The Insubric line belongs to the Peri-adriatic fault array. Location of this section on Fig. 4.4. *Simplified from Doglioni 1987, Fig. 2, p. 182.*

nappe complex corresponds to a lower Austro-alpine thrust slice derived from the South-alpine zone, as noted earlier.

The Southern Alps are bounded to the south by the foreland basin of the Po plain and to the north by the Peri-adriatic fault array (Figs. 2.2–2.4). To the east, they pass into the Dinarides (Figs. 2.1 and 2.2). With the exception of restricted areas surrounding the Peri-adriatic fault array and decollement surfaces, Alpine folding is weak but thrust displacement is at a scale of tens of kilometres (Fig. 2.9). The weak deformation at the micro- and mesoscopic scale may be related to the rigidity of the 2000 m total thickness of Permian ignimbrites and rhyolites which comprise a significant part of the succession of the Southern Alps.

3. ALPINE METAMORPHISM

Successive metamorphic events can be distinguished in the Alps and correspond, respectively, to ocean–continent subduction then to the formation of a collisional prism (Fig. 2.15 and Chapter 14).

The rocks subjected to subduction were affected by HP–LT metamorphism of blueschist and eclogite facies grade (trajectory 3 of Fig. 2.10). The local

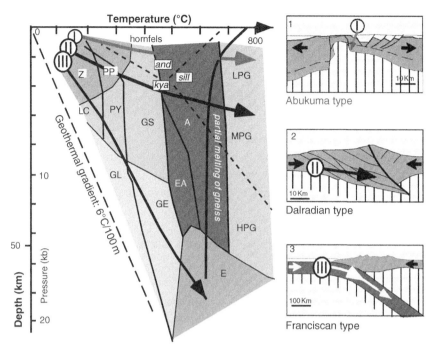

Figure 2.10 *Position of different metamorphic facies in temperature–pressure space and the paths of different cases of prograde metamorphism.* And: andalusite; kya: kyanite; sill: sillimanite. Metamorphic facies – Z: zeolite; PP: prehnite–pumpellyite; LC: lawsonite–chlorite, lawsonitic greenschist facies; PY: pumpellyite–actinolite; GS: greenschists; GL: glaucophane–lawsonite, low-temperature blueschists; GE: glaucophane–epidote, high-temperature blueschists; E: eclogites; EA: epidote amphibolites; A: amphibolites; LPG, MPG, HPG: low-, middle- and high-pressure granulites, respectively. (1) Prograde metamorphism at low pressure: Abukama type gradient; example of continental crust undergoing stretching. (2) Prograde metamorphism at intermediate pressure: Dalradian type gradient: thickening of continental crust during collision at medium speed. (3) Prograde metamorphism at high pressure: Franciscan type gradient: Oceanic lithosphere undergoing subduction. *Adapted after Kornprobst 1994 and Lemoine et al. 2000, Fig. 4.9, p. 53.*

presence of coesite-bearing whiteschists demonstrates Ultra-High-Pressure (UHP) metamorphism implying rapid burial to at least 100 km.

The rocks involved in the crustal or lithospheric collisional prism were then subjected to greenschist or amphibolite facies metamorphism (trajectory 2 on Fig. 2.10).

In the external, west, north or northwestern part of the Alps, the intensity of metamorphism decreases from greenschist metamorphism through low-grade metamorphism (prehnite–pumpellyite facies) to finally disappear altogether.

South of the Peri-adriatic line, the rocks of the Southern Alps are folded but almost unmetamorphosed. The Alpine thermal overprint is less than 200°C. Brittle thrusts imply an overall N–S shortening of less than 100 km.

4. THE ARC OF THE WESTERN ALPS COMPARED TO THE CENTRAL ALPS

The arc-shaped Western Alps can be followed from the Dent Blanche–Cervin (Matterhorn) massif as far south as the Maritime and Ligurian Alps on the Mediterranean coast. The palaeogeographical and structural characteristics of the Western Alps differ from those of the Central Alps.

4.1 Absence of a continuous foreland basin in the Western Alps

The flexural foreland basin located in front of the Central and Eastern Alps disappears in Savoy between the Sub-alpine chains and the southern folds of the Jura (Fig. 2.4). Towards the SW and S, Neogene molasse is locally well exposed on all the outer perimeter of the arc though relatively thin. In several parts of the Lower Dauphine and Digne-Valensole Basins, the sedimentary thickness reaches several kilometres (Fig. 13.5). Although these small molasse basins do not represent an originally continuous basin, they also originate from flexural loading.

4.2 External zone: north/south palaeogeographic evolution

Jurassic and Cretaceous palaeogeographic boundaries trend in part obliquely to Western Alpine structural trends reflecting a subsequent rift phase during the Late Jurassic–Early Cretaceous. The Urgonian carbonate platform of Barremian–Early Aptian age can be followed to the south where it comprises the northern Sub-alpine area (Bauges, Bornes, Chartreuse and Vercors; Fig. 13.5). The coeval, pelagic basin extended to the Ultrahelvetic zone in the north. The so-called Vocontian domain, south of the Vercors, was a western continuation of the Ultrahelvetic facies tract but without the Urgonian facies. The horseshoe-shaped Vocontian pelagic basin opened eastward and its boundaries are transected by Alpine lineaments (see below, Figs. 7.10 and 13.5). In addition, the Jurassic–Cretaceous carbonate platform of Provence, present to the south, encompasses the southern part of the southern Sub-alpine chains, the Provence chains and the Corsica-Sardinia block, which was then attached to Provence.

4.3 External zone: the Sub-alpine chains

Southward from the Mont Blanc transect, the system of large Helvetic nappes gives way to the folded Sub-alpine Chains which curve along the outer edge of the External Crystalline Massifs. The curvature of the Alpine arc is outlined by structural features that involve both basement and the post-Hercynian cover. The External Crystalline Massifs from Mont Blanc to the Pelvoux show a rectilinear pattern inherited from Jurassic syn-rift fault trends as shown below (Chapter 13 and Fig. 13.5).

The overall trend of the folds of the Sub-alpine chains deviates progressively from NE–SW in northern Savoy to N–S near Grenoble, then to NW–SE and again to E–W in the southern Sub-alpine and Provence chains. However, the last trend largely reflects interference between the E–W Pyrenean–Provence fold system and subsequent NW–SE Alpine folds (Chapter 13, Fig. 13.5).

4.4 Internal zones: absent Valais zone, lack of Austro-alpine at outcrop in the Western Alps

The Internal Alpine Arc is a characteristic of the Western Alps. However, several main structural units disappear from north to south along strike. Valais units do not exist further south than the Tarentaise valley near Moutiers. In the Sub-alpine and Provence chains (external Western Alps), the Early Cretaceous rift shares a common age with the Valais domain. Austro-alpine units do not outcrop southwest of the Dent Blanche massif and its surrounding small thrust slivers.

4.5 Flysch nappes

The Flysch nappes outcrop in part of the Embrun and Ubaye areas, between the crystalline cores of Pelvoux and Argentera and elsewhere in the Maritime Alps towards the Mediterranean coast. Within the interior, they are confined by the Brianconnais front and rest on the external Dauphine or Provence zone. The two nappes named as Parpaillon and Autapie are mainly composed of Late Cretaceous Helminthoid Flysch and slivers dislodged from the sedimentary cover of Subbrianconnais and Brianconnais.

5. DEEP STRUCTURE OF THE ALPS FROM GEOPHYSICAL STUDIES

Industry seismic profiles as well as deep scientific seismic profiles acquired by the ECORS–CROP, NFP and TRANSALP programmes (Figs. 2.11, 2.12, 2.16, 2.17) have provided new insights into the shallow

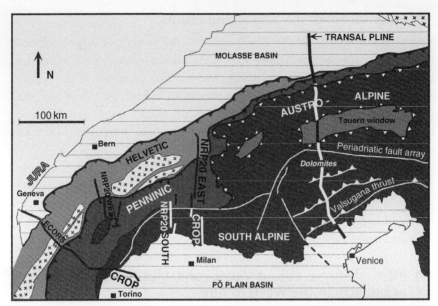

Figure 2.11 *Location sketch map for the deep reflection seismic ECORS, CROP, NRP 20 and TRANSALP lines.*

and deep structure of the Western, Central and Eastern Alps especially when complemented with surface structural geology studies, seismic refraction, seismic tomography and gravity surveys (Fig. 2.13).

The gravity map (Fig. 2.13) shows two associated positive and negative Bouguer gravity anomalies that are typical of collisional fold belts. The negative anomaly increases from −30 mgal beneath the Rhone valley to −150 mgal beneath the Brianconnais zone (Masson et al. 1999). The change is attributed to thickening of the continental crust, which increases eastward to exceed 50 km in the middle of the fold belt. Immediately to the east, the positive anomaly of Ivrea reaches +80 mgal.

The 'Ivrea body' has been identified on the ECORS–CROP profile. It has been interpreted in the past as a flake of Apulia mantle elevated to shallow depths of 10–12 km (Fig. 2.12). A more recent interpretation proposes that it may be a flake derived from the lower crust, but this interpretation has not been followed here.

The 'Ivrea body' is characterized by high seismic velocities that allow it to be mapped by means of seismic tomography. It does not continue south of Cuneo. It is sub-vertical and has a north–south orientation. Shorter in length than width, it widens from south to north. In contrast to adjacent structures visible at outcrop, it is not curved. In terms of the hypothesis of a

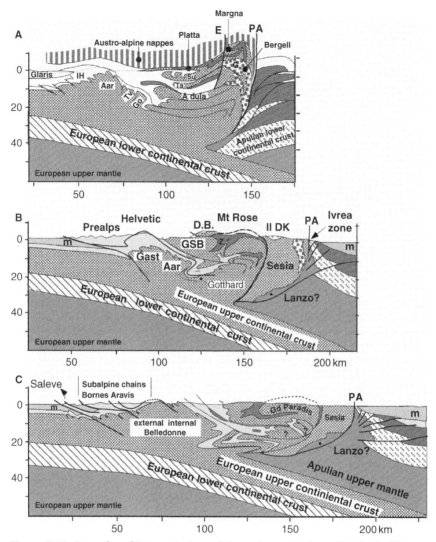

Figure 2.12 *Examples of interpretations of three deep seismic profiles in the Western Alps (ECORS–CROP) and the Central Alps (West and Eastern Switzerland).* (A) Eastern Switzerland (Grisons). (B) Western Switzerland (Matterhorn transect). (C) Western Alps (Savoy, Po basin). E: Engadine fault. PA: Peri-adriatic lineament. m: molasse. IH: Infra Helvetic.- Crystalline Massifs: Tv: Tavetsch; Gast: Gastern; Go: Gotthard.- Penninic basement nappes: Su: Suretta, Ta: Tambo, GSB: Grand Saint Bernard. - G: Late orogenic granites (N. Oligocene) of Bergell etc. DB: Dent Blanche nappe. IIDK: Second diorite-kinzigite zone (the first is the Ivrea zone). Same colours have been used for the upper and lower European and Apulia crust; however, the continental basement of Sesia, whose attribution is controversial, is not represented by a specific colour. (See colour plate 5) *Simplified from Marchant 1993. Adapted from Lemoine et al. 2000, Fig. 5.8, Table V.*

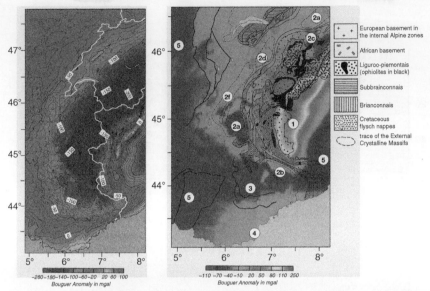

Figure 2.13 *Gravity results.* The negative gravity anomalies indicate a mass deficit due to the presence of less dense rocks at abnormal depths apart from the mountain belt. These less dense rocks are interpreted as belonging to continental crust pushed down into the mantle as far as 55 km depth at the time of crustal thickening engendered by the collision. They form the crustal root of the fold belt. Conversely, the strong positive, narrow and arcuate anomaly located in the heart of the arc is attributed to the 'Ivrea body' classically interpreted as a slice of mantle elevated to shallow depths through the crust. (1) the 'Ivrea body'; (2a): Pelvoux; (2b): Argentera; (2c): Simplon-Ticino Nappes (Valais zone); (2d): Mont Blanc; (2e): Aar; (2f): Belledonne; (3): Southwestern Sub-alpine units; (4): Maures and Esterel; (5): Molasse basins. 2a, b, d, f and 4 correspond to the External Crystalline Massifs. Bouguer gravity anomalies in milligals. (See colour plate 6) *From Masson et al. 1999, Fig. 2, p. 868 and Fig. 4, p. 870.*

mantle indentor responsible for the formation of the internal arc of the Western Alps, it is necessary to invoke one or more horizontal decoupling surfaces separating the rectilinear structures below from the curved structures above which form the Alpine arc.

The interpretation of the available geophysical data remains controversial. Major uncertainties in the deep geology and structure are reflected in varied and speculative models of Alpine subduction and collision. It is not our intent to review this debate in detail, and we emphasize those features which are well established from the data before commenting on aspects of the various models. Deep seismic data have provided an invaluable insight into basement-involved tectonics, the flexural foreland basin, the role of detachments in the thin-skinned tectonics of the Sub-alpine chains as well as the uplift of the Mont Blanc and Belledonne Massifs by thrusting along crustal ramps. The structure of the internal zones remains equivocal and thus open to a wider range of interpretations.

Notably, section-balancing techniques work successfully and accurately measure shortening west of the Penninic Frontal Thrust. However, the amount of subduction and thus the amount of crustal shortening within the internal zones remain highly conjectural, thus accounting for the variety of models that attempt to accommodate the wide range of shortening estimates made from a variety of different studies.

The 320-km long ECORS–CROP profile (Fig. 2.12C) extends from the Bresse basin in the northwest through the Jura, the Molasse basin, the Sub-alpine Borne massif, Belledonne Crystalline Massif, Penninic front and internal Alpine zones to terminate in the Po plain. The line crosses directly through the heartland of the Alps described in this book, and its interpretation has provided useful insights into the deep structure.

The ECORS–CROP line (Fig. 2.14) shows that the layered European crust dips eastward to plunge beneath the internal Alpine zones. This flexure

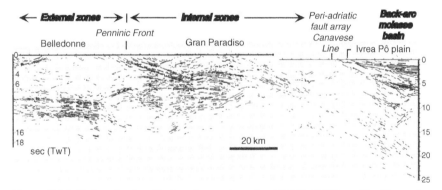

Figure 2.14 *Main seismic reflectors imaged on the ECORS–CROP profile.* The major feature is the layered European crust dipping eastward to plunge beneath the internal Alpine zones. In the Penninic zone, high amplitude reflections down to 8–10 s contrast with the seismically transparent crust that underlies the External Crystalline Massifs and the roots of the Penninic nappes in the Ivrea zone. The Penninic zone appears to be characterized by low angle or slightly curved horizons whereas steeply dipping structures characterize the surrounding domains. The well-documented rise of the Moho to 10–12 km beneath the Ivrea zone seems to comprise the hanging wall to a stepped descent of the Moho beneath the South Alpine crust of the Po plain. Other mid-crustal reflectors may represent the Moho in flakes of lithosphere caught up in crust thickened by the Alpine collision. Integrated seismic and gravity data also suggest the presence of a high-density peridotite or eclogite body beneath the Gran Paradis massif (Fig. 2.12C). *From A. Nicolas et al. 1990.*

of the European plate is clearly the response to the lithospheric loading caused by the Alpine collision. East of the well-imaged Penninic Frontal Thrust, most reflections dip eastward, indicating a fundamental asymmetry in Alpine structure whose explanation is a requirement of any valid model.

Many models that might fit the observed crustal geometries can be proposed to account for the evolution of the Alpine collisional prism, but all are necessarily based on a limited number of starting scenarios. Expressed in very simple terms, once subduction and continent–continent collision had begun, the most influential parameter was the location of the main decoupling surface within the subducting lithosphere. Modern accretionary prisms associated with oceanic subduction zones provide a useful analogy showing the location of the decoupling surface within or at the base of the sedimentary section overlying the oceanic crust (Fig. 2.15).

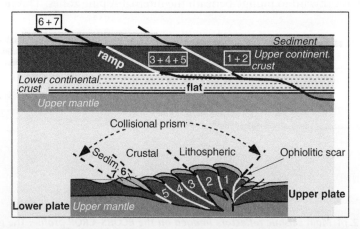

Figure 2.15 *Simplified model of the construction of a collisional prism.* This model shows how a system of ramps and flats in a series of thrusts, numbered here from 1 to 7, and organized in sequence in time and space allow interpretation of the formation of the prism and its three structural subdivisions (when present). These are represented, from the internal to the external part of the folded belt, by the lithospheric, crustal and sedimentary parts. This model is simplified and not to scale. It does not take into account the common detachment of the dense lower crust from the upper crust to be consumed at depth. In this case, it will not appear in the prism. For simplification, the model takes into account the absence of dense lower continental crust in the wedge since it was detached by crustal delamination from the upper crust to be consumed by subduction. *Adapted from Lemoine et al. 2000, Fig. 4.6, p. 50.*

In fact, the major decoupling surface in the Alps seems to climb from east to west from the internal to the external zones via a system of ramps and flats. In the east, the decoupling surface lies close to the crust/mantle boundary but then moves upward to the boundary between the upper brittle and ductile lower crust to then emerge at the base or within the sediment pile. In the west, within the Alpine foredeep and Bresse basin, detachment surfaces follow Triassic and Oligocene evaporite beds. In the case of the thin-skinned Jura fold belt, the underlying master detachment lies in Triassic evaporites and roots eastward into the Penninic Frontal Thrust (Fig. 2.16).

In a collisional belt, depending on whether the decoupling level lies within the mantle, at the crust/mantle interface or within the crust, collision can involve accretionary wedging or lithospheric slabs involving either the entire crust or only the upper crust.

The most striking observation made for the ECORS–CROP profile is the identification of a slab of mantle inserted within the thickened crust of the internal domain. The thickened crust of the latter domain suggests that the primary decoupling level was located within the mantle suggesting a likely mechanism of lithosphere scale accretion (Fig. 2.16).

The latter would most probably have occurred after complete subduction of the Liguro-piemontais Ocean. The subsequent collision of Apulia and Europe was responsible for uplift of the deep crust and upper

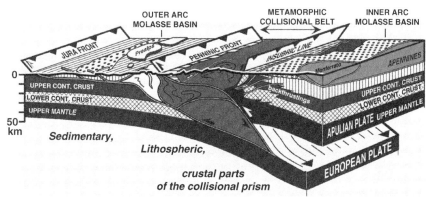

Figure 2.16 **Block diagram for the Alpine collisional prism, drawn from the ECORS–CROP seismic lines.** Simplified from Roure et al. 1996, cover page of the volume.

mantle of the Apulia promontory to form the Ivrea body. In this model, the foreland of the European plate was cut by a major lithosphere scale thrust that rooted eastward and effectively doubled the thickness of the lithosphere below the internal Alps. A reasonable speculation is that this shear was guided by the existence of the weak Valais Trough so that the thrust would have emerged along the locus of this former Trough and thus in the vicinity of the Penninic Frontal Thrust. A single headed 'crocodile' model has been invoked to depict the indentation by Apulia upper mantle (Fig. 2.12).

It is useful to compare and contrast the results from the Swiss and Austrian seismic lines with those discussed above. Although the Swiss NFP 20 line does not extend as far south as the southerly verging thrusts of the Southern Alps, both the ECORS–CROP and the NFP20 lines show a similar southward dip of the European Moho and crustal geometry of the External Crystalline Massifs. Similar reflectors outlining the base of the upper Pennine nappes are present. However, no evidence of imbricated mantle slices was found east of the Ivrea body. The ECORS–CROP profile therefore reveals a more complex crust mantle structure imaging two independent Mohos represented by the Ivrea body and a mid-crustal marker.

The recent 300-km long TRANSALP profile (Fig. 2.17) extends southward from the Molasse basin near Munich over the Eastern Alps to end on the Po plain near Venice. In common with the ECORS–CROP and NFP-20 sections, the European plate dips southward below the Northern Calcareous Alps and the Hohe Tauern Window. South of the Hohe Tauern Window, a giant bi-verging pattern of sub-parallel reflectors extends to the surface from the crustal root at 22 km depth. Northward dipping reflections beneath the Dolomites can be interpreted as a system of backthrust ramps.

Two alternate crustal models (Figs. 2.17A: 'crocodile model' and 2.17B: 'lateral extrusion model') have been proposed to explain the observed geometries though both share in common a major crustal shear dipping south for 80 km from the Inn valley to beneath the Hohe Tauern window. The 'crocodile' model (Fig. 2.17A), which focuses on the bi-vergent character, provides a better way of understanding the way present structure has evolved from the start of continental collision. In the Western Alps, a single-headed crocodile model has been invoked to explain the reflector character caused by the presumed indentation of the Apulia upper crust and mantle (Fig. 2.12).

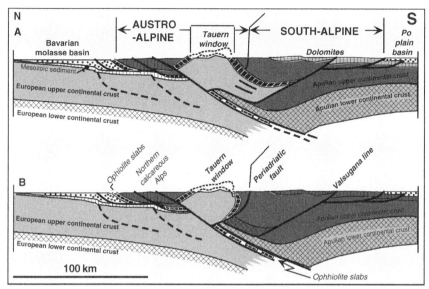

Figure 2.17 *The TRANSALP deep seismic line, from the Bavarian molasse basin to the Po plain: two proposed models.* (A) 'Crocodile model'. This model considers the prominent reflector to image Alpine brittle and ductile shear zones. In this model the edge of the Tauern window, supposedly representing the former European passive plate boundary, wedges deep into the Apulia plate splitting up the upper and lower crust at the mid-crustal brittle–ductile boundary thereby displacing the Dolomite block to the south. The upper crust of the Apulia plate was thrust as the Austro-alpine nappes over the Hohe Tauern Window. In this model, the seismic interfaces resemble a double-headed crocodile. In this respect it differs from that of the Western Alps (single crocodile model) and implies that the Peri-adriatic fault array does not directly correspond to the Insubric line. (B) The second model, 'lateral extrusion', describes the Eastern Alps in terms of an indentor model similar to that developed form the NFP20 lines in Switzerland (Fig. 2.12). Location: see Fig. 2.11. Line is close to section shown on Fig. 2.6A. *Redrawn and simplified from TRANSALP Working Group 2002, Copyright 2002 American Geophysical Union; Fig. 3 of p. 92.4 modified by permission of American Geophysical Union.*

FURTHER READINGS

Bigi et al. (1990); Bucher and Bousquet (2007); Frey et al. (1999); Goffé et al. (2004); Lardeaux et al. (2006); Lippitsch et al. (2003); Nicolas et al. (1996); Paul et al. (2001); Pfiffner et al. (1997); Roure et al. (1996); Schmid et al. (2000); Schmid et al. (2004); Thouvenot et al. (2007); TRANSALP Working Group (2002).

CHAPTER THREE

On the Origin of the Alps: The Vanished Oceans

Contents

Summary

Ophiolites in the nappe pile of the Alps originate from a now-vanished oceanic basin, called the Liguro–piemontais Ocean.

The Western Alps, from Rift to Passive Margin to Orogenic Belt, Volume 14
ISSN 0928-2025, DOI 10.1016/S0928-2025(11)14003-1
© 2011 Elsevier B.V.
All rights reserved.

The opening of the Liguro-piemontais Ocean (or Ligurian Tethys) was one of a series of multiple events, within the east–west Tethys Ocean *sensu lato*, which resulted in the separation of Gondwana from Eurasia between the Palaeozoic and the Tertiary. The opening of the Tethys cut across the Pangaea supercontinent assembled from all the continental plates at the end of the Palaeozoic. Most of the Hercynian and Alpine fold belts were derived from this ocean and its margins. The Liguro-piemontais segment of the Tethys existed for around only 100 Ma of the total, 350 Ma, duration of the Tethys. Its opening, like that of the Central Atlantic at the end of the Lower Jurassic, led to the separation of Gondwana from the then single Eurasia–North America plate. The Liguro-piemontais Ocean separated the European and Iberian margins to the north and northwest from the African or more precisely, the Apulo-african margin to the south and east. Apulia (or Adria) was then a microcontinent, satellite or northern spur of Africa represented today by the basement that underlies the Adriatic and Po basins as well as part of the Italian peninsula.

1. NAPPE STRATIGRAPHY AND THE EXISTENCE OF THE PRE-ALPINE TETHYAN OCEANIC DOMAIN

In the Alps, the precise stacking order of the nappes reflects their original disposition (Figs. 2.4–2.6). The lower and external group of nappes is of European origin. In the Central Alps, this lower group comprises three subgroups: the lowest derives from the European continental margin *sensu stricto*, the middle subgroup originates from the Valais domain while the upper subgroup is the Grand-Saint-Bernard–Monte Rosa ('SBR') Block. The SBR block consisted mainly of pre-Mesozoic continental crust with a thin Mesozoic sedimentary cover. Before Alpine thrusting, the SBR block separated the small Valais basin from the Ligurian Tethys.

The middle group is derived from the Liguro-piemontais Ocean and consists of ophiolites covered by pelagic Jurassic and Cretaceous sediments ('Schistes Lustrés'). As noted earlier, the Liguro-piemontais Ocean was a segment of the Mesozoic Tethys formed at the same time as the Central North Atlantic during the breakup of Pangaea (Fig. 3.1).

The upper and internal group of Apulia-african origin corresponds to Austroalpine and Southalpine units. These units (Figs. 2.2–2.5) comprise most of the Eastern and Central Alps of Austria, and Northern Italy (about half the area of the Alps). They thus represent a large fragment of Africa, or

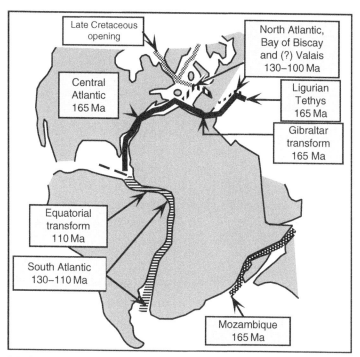

Figure 3.1 *Successive oceanic spreading phases cutting through Pangaea since the start of the Jurassic.* Opening of the Central Atlantic Ocean crust at about 165 Ma (end of the Middle Jurassic) is recorded by the oldest observed oceanic magnetic anomalies. At this time, the Africa–South America plate moved ESE (actual direction) along the Newfoundland–Gibraltar transform zone. The displacement led to the opening of the Liguro-piemontais Ocean (or Ligurian Tethys). Later, the South Atlantic opened during the Early Cretaceous. During the Early Cretaceous, the opening of the North Atlantic between Iberia and Europe as well as the Bay of Biscay led to the separation of Iberia which became a relatively independent microplate. Iberia accompanied the movement of Africa to the SE almost up to the time of opening of the small Valais Ocean, according to present-day classical models. *Modified from Lemoine et al. 2000, Fig. 6.1, p. 76.*

more precisely Apulia, that has been thrust onto both the 'SBR' block and the southern edge of the European continental plate.

In contrast, most of the outcrop of the Western and Central Alps is composed of units of both European continental crust and Tethyan oceanic origin.

Thus the high peaks of the Eiger, Mont Blanc, Monte Rosa and the Barre des Ecrins are of European origin, while the Monviso, Piz Platta and Gross Glockner represent the remnants of the oceanic crust of the former

Ligurian ocean. The exception, in the Central Alps, is the Matterhorn (Cervin) which is a thrust fragment of Africa or, more precisely, Apulia (see Fig. 16.9).

2. ORIGINS OF THE TETHYS

The Liguro-piemontais Ocean, which opened in the Jurassic, is a small part of the much larger Tethys, whose history was long and complex. The Tethys, distant ancestor of the Mediterranean, was an east–west ocean that separated Gondwana and Eurasia from Palaeozoic times. Its evolution is marked by a series of opening phases followed by closures which created successive folded belts. Multiple phases of crustal accretion of fold belts increased the surface area of Eurasia towards the south.

2.1 Palaeotethys and the Hercynian orogeny

The Palaeozoic Tethys (Fig. 3.2) opened progressively during the Ordovician from Australia to Western Europe across the northern edge of Gondwana. The opening led to the detachment of an array of continental microplates, which later amalgamated with Eurasia during the Devono-Carboniferous Variscan orogeny. These microcontinents now constitute a large part of the Hercynian basement of Europe from Iberia to the Pontides and even further east. Later subduction and resorption of the Palaeotethys oceanic remnants beneath the Eurasian craton led to magmatic and volcanic activity typical of the fragments of the Variscan chain ultimately incorporated into the Alpine domain. The majority of Hercynian magmas, including granodioritic intrusions, prove partial fusion at depth in the remains of Pre-Cambrian continental crust. The Hercynian fold belt differs from those fold belts associated with arc-continent or continent–ocean collision and superposed on long-lived subduction zones like the Andean belt in that the contribution of juvenile magma of mantle origin was particularly minor or modest.

The Palaeotethys is reputed to have begun progressive west to east closure from North Africa at the end of the Carboniferous (post-Namurian). This was equally the case in the Pyrenees where the thrust sheets were progressively emplaced from the Namurian to the Westphalian in an east to west sense from the Eastern Pyrenees to the Basque country. In Iran, closure only took place during the Triassic as shown by the age of early folding, palaeomagnetic and palaeobotanical data. At this time, Pangaea was consolidated as a supercontinent.

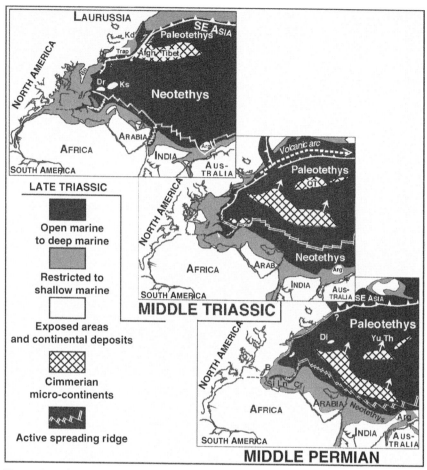

Figure 3.2 *The origins of the Tethys.* The successive phases shown here are (A) opening of the Neotethys Ocean on the northern border of Gondwana, (B) progressive subduction of the remains of Palaeotethys, (C) northward migration of the group of 'Cimmerian' microcontinents, which generated a volcanic arc of Triassic age, then (D) their incorporation into the southern margin of Laurussia during the Eocimmerian (or Indosinian) orogeny. The area affected by the Cimmerian folding is shown in Fig. 4.1 and its northern boundary is named as the Scythian front in Fig. 2.1. Afgh: Afghanistan; Arg: Argo; Bo: Bosnia; Cr: Crete; Di: Dizi; Dr: Drama; Kd: Kopet Dagh; Ks: Kirsehir (Asia Minor); L: Lagonegro Trough; Si: Sicanian Basin; QT: Quanc-Tang in China; Yu: Yunan; Th: Thailand. *Simplified after Marcoux et al., in Dercourt, Ricou and Vrielinck, Eds., 1993* (Baud et al., 1993).

2.2 The Neotethys and the Eocimmerian or Indosinian orogeny

The succeeding Neotethys (Fig. 3.2) phase of Tethyan evolution was marked by renewed opening from east to west that again cut across the northern edge of Gondwana. The onset of opening is variously dated from the Permian age (265 Ma) of MORB pillow basalts in Oman and the overlying radiolarites as well as the basal transgression across the rift shoulders on the Arabian plate.

The opening led to the detachment of the array of 'Cimmerian' microcontinents such as Apulia (Italy), southern Asia Minor (Kirsehir), central Iran (Lut block), central Afghanistan and southern Tibet. The broad topographically complex Tethys Ocean widened eastward between the north margin of Gondwana (i.e. present day North Africa, Libya, Arabia) and southern Laurussia (Fig. 3.2). As shown on available reconstructions, it consisted of two main plates:

- A northern composite plate comprising (a) the remains of Palaeotethys oceanic crust being subducted to the north beneath eastern Eurasia, (b) the 'Cimmerian' microcontinent(s) separating Palaeotethys from Neotethys, and (c) new oceanic crust in the South (the northern half of Neotethys) including an active, southward spreading ridge.
- A southern plate, comprising the southern part of Neotethys oceanic crust attached to Gondwana continental crust. At this stage, the western part of this southern plate was contiguous with the Laurussia plate then consisting of Eurasia and North America (Fig. 3.2).

Spreading of Neotethys to the south, combined with subduction beneath Eurasia to the north, led to the progressive migration, then collision, of the 'Cimmerian' group of microcontinents against southern Eurasia in Late Triassic to Liassic times. This orogeny is known as the Eocimmerian in Romania (Dobrodgea), the Crimea and the Pontide Alps (northern Asia Minor; Fig. 2.1, Scythian thrust front and Fig. 4.1) and further east as the Indosinian. This new fold belt thus separated the Variscan orogenic domain to the north from the area unaffected by the Hercynian orogeny to the south and east (Fig. 3.2).

2.3 The aborted Permo-Triassic rift phase in the Western Mediterranean area: relationship with the Neotethys

During the Permo-Triassic, a network of active extensional faults defined a multidirectional rift system of complex geometry. The rifts appear as localized sedimentary basins exhibiting contrasting differential subsidence (Fig. 3.3). In the Alpine domain, the rifting was punctuated by volcanic activity in many places, notably the Hellenides, Dinarides, Carnic and Julienne Alps, Dolomites and the Austroalpine domain as well as in the Taurides and Syria. Rifting of this age was equally prevalent on the European side of the future Ligurian Tethys, in particular, in the Pyrenees and

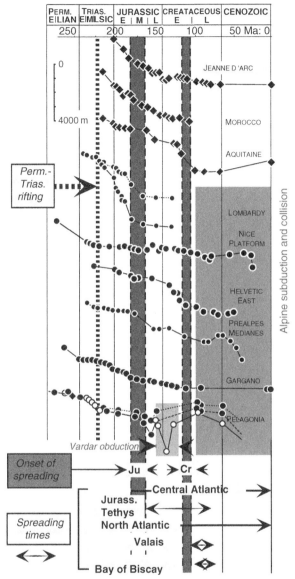

Figure 3.3 *Synthetic subsidence curves for selected circum-Iberia domains.* Subsidence has been continuous but not uniform in the Jeanne d'Arc basin and Morocco. The subsidence curve for the Aquitaine basin shows two phases of rapid subsidence. That of the Jurassic records the opening of the Central Atlantic while the more rapid Early Cretaceous subsidence is linked to the opening of the immediately adjacent Bay of Biscay. The reversal of subsidence in the Cenozoic records the Pyrenean orogeny. The Dauphinois–Vocontian domain shows a comparable subsidence curve. The Nice area is located between the Ligurian Tethys Ocean and the Vocontian 'rift'. The Liassic and Cretaceous subsidence phases record Tethyan rifting and that in the Valais–Vocontian corridor, respectively. The oddity is the reversal of subsidence in the Late Jurassic. *From Stampfli 2001, Fig. 7, p. 17.*

the Corsica–Sardinia block. Volcanism ceased almost everywhere by the Late Triassic.

The rifting affected a large part of the southern and western borders of the European craton and part of the northern edge of the African craton.

Nonetheless, the palaeogeography of the western Tethys shows little difference between the Middle Permian and Middle Triassic (Fig. 3.2). The Permo-Triassic rift system thus does not seem to mark a significant westward propagation of oceanic spreading. In contrast, the extensional fault patterns of the Late Triassic truly herald the subsequent pattern of Jurassic extension (Fig. 3.4). On a wider scale, the main fault trends also herald those of future rifts that became precursors to subsequent oceanic opening, mainly in the future North Atlantic and Bay of Biscay.

While the Permo-Triassic extensional event thus represents the first tentative dislocation of Pangaea, it should be regarded as an aborted phase of rifting in the Western Alpine area. Not until the Late Triassic and early Jurassic would renewed rifting result in the middle Jurassic opening of the Tethys and Central North Atlantic, and, later, the North Atlantic in the Early Cretaceous (Fig. 3.4).

However, the Triassic of the Sub-Pelagonian and Pindos—Olonos units of the Hellenides consists of pelagic sediments. Though not proven, these sediments may possibly have rested on oceanic crust that is now subducted or completely concealed in the overthrust belt. We have therefore not retained the hypothesis of the Lausanne School of Geology, which proposes a complex, multibranched spreading system called the Meliatta-Hallstatt-Maliac Ocean cutting the South European continental margin and the area between Iberia and North Africa.

2.4 An unresolved problem: the development of shoshonites in the Dolomites

The period of major transgression, rapid subsidence and accelerated extension of Ladinian age is contemporaneous with shoshonitic magmatism in the Dolomites at this time. Various authors have interpreted the shoshonitic magmatism, which belongs to the calc-alkaline field, as typical of back-arc basin settings. In this case, the back-arc basin would be associated with a westward plunging subduction zone to the east of Northern Italy in Yugoslavia or Hungary (Carlo Doglioni, pers. comm., August 2002). If such an interpretation is adopted, the extensional tectonism of Middle Triassic age in the Dolomites must be clearly separate from that caused by the general network of extensional faults which affected most of Europe at this time. It would then be the westernmost expression of the Eocimmerian orogeny which closed the Palaeotethys to the east.

However, an interpretation which proposes the notion of a subduction zone in the Southern Alps is not compatible with rift-related magmatism, even in aborted rifts like that of the Dolomites (see Chapter 4). It is also true that there is no expression of the Eocimmerian orogeny in the classic sense further west than the Dobrodgea of Romania. Western Europe, from north to south, clearly appears to have been subject to extension during Triassic times.

2.5 The Jurassic Tethys

A major palaeogeographic and tectonic event which began towards the end of the Triassic marks the separation of Gondwana from Europe–North America, then still a single plate (Fig. 3.2). The ensuing westward extension of the Tethys Ocean, which previously did not pass west of the Triassic domain, resulted in this new rupture, which would commence spreading from the middle Jurassic (Fig. 3.4; Callovian).

The separation resulted from the propagation and then the conjunction of two rift systems (Fig. 3.4; Toarcian to Tithonian). One rift system marked the future Central Atlantic and initiated the separation of North America and northwestern Africa. The second propagated from the northern branch of Neotethys to create the Ligurian Tethys Ocean between Southern Europe and North Africa. The two rifts and their later spreading axes were linked by the Newfoundland–Gibraltar transform fault system (Fig. 3.4; Toarcian).

The various segments of the Jurassic Tethys differ in their subsequent evolution as a result of successive changes in the movement of Africa with respect to Europe (Fig. 3.5; see Wortmann et al., 2001). In the Tethyan part of the Central Atlantic (Figs. 3.1 and 3.3), spreading has continued to the present day. The prolongation of the Newfoundland–Gibraltar plate boundary south of Iberia remained in general transtension until during the Miocene. During the Early Tertiary, inversion along the boundary resulted in the Betic Cordillera of Southern Spain and the Rif and Tellian fold belts of North Africa. However, the Liguro-piemontais Ocean began to close by the end of the Cretaceous thereby initiating the formation of the Alps. The

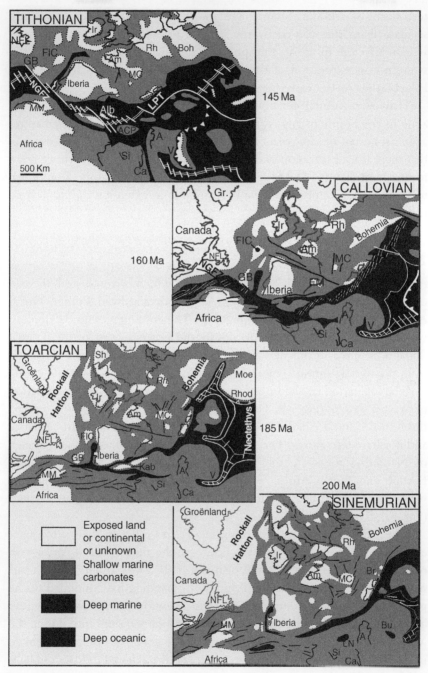

Figure 3.4 *(Continued)*

Tethyan Jurassic Ocean remaining in the confines of the Neotethys began to close from the end of the Jurassic to form the Dinaric–Hellenide fold belt (Fig. 3.3; Vardar) and the western Carpathian virgation.

3. THE DEVELOPMENT OF THE TETHYS IN THE CRETACEOUS

3.1 Overview

From the beginning of the Cretaceous, the main role of the Tethys in successive extensional events in the fragmentation of Pangaea came to an end. This role was from then on played by the Atlantic (Fig. 3.1). Nonetheless, a new system of rifts forming two trends cut across the north margin of the Tethys from the end of the Jurassic. The westernmost rift cut across the future Bay of Biscay, which would later open by spreading between Iberia and Europe between the end of the Aptian and the Campanian (Fig. 3.1). The second, eastern rift defined the Valais zone and its continuation to the southwest in the Helvetic and Dauphinois domains, and notably the Vocontian basin of the Sub-alpine chains. According to traditionally accepted interpretations, the Valais rift would have been transformed into a narrow and ephemeral oceanic basin at the start of the Cretaceous then closed by subduction by the end of the Cretaceous. But if the ophiolites intercalated in the Valais units are truly Carboniferous in age as documented by Masson (2002); Masson et al. (2008), it must place in question the existence of a branch of the Tethys Ocean cutting the south part of the European craton during Cretaceous times. However, recently obtained middle Jurassic dates from Valais ophiolites (Manatschal et al. 2004–2006) suggest an alternative possibility that both the Ligurian and Valais branches of Tethys might have opened simultaneously during middle Jurassic times (for discussion, see Chapter 12). This is the subject of current research which is under review without, as yet, any consensus.

Figure 3.4 *Development of the Jurassic Tethys of the Central Atlantic from the Neotethys between Western Europe, North Africa and North America.* A: Adria; ACP: Apennine carbonate platform; Alb: Alboran; Am: Brittany; Br: Brianconnais; Bu: Budva; Ca: Calabria; FIC: Flemish Cap; GB: Grand Banks; Ir: Irish massif; K, Kab: Kabylies; LN: Lago Negro; MC: Massif Central; MM: Moroccan Meseta; NFL: Newfoundland; NGFZ: Newfoundland Gibraltar Fracture Zone; Rh: Renish massif; Sh: Shetland Platform; Si: Sicily; V: Vardar. *Simplified after Dercourt, Ricou and Vrielink 1993 and Dercourt et al. 2000.*

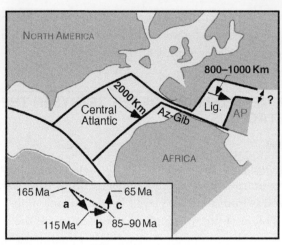

Figure 3.5 *Sketch showing the difficulty in reconstructing the pre-Jurassic.* The three stages of displacement of Africa with respect to Europe between the Middle Jurassic and the present are schematically presented here (commentary: Box 3.1). **a:** Displacement towards the SE between 165 and 115–100 Ma, around 2000 km at the level of Gibraltar and 800–1000 km in the Ligurian Tethys. **b:** Displacement (poorly constrained) between 115–100 Ma and 85 Ma probably about 500 km towards the East. **c:** Displacement towards the North of about 600–700 km between 85 and 65 Ma, date of the onset of collision in the Central and Western Alps. Direction and extent of the displacement of Africa with respect to Europe, trajectories **a, b** and **c,** determined after *Dewey et al. (1989), Olivet (1996) and Olivet oral communication (1999) and Patriat et al. (1982). Modified from Lemoine and others 2000, Fig. 3.5, p. 81.*

3.2 The traditional view: two branches of the Tethys Ocean: the Liguro-piemontais branch and the Valais appendix

Two oceanic domains are classically represented in the Central Alps compared to the Western Alps. Figures 2.5–2.7 show that in the Central Alps, the units comprising the European basement can be divided into three superposed groups:

- To the north, the lower group consists of the granite-gneisses forming the basement of the external crystalline massifs of the Aiguilles Rouges, Mont Blanc, Aar and Gotthard.
- The middle units consist of variably metamorphosed sediments, i.e. the *Schistes Lustrés* of mainly Late Cretaceous age and cover of Late Cretaceous to Early Eocene flysch. These sediments are associated with rare, small ophiolite bodies (basalts, gabbros, serpentinites) as well as with the continental basement of the Simplon–Ticino massifs that constitute the external Valais units.

– In the south, the upper group comprises basement nappes of the Grand-Saint-Bernard and Monte Rosa that are overthrust in turn by oceanic units of Liguro-piemontais origin.

In contrast, ophiolites related to the Valais are *absent* in the Western Alps southwest of the Petit Saint-Bernard Pass. In this area, there is no tectonic suture to demonstrate the former presence of the Cretaceous Valais Ocean if it existed.

3.3 The intermediate SBR continental block (Grand-Saint-Bernard–Monte Rosa)

While the Liguro-piemontais Ocean was a true branch of the Tethys, the Valais Ocean is classically said to intersect the southern part of the European margin *sensu lato*. These two oceanic domains were separated by a rather large continental block (c. several hundred kilometres in width) called here the SBR block. This block corresponds to the basement of the Grand-Saint-Bernard complex comprising the Pontis, Siviez-Mischabel and Mont Fort units (Fig. 2.7), plus that of the Monte Rosa. In a palaeogeographic context, the SBR block also includes the Subbrianconnais and Brianconnais domains as well as the Piemontais in the sense of the usage adopted here.

3.4 Difference in age of the Valais and Liguro-piemontais basins

The Liguro-piemontais oceanic domain and the Central North Atlantic form part of the same spreading system initiated at the end of the middle Jurassic (about 165 Ma) after a period of rifting between the Late Triassic or earliest Jurassic and early middle Jurassic (200–220 to 165 Ma). During rifting, subsiding basins developed at a scale of about a hundred kilometres. One example is the Helvetic–Dauphine basin, which would later be part of the European continental margin of the Ligurian Tethys.

The Valais domain is considered classically as belonging to the Biscay–Valais–Grisons branch of an intra-continental rift system of Late Jurassic–Early Cretaceous age (see Fig. 7.10). The only sectors of this rift that developed true spreading were the Bay of Biscay to the west and putatively the Valais domain to the east.

In the Central Alps, the Valais 'oceanic' domain extends from Versoyen, near the Petit Saint-Bernard Pass in the west as far as Graubünden. It continues into the Eastern Alps where it probably joined the Liguro-piemontais Ocean as shown by the disappearance of the SBR block at outcrop.

3.5 Did the Valais branch of the Liguro-piemontais Ocean really exist?

Recent observations have put in doubt the existence of the Valais Ocean in the Cretaceous at least in the north part of the Western Alps and in the transition to the Central Alps. These observations were made in the Versoyen area near the Petit Saint-Bernard Pass. There, gabbros previously assigned to the Valais have been dated as Early Carboniferous. In addition, the same ophiolites form the normal basement to the Triassic and Jurassic succession of the Petit Saint-Bernard nappe. Consequently, the corresponding ophiolitic suture must be assigned to the evolution of the Hercynian basement of the Alps and not to the Mesozoic evolution of the Alpine Tethys.

The development of this system of Early Cretaceous rifts took place during the post-rift phase of the formation of the Liguro-piemontais Ocean *sensu stricto*. But this point of view is still controversial (see discussion in Chapter 12; Fig. 12.8). A recent – but not traditional – point of view is to consider that the oceanic opening was simultaneous in both Ligurian and Valais branches of the Tethys (see discussion in Chapters 7 and 12). This view is based in part on modelling (Manatschal et al. 2004–2006).

These conclusions contradict traditional palaeogeographic reconstructions and are too new for a consensus to emerge. Future work will be enabled notably by new chronological studies on the different ophiolite bodies.

Several models, which accept the existence of the Valais Ocean, have been proposed for the westward continuation of the supposed Valais Ocean (Fig. 12.8). One model suggests that the Valais Ocean continued longitudinally between Provence and Corsica, an area now under the Mediterranean and thus unknown, and finally via the Pyrenees and Aquitaine Basin into the Bay of Biscay. In the model adopted in this book, which assumes that the Valais is a rift zone, it crosses a transform fault (Chapter 12) and is represented in the Southern Sub-alpine and Provencal chains as an intracontinental rift of Late Jurassic–Early Cretaceous age (see Chapter 7) that has not evolved to spreading (Fig. 7.10).

The short life of the Valais Ocean, if it existed at all, during Jurassic (?) and Cretaceous times suggests it was probably very narrow and possibly discontinuous. If it was 'oceanic', the Valais rift and ocean may have been comparable to the present northern Red Sea. There, the narrow and discontinuous spreading centres terminate at the Dead Sea transform and do not continue into the wholly intra-continental Gulf of Suez rift.

In summary, in the Western and Central Alps from the external to the internal parts of the fold belt, the following elements can be recognized according to the traditional views (Fig. 3.6):

* the European margin *sensu stricto*,
* the narrow and discontinuous Valais 'oceanic basin' that is unknown in the southern part of the Western Alps,

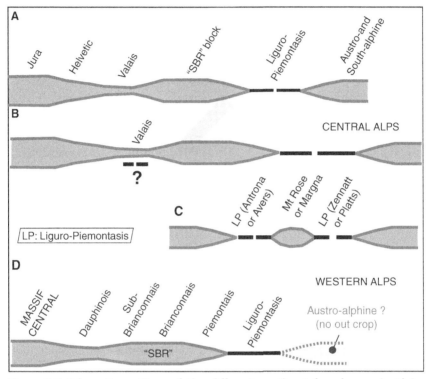

Figure 3.6 *Schematic sections of the different continental and oceanic Alpine palaeogeographic domains across different transects.* Section A shows the possible disposition of the palaeogeographic domains of the Central Alps in early Late Jurassic times. This sketch section implies that the Valais post-dates the Liguro-piemontais continental breakup. Section B shows the dispositions in the Cretaceous, for the case where the Valais Ocean is truly open, a view doubted by Henri Masson *et al.* (2002–2008). Section C is a variant illustrating the possibility of 'microcontinents' in the form of slivers if present. For the Monte Rosa, this hypothesis has not been used in this work. Section D concerns the Western Alps where neither the Valais oceanic domain nor the Austroalpine nappes outcrop. If the Austroalpine nappes existed on this transect, they can only have been eroded totally and/or buried under recent sediments. It is probably shown by the presence of the Ivrea zone on one hand (Fig. 2.12B), and by geophysics on the other, i.e. by a double Moho beneath the Po plain. *Modified from Lemoine and others 2000, Fig. 6.3, p. 77.*

* the several hundred kilometre wide 'SBR' (Saint-Bernard–Monte-Rosa) continental block,
* the Liguro-piemontais Ocean, a true segment of Tethys, and lastly
* only in the Central Alps, the Apulia-African margin.

3.6 Are other small narrow oceanic domains separated by small continental fragments?

On certain transects, the present disposition of the nappes (Figs. 2.6 and 2.7) seems to suggest that the European side of the Tethyan ocean might have been subdivided by blocks of continental crust (Fig. 3.6). This configuration may be comparable to the Southern Red Sea, which is separated from the Afar subaerial spreading centre by the Danakil continental fragment. In this manner, the Monte Rosa continental basement might have separated the ophiolites of Zermatt–Saas Fee and Antrona. However, many models are possible (Fig. 3.6, B and C; see also Chapter 12; Fig. 12.8). As noted above, it can be proposed that the Monte Rosa block (Fig. 2.6) was a microcontinent separating two oceanic areas (Fig. 3.6C). In a preferred model (Chapters 2 and 11), the ophiolites of Zermatt–Saas Fee and Antrona would then have been initially thrust on the future normal and inverted flanks of the recumbent fold nappe of Monte Rosa.

A comparable disposition exists in the Graubünden area, implying also the choice between several models. There, the continental basement of Margna lies in a comparable position between the Avers and Platta ophiolite units (Fig. 12.10).

Similarly, the Sesia zone (Fig. 2.12) is considered to be either European or African, or intermediate in origin. In the latter case, it may separate two oceanic domains, one of which remains between the Monte Rose and Sesia.

4. EVOLUTION OF THE LIGURO-PIEMONTAIS TETHYS: TETHYAN AND ALPINE PHASES

The Liguro-piemontais Ocean opened between the Apulo-africa continent and Europe–Iberia. To a first approximation, the movements of Apulia differed little from Africa (discussion: Box 3.1). The movement of Africa with respect to Europe can be reconstructed using Central North Atlantic magnetic anomalies. Between 165 and 115 Ma and probably up to 80–85 Ma, the relative motion was to the ESE and E (actual direction) but after 80–85 Ma, the subsequent motion was to the north.

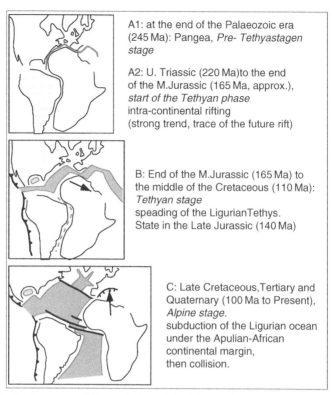

A1: at the end of the Palaeozoic era (245 Ma): Pangea, *Pre- Tethyastagen stage*

A2: U. Triassic (220 Ma)to the end of the M.Jurassic (165 Ma, approx.), *start of the Tethyan phase* intra-continental rifting (strong trend, trace of the future rift)

B: End of the M.Jurassic (165 Ma) to the middle of the Cretaceous (110 Ma): *Tethyan stage* speading of the LigurianTethys. State in the Late Jurassic (140 Ma)

C: Late Cretaceous,Tertiary and Quaternary (100 Ma to Present), *Alpine stage.* subduction of the Ligurian ocean under the Apulian-African continental margin, then collision.

Figure 3.7 *Main stages of the evolution of the Liguro-piemontais Tethys.* Refer to the Wilson cycle, Fig. 1.15. *Redrawn and completed from Lemoine et al. 2000, Fig. 6.4, p. 80.*

These quite different episodes of relative movement between Apulia (Africa) and Europe (Fig. 3.7) define the two major evolutionary phases of the Liguro–piemontais Ocean and the Alps.

Box 3.1 DISPLACEMENTS OF THE APULO-AFRICA BLOCK

SIZE OF THE LIGURO-PIEMONTAIS OCEAN: A DOUBLE UNKNOWN

Apulia, and not strictly Africa, comprised the southern margin of the Liguro-piemontais Ocean. While the movement of Africa can be reconstructed using Atlantic magnetic anomalies, the movement of Apulia cannot be reconstructed because subduction of the Ligurian Tethys has destroyed the magnetic anomaly evidence. The oceanic crust, located in the Eastern Mediterranean between the Cretan arc and North Africa (Libya, Egypt), is thought to represent the remains of the Neotethys (Fig. 2.1). In fact, any reconstruction must therefore make the simplifying assumption that Apulia was a simple northern promontory of Africa and not an independent microplate.

A key question is therefore whether the Apulia microcontinent moved independently of Africa and especially whether it was subjected to rotation. However, serious difficulties are encountered if Apulia was joined to Africa during the Mesozoic. In effect, between the onset of spreading (165 Ma) and collision (at the latest 65 Ma), the trajectory of North Africa with respect to North America has to be decomposed into three directions (Fig. 3.5, trajectories **a, b** and **c**).

The first trajectory **a,** between 165 and 110 Ma, is well known from Atlantic magnetic anomalies and represents an ESE displacement of about 1500 km at the latitude of Gibraltar. The second, **b,** is less well known (though fracture zones can be mapped from Seasat images) because there are no magnetic anomalies during the Cretaceous quiet zone (Chron M0-115 Ma to Chron 33–85 Ma) that would allow precise reconstruction. However, this interval probably represents a WNW–ESE displacement of about 500 km (J.L. Olivet, pers. comm.). By the Late Cretaceous (c. 85 Ma), the total width of the Atlantic Ocean at the latitude of Gibraltar was probably about 2000 km. The position of the northern Ligurian ocean (site of the future Alps), roughly halfway between North Africa and the Europe–Africa rotation pole located in Southwest Finland (Fig. 6.16), suggests that its ENE–WSW width was between 800 and 1000 km at most, and certainly so if Apulia was part of Africa.

The third displacement, **c,** between 85 and 65 Ma (start of the collision phase in the Western Alps) corresponds to a new movement of Africa and Apulia to the north of some 400 km only.

Whatever the reconstruction, this north–south displacement is far from sufficient to close the Liguro-piemontais Ocean which had a NNW–SSE width of between 800 and 1000 km.

The two following conclusions are therefore inescapable. First, Apulia and Africa, known to be displaced in the same sense with respect to Europe, were separated at the start of or during spreading. Second, the magnitude of the displacement of Apulia was necessarily less than that of Africa indicating that the Liguro-piemontais Ocean was quite narrow, perhaps only a few hundred kilometres in width. The narrow dimension of the small palaeo-ocean implies very slow spreading rates of less than 5 mm/yr which are in reasonable agreement with the rates inferred from the Alpine and Apennine ophiolites (see Chapter 11).

4.1 Initiation of the Atlantic Tethyan phase began with the rupture of Western Pangaea at about 165 Ma

The first Late Triassic to Middle Jurassic rift and subsequent ocean that cut across Western Pangaea (Figs. 3.1–3.2, 3.4) can be divided into two sectors. In the Central Atlantic and the Liguro-piemontais Tethys, simple spreading resulted in divergence of Europe–North America and Africa (Apulia). However, other sectors were affected by strike-slip displacement along continent-to-ocean transform faults that linked the Liguro-piemontais and Central North Atlantic Oceans during spreading. The Newfoundland–Azores–Gibraltar line was one such transform. To the east, a similar transform may have linked the Grisons(= Graubünden)–Tauern oceanic segment (Fig. 3.5; also see Figs. 12.8 and 12.10). The dimensions of this segment are unknown, though it was situated along the future eastern part of the Central Alps, Eastern Alps and Southern Carpathians.

4.2 Onset of the Alpine phase began between 115 and 80 Ma

The transition to the Alpine phase began in the Late Cretaceous though it is not possible to be more precise (Box 3.1). During this interval, Africa changed direction to converge northward with Europe (Fig. 3.5) as a result of the progressive opening of the South Atlantic. This motion resulted in shortening of the Liguro-piemontais Ocean by subduction prior to continental collision.

4.3 Two important dates

The first continental scale rifting of Western Pangaea took place in Late Triassic times leading to complete rupture and spreading, by about the end of the middle Jurassic (165 Ma Bathonian–Callovian), of the Central Atlantic and Liguro-piemontais Oceans. The second key event took place sometime between 115 and 80 Ma (Box 3.1) and marks the onset of subduction of the Liguro-piemontais Ocean which rapidly led to the collision of Europe and Apulo-Africa.

5. ALPINE PALAEOGEOGRAPHIC DOMAINS IN THE CONTEXT OF THE LIGURO-PIEMONTAIS OCEAN AND CONTINENTAL MARGIN

5.1 Limits to palaeogeographic reconstructions: the negative effects of subduction

Much palaeogeographic information has been irretrievably lost for two main reasons. Obviously, almost all the oceanic crust and some continental margin units have been lost by subduction. In addition, important parts of the initial palaeogeographic domains have been deeply eroded and the eroded products are now represented by the sediments infilling the flysch and molasse basins.

5.2 Structural zones and palaeogeographic domains: the heritage of the Tethyan phase

In Chapter 2, the Central and Western Alps were described in terms of structural zones or units. These zones were derived from different palaeogeographic domains differentiated both during the Tethyan spreading phase and more especially during the course of rifting between 210 and 165 Ma.

This nomenclature, which dates back to the first quarter of the 20th century, has nearly always given the same names to palaeogeographic domains and to structural zones. As a result of this semi-general rule, each major zone has been characterized by a particular type of stratigraphy which, though variable, is generally different from neighbouring zones. Each of the zones has been independently described in terms of its own palaeogeographic evolution. The major structural limits and thrusts were in fact located, prior to the collision, on palaeogeographic boundaries. It will be later shown (Chapters 12 and 16) that these types of boundaries coincide mostly with Tethyan rift phase faults or normal fault arrays, which partition the margin with each defining different sedimentary basins characterized by a unique rheological stratigraphy. These differences in rheological stratigraphy between the basement and sediments and within the sediments comprising the cover strongly influenced the development of Alpine thrusts.

6. THE ALPINE PHASE, LATE CRETACEOUS AND TERTIARY

From the middle of Cretaceous (c. 100 Ma), the palaeogeography began to change and was then totally disrupted by subduction and the Alpine collision. Oceanic subduction took place first beneath the Apulia margin. Nappes derived from Apulia continental basement and the overlying sediments were thrust over the European margin. Starting in the most internal zones of Apulo-Africa, thrusts and folds, which developed on the Apulia-African margin, defined a subducting active margin that consumed the Liguro-piemontais Ocean and progressively encroached across the external zones of the European margin (see Chapter 13; Fig. 13.5).

Henceforth, for each phase in the evolution of the Alpine chain the following main domains can be recognized from north to south in the Central Alps and from west to east in the Western Alps:
- An external domain, without further deformation in this phase.
- A foredeep basin, or flexural basin, infilled by clastic sediments composed of turbidites, flysch or continental molasse and very shallow marine deposits in the outer of these basins in front of the Alps or between the Alps and the Jura.
- An internal domain, already deformed, uplifted and undergoing erosion. Erosion of the Alps during contemporaneous deformation produced shales, sands and conglomerates that accumulated in successive foredeep basins as well as in the Po basin behind the fold belt. This palaeogeography was progressively displaced outward across the external zones in three main phases.

The first deep basins were filled with Late Cretaceous flysch between 110 and 65 Ma, represented in one area by the Liguro-piemontais flysch, notably the Helminthoid flysch, and in the other by the Valais flysch (Fig. 7.3). These basins are more or less superposed and can be related to the two subduction phases that correspond to the disappearance of the Valais palaeo-geographic domain and Liguro-piemontais Oceans. Subsequently, flexural foredeep basins, developed initially on the Brianconnais and Subbriancon-nais zones and later on the external Helvetic–Dauphinois zone, were infilled by Tertiary flysch of Eocene and Early Oligocene age (45–30 Ma; Figs. 14.2–14.4). Finally, at 35 Ma, a subsiding, very shallow marine or fluvial basin formed in front of the fold belt to be infilled by Oligocene and Neogene molasse.

FURTHER READINGS

Keppie (1994); Stampfli and Borel (2002).

CHAPTER FOUR

Hercynian Inheritance, Tethyan Rifting and Alpine Nappes

Contents

Summary

The pre-Mesozoic basement composition of more than half the 1000-km long outcrop of the Alps from Italy to Austria implies that the Alpine fold belt is partly composed of remobilized Hercynian and older structures that record pre-Alpine orogenic cycles. Therefore, the structural history of the Tethys and the later Alps cannot be completely understood without knowledge of earlier deformation and structure. Examination of basement types and structures involved in Alpine folding allows analysis of how the earlier basement structure has, or has not, controlled the rift fabric and shaped the continent–ocean boundary and later the thrusts separating the main Alpine units.

This chapter will show that the position of the Ligurian and Valais lines of rupture was already predetermined by a major discontinuity between pre-Mesozoic basement blocks that are now adjacent. This was also the case for the boundaries of the nappes constituting the assemblage of the 'Zone Houillere'

The Western Alps, from Rift to Passive Margin to Orogenic Belt, Volume 14
ISSN 0928-2025, DOI 10.1016/S0928-2025(11)14004-3
© 2011 Elsevier B.V.
All rights reserved.

(Coal zone) and the Saint Bernard–Monte Rosa (SBR) block, which are respectively characterized by distinctly different Paleozoic successions. The relative permanence and role of specific structural discontinuities can therefore be inferred as well as from the frequent coincidence of Alpine nappe boundaries with those defining tilted fault blocks created by Tethyan rifting.

1. THE ALPINE FOLDBELT INCORPORATES PART OF THE HERCYNIAN FOLDBELT

The geographic area affected by Alpine deformation cuts partly and obliquely across the region affected by Hercynian deformation (Fig. 4.1).

The northern limit of the Hercynian deformation is situated largely to the north of the North Alpine Front.

The southern limit of the Hercynian deformation crosses Morocco. Further to the east, it disappears under the southern front of the Alpine fold belts of North Africa. In southern Europe, the South Hercynian front cannot be located precisely in the interior of the Alpine domain. Nonetheless,

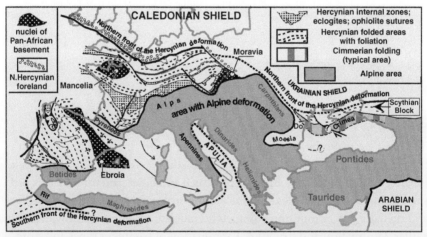

Figure 4.1 *Superposition of the folded Alpine domain over the southern part of the Hercynian domain.* The distribution of Hercynian structures shows the existence of blocks or nuclei derived from Pan-African basement whose presence is partly responsible for the pattern of Hercynian virgation. The best interpretation of these structures requires that Corsica, Sardinia and Calabria are restored to their pre-Miocene positions prior to their displacement to the SW during the oceanic opening of the Western Mediterranean (arrows). Alpine deformation in Europe cuts only a part of the southern branch of the fan described by Hercynian structures. *Inspired and simplified after Matte (1986, 1998); Autran et al. (1995).*

remnants of the Hercynian fold belt are known in Italy in the Apuane Alps, in the Balkans, the internal (Tatras) and external zones of the Carpathians, the basement of the Moesian platform (region of Bucharest), Pelagonia and the Rhodope massif in Greece, the Pontides along the southern side of the Black Sea and probably in the Karakaya massif in Asia Minor. Major Mesozoic and Tertiary structures intersect Hercynian structural trends of which the best large-scale examples are the Rif and Betic cordilleras. In contrast, the location of the Mesozoic rift of the Bay of Biscay and the Tertiary structure of the Pyrenees was controlled by a system of transcurrent faults of Carboniferous age. Apart from these obvious large-scale regional trends, the question and problem of the detailed relationship between Hercynian, Tethyan and Alpine lineaments cannot be oversimplified. In this chapter, different relationships will be illustrated using several case studies. However, it must be emphasized that the explanations offered below are by no means a complete overview of a complex and incompletely understood subject.

1.1 Major common phases in the pre-Mesozoic history of the Alpine basement and foreland

A key element of the pre-Mesozoic history of the Alpine basement is the succession of several orogenic cycles from late Precambrian time. The Variscan or Hercynian orogeny of Late Ordovician (or Silurian) to Carboniferous age is the youngest of these pre-Alpine cycles. These cycles are all characterized by:

(i) Successive formation of oceanic basins across the north margin of Gondwana between the Late Proterozoic and the Devonian.

(ii) Drift towards the south margin of Laurussia of microcontinents rifted off north Gondwana by the successive phases of rifting and spreading that formed the Rheic Ocean and its successor, the Paleotethys. In the north, the immense continental block of Laurussia, which extends as far as the Urals, resulted from the collision of the Laurentides with the Scandinavian–Russian platform. The latter formed as a result of the Grenville (1000–900 Ma) and later Caledonian orogenies (500–420 Ma). In the south, the Gondwana plate or continent was consolidated following the Pan-African orogeny.

(iii) Collision, followed by subduction of the microcontinents against the south margin of Laurussia, was accompanied by magmatism and voluminous volcanism; the geochemical signatures of these magmatic rocks provide their geodynamic context.

Successive phases of oceanic basin formation in Late Proterozoic, Cambro-Ordovician and Devonian times detached several microcontinents of north Gondwana origin, which were then separated by oceanic basins. These elements were first involved in the Hercynian deformation, then *a fortiori* in the basement of the Alps and Alpine foreland (see above, Fig. 3.2). The same type of event is moreover duplicated several times during the Mesozoic.

1.2 Main structural elements of the Hercynian fold belt in map view

In the whole Hercynian fold belt, old basement blocks, detached from the north Gondwana margin and overlain by Palaeozoic sediments have remained more or less preserved from Hercynian deformation. The most reliable basement age dates are from U/Pb analyses of single, *in situ* zircon grains. They show Pan-African ages reflecting many arc and collision events ranging in age from 650 to 500 Ma. The basement is covered by thin epicontinental sediments of Devonian and Carboniferous age with minor deformation. Examples (Fig. 4.1) include the partly hidden Ebroia Massif in Spain, Mancelia in Central Brittany and the Moravian craton around the city of Brno ('Moravia' or 'Brnoia') in East Bohemia.

The presence of nuclei of Pan-African basement is partly responsible for the geometry of structures, which were moulded around them during Hercynian orogenesis. The end result was a complicated structural pattern in map view made up of large-scale festoons and opposing virgations. Such induced structures are exemplified by those which envelop the Ebro block ('Ebroia') over which the Pyrenean nappes, Hercynian then Alpine, were thrust in a SW direction on one side while the major nappes of Galicia and the Iberian meseta were thrust towards the east and NE on the other (Fig. 4.1) forming the Cantabrian orocline.

1.3 Deformation of the south branches of the Hercynian fan by Alpine structures

The subduction of Gondwana-related microcontinents began during Silurian times as shown by the crystallization ages of eclogites. Hercynian closure of the Palaeozoic Ocean and the collision between Laurussia, Gondwana and detached Gondwana microcontinents began in the early to middle Devonian and ended by Early Carboniferous times (385–325 Ma). It led to the northward subduction of the oceanic crust below the internal zones of the Hercynian belt and consequently to the high-pressure Hercynian metamorphism found in the Moldanubian units of the Alpine foreland, in the External Alpine Crystalline Massifs, and in lower and

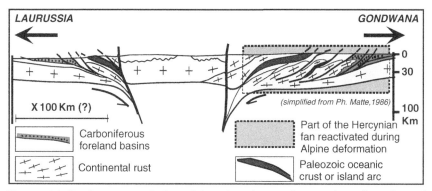

Figure 4.2 *Emplacement of the Alpine domain on the southern branches of the Hercynian fan in Western Europe.* Hercynian features, simplified after Matte (1986, 1998).

middle Austro-alpine nappes. The collision resulted in the very large, overall fan structure of the Hercynian fold belt with double vergence northward to Laurussia and southward to Gondwana (Fig. 4.2). The axis of the fan is occupied by the highest grade metamorphic zones, i.e., the internal Hercynian zones. Ages of terranes affected by the Hercynian orogeny young northward (Laurussia) and southward (Gondwana) from the axis of the fan. Only the south-verging part of the Hercynian fan was subsequently affected by the Alpine orogeny in the Central and Western Alps (Fig. 4.2).

 ## 2. RELATIONS BETWEEN HERCYNIAN, TETHYAN AND ALPINE STRUCTURAL UNITS

A large part of the outcrop of the Alpine domain consists of pre-Permian, Hercynian or older basement present in both the internal and the external Alpine zones (Fig. 4.3). This importantly emphasizes the role played by pre-Alpine basement(s) during the course of Tethyan and then later Alpine history.

2.1 The external Crystalline Massifs: a shared history with the basement of the Alpine foreland

The pre-Mesozoic basement underlying the Alpine foreland outcrops in Bohemia, the Vosges, Black Forest and part of the Massif Central. It consists of the so-called Moldanubian part of the Hercynian internal zones, which is characterized by high-grade metamorphism. The Maures massif of

Provence (Fig. 4.3) and parts of Corsica–Sardinia also comprise part of the European basement. Within the Alpine domain proper, outcrops of the internal Hercynian zones comprise the External Crystalline Massifs (Figs. 4.3 and 4.4).

The Hercynian history of the Alpine foreland is comparable with that of the External Alpine Crystalline Massifs as well as the Simplon–Ticino

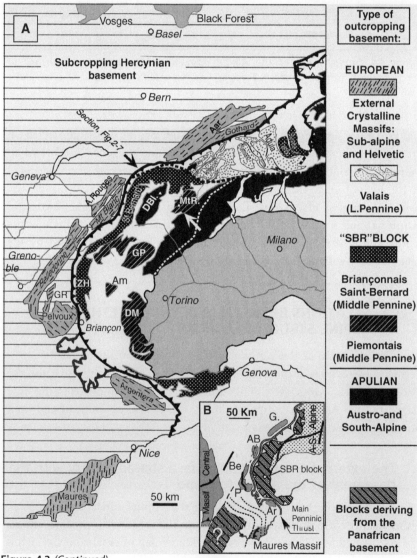

Figure 4.3 *(Continued)*

nappes, which were derived from the European side of the small Valais basin. By the way of example, the External Crystalline Massifs of Belledonne–Pelvoux–Argentera provide a resume of the Hercynian history of the Alpine foreland (Fig. 4.4).

The main common phases of this shared history are as follows:

– A pre-Hercynian crustal extensional phase took place in Cambrian times on the southern part of the mid-European margin. The event marks the appearance of Cambrian oceanic crust in mid-Europe. It gave birth to the future Chamrousse ophiolite complex of the Belledonne massif dated (U-Pb) at 496 Ma (Fig. 4.4) analogous to remnants found in the Massif Central, the Bohemian massif and the Eastern Alps ('*Plankogel terrane*'; Figs. 2.1 and 4.1). Other fragments of oceanic terranes in the Pennine basement, e.g., in the Tambo and Suretta nappes of the Brianconnais zone in eastern Switzerland (Fig. 12.10) have been identified from their petrographic and geochemical character.

– A major subduction–collision event is marked by the emplacement of a first system of nappes that exhumed 'relict' Cambro-Ordovician ophiolites as well as eclogites. Thrusting was accompanied by high-temperature, middle-pressure metamorphism with associated anatexis. The rocks correspond to the 'Leptinite and Amphibolite' group of Hercynian Western Europe, which corresponds to protoliths dated to end Cambrian and Early Ordovician. However, in the Belledonne massif, the Chamrousse ophiolite complex remained undeformed until as late as the Early Carboniferous

---

Figure 4.3 (A) Location of pre-Mesozoic basement in the Western and Central Alps. Compare with Figure 2.4. *Simplified after Marthaler (2001), p. 70.* **(B) A hypothesis concerning Hercynian structures in the Internal Alpine zones and External Crystalline Massifs.** In the basement of the middle Pennine of the Eastern and Central Alps ('SBR' block), the absence of any Hercynian deformation suggests the idea that this block might have been a nucleus of Pan-African basement comparable to Ebroia in Iberia and Mancelia in Normandy. The difference in present day structure is the result of Alpine thrusting. The Late Penninic corresponds to ophiolites derived from Tethyan oceanic crust. In the External Crystalline Massifs including the Maures, Hercynian structures have the form of an S-shaped curve (see Fig. 4.4 also). The lower part of the S matches perhaps the shape of another block of Pan-African basement hidden beneath the Mesozoic Provencal basin comparable also to the 'Ebroia' block of Iberia. The festoon can result equally from late strike-slip movements of Carboniferous age, which are characteristic of the Hercynian fold belt. *Inspired from Matte (1998) and Autran et al. (1995). Redrawn and completed from Neubauer and von Raumer (1993); Autran et al. (1995).*

- Intrusion of the first major Hercynian granite is dated to end Devonian. Such granites occur in Brittany, the Vosges and the NE Massif Central and the External Crystalline Massifs including the Maures Massif.
- A new phase of crustal extension occurred from latest Devonian to Visean times (c. 350 Ma) resulting in regional anatexis and granitic intrusion in the back-arc of the subduction zone. Calc-alkaline

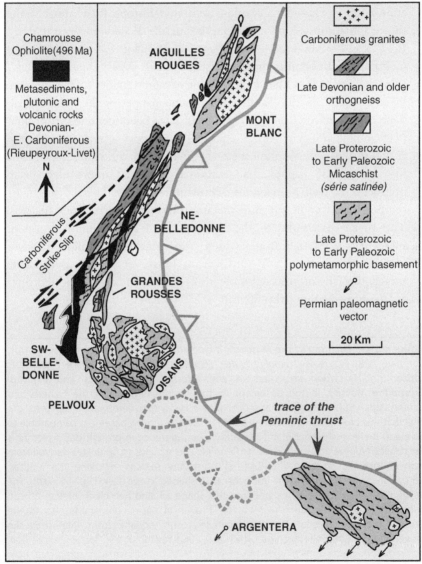

Figure 4.4 *(Continued)*

magmas, consanguineous over large distances, known from the Hohe Tauern (Austria), the External Crystalline Massifs and the Massif Central (France), may be an early expression of crustal thinning. Their age (360–350 Ma) is not later than basal Visean. No subduction-related magmatism is known after the Visean.

- The Devonian to Late Dinantian crustal extension resulted also in the formation and infill of transtensional (? back-arc) basins. It is characterized by mesozonal metamorphism but without granites and eclogites. Metasediments and associated magmatic rocks are known in the southern Belledonne Massif and called the 'Rieupeyroux-Livet and Taillefer formations' (Fig. 4.4). These are probably the time equivalent of the 'Brevenne Schists' of the NE Massif Central. This rock suite is similar to the 'Schistes des Cevennes' of the Early Palaeozoic SE Massif Central. Equivalents are known in Vosges and Black Forest.

- A second and brief phase of low-pressure metamorphism postdates the Early Carboniferous extension phase and the emplacement of the Rieupeyroux–Livet–Taillefer formations. The granitic plutons emplaced after the Visean during Carboniferous times formed by crustal anatexis.

Figure 4.4 *The Belledonne, Oisans and Argentera Massifs: the Hercynian grain and Alpine deformation* The Belledonne Massif is composed of two different parts. The north-eastern part is characteristic of the Hercynian internal zones at the north. The southwest sector of the Belledonne Massif belongs to the external Hercynian zones. During the Alpine cycle, the entire External Crystalline Massifs were transported to the NW along a deep detachment surface shown by geometric constraints and deep seismic profiles (ECORS: Figs. 2.16, 13.1). The Crystalline Massifs of Belledonne, Pelvoux and Argentera are separated from the Dauphinois basin by a lateral ramp that terminates the surface of the deep detachment to the SW. In consequence the Carboniferous strike-slip faults shown on this figure do not continue those which were reactivated in the Jurassic during Tethyan extension such as the Nimes and Durance faults and which defined the tilted blocks of the Dauphine and Provencal domains. However, the whole system of faults belongs to the same family. Note that the map trace of the frontal Penninic thrust follows approximately the envelope of Hercynian structures (Chapter 6) The Hercynian grain has not been reoriented by later Alpine deformation. In effect, the directions of Permian palaeomagnetic vectors, determined on sediments resting unconformably on the basement of the Crystalline Massif, are congruent with those observed in the extra-Alpine domain in this part of Europe. In contrast to the other main Hercynian lineaments shown below, the Hercynian suture separating the internal and external zones within the Belledonne Massif was not reactivated during Mesozoic extension and later Alpine deformation. *Simplified from Pfeifer et al. (1993), Fig. 2, p. 120; from Ménot et al. (1994), Fig. 1, p. 459; Bogdanoff and Schott(1977).*

Emplacement of such granitoids occurred until the Late Westphalian (310–305 Ma) in the Aar and Aiguilles Rouges massifs (Vallorcine granite), in the Eastern Pelvoux in the Maures Massif, in Corsica and in the Pyrenees.

The Chamrousse ophiolites in the Belledonne massif were thrust westwards, i.e. above the Rieupeyroux–Livet formation not earlier than Dinantian time (Fig. 4.4). This event was in response to the collision of the Belledonne massif with the Grandes Rousses and Oisans. Thrusting developed under mesozonal metamorphic conditions. This late orogenic development is closely comparable to that of Central Brittany (Late Devonian onset of folding).

For these reasons, the evolution of the segment of the internal zones of the Hercynian fold incorporated in the Alpine External Crystalline Massifs is somewhat different from that of the Massif Central and the Moldanubian Zone of Central Europe. This Hercynian segment of the Western and Central Alps probably also represents an eastern part of the Hercynian internal zones displaced in the Carboniferous by the advance of the Gondwana indentor (Fig. 4.4).

–　Erosion, commencing in the Namurian (from about 325 Ma), which exposed the deepest parts of the Hercynian fold belt, resulted in the deposition of Carboniferous sediments along the periphery and the interior of the fold belt. In the marine domain south of the orogen, the early, flexural, Hercynian foreland basins, infilled by *flysch* eroded from the rising fold belt, are largely distributed along the south border of the Hercynian fan. Their age ranges between Namurian and Westphalian depending on locality. Such basins are known in the Pyrenees, the Montagne Noire (south of the French Massif Central), in the Rif and Betic Cordilleras, the Kabylies, Balearic Islands, Sardinia, the upper Austro-alpine nappes, the basement of parts of the Southern Alps, Attica (Greece) and the island of Chios off Western Turkey.

–　In the continental domain, Stephanian to Autunian age sediments (320–270 Ma) were deposited in discontinuous, transtensional basins in fluviatile to lacustrine environments. These sediments were unaffected by Hercynian metamorphism and rest discordantly on metamorphosed Early Carboniferous or older rocks. The underlying hiatus may represent an interval of 30–40 Ma or more. Examples include the coal basins of La Mure in the Alps and Saint Etienne and Cevennes in the eastern Massif Central. In the Bigorre and Cadi basins of the Pyrenees, the sedimentary infill is of Autunian age and includes thick volcanics dated at 270 Ma.

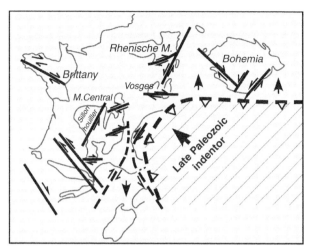

Figure 4.5 *A hypothetical model for Late Carboniferous to Early Permian wrench tectonics in the Alpine foreland.* This model, among others, represents the network of late Hercynian strike-slip faults in the Alpine foreland as being generated by the advance of an Apulia–african promontory at the end of continent–continent collision. The main directions of expulsion of blocks at the periphery are shown. Note the leading edge of the indentor is occupied by the Saint Bernard–Monte Rosa block ('SBR' block). *From Neubauer and Raumer (1993), Fig. 14, p. 641.*

– Dissection of the Hercynian fold belt by a system of Late Carboniferous and Permian strike-slip faults that affected the whole of Western Europe (Fig. 4.5). Activity along this system of faults led to the formation of narrow elongate basins along transtensional fault segments. The type example is the 'Sillon Houiller' (= Coal furrow) of the Massif Central. In addition, the massifs form a virgation, which suggests the possible presence of a nucleus of Pan-African basement under Provence (Fig. 4.3B) whose significance may be comparable to the 'Ebroia' block of the Ibero-amorican arc (Fig. 4.1). This nucleus, whose identification is indirect, resulted from the organization of Hercynian structures in the External Crystalline Massifs and the Corsica–Sardinia block, which defined an envelope around the supposed nucleus, before the opening of the Gulf of Lions.

2.2 The Palaeozoic development of the South-alpine and Austro-alpine basement

The basement(s) of lower and middle Austro-alpine units must be distinguished from those of South-alpine units *pro parte* and upper Austro-alpine units.

Lower and middle Austro-alpine units, as well as the Ivrea zone, are highly deformed, metamorphosed and intruded by huge Hercynian granites. In places, there are traces of deformation phases of Early Palaeozoic or older ages. However, the Palaeozoic, and especially the Proterozoic history, is difficult to document because of the later Alpine overprint.

The Upper Austro-alpine nappes together with the lower part of the pre-Mesozoic sequence in the Southern Alps originated from the north Gondwana passive margin and also from the south Hercynian foreland (Figs. 2.9, 4.2, 4.6, right part). The typical stratigraphic succession consists of thick quartz phyllites dated to the Late Cambrian to Early Ordovician from acritarchs. The succeeding sediments are platform carbonates of Late Ordovician (Ashgill) age and then reefal limestones of Devonian age. The carbonates are contemporaneous with shales deposited in adjacent basinal environments. The palaeogeography of platform and basin became uniform from the Frasnian (Late Devonian) with deposition of flysch *(Hochwipfelflysch)* until Westphalian B time. The succession was unconformably covered first by molasse-type deposits from Westphalian D to Stephanian (Auernig beds) times, then by clastic and volcano-clastic deposits of Late Stephanian and Permian age, which were undeformed during the Hercynian cycle. The best sections are located in the Carnic Alps, though a similar succession, with some minor differences, is characteristic of the external Hercynian domain of the Pyrenees, southern Spain, Rif and Betic cordilleras, the Balearics, Sardinia, Greece and western Asia Minor, as mentioned above.

The basement of the Southern Alps is known only from relatively small and widely dispersed outcrops within post-Hercynian structures (Fig. 4.7). The succession comprises shales and psammites that are poorly dated except at the base where Cambrian acritarchs are found. It is divided into two parts by very widespread acid volcanics, which record an episode of Late Ordovician extension. Metamorphism is of greenschist grade. An approximately Visean age (350–320 Ma) is given by Rb/Sr whole rock and individual mineral ages. Granitoids of Permo-Carboniferous age intrude the succession.

2.3 The Penninic domain: unique basement compared to the External Crystalline Massifs

Published interpretations dispute the presence or absence of Hercynian metamorphism in the Penninic domain, namely in the Pontis and Siviez-Mischabel nappes (see Chapter 2 and Fig. 2.7). Some authors consider that the basement of these nappes must be integrated in a consistent manner in

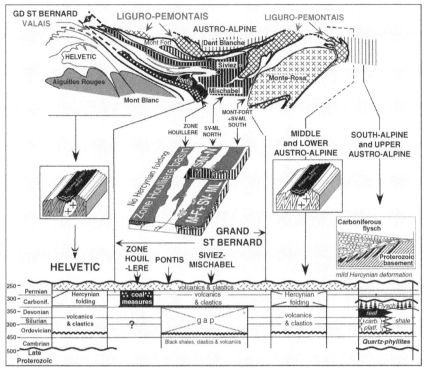

Figure 4.6 *Correlation between Alpine structural units and Late Palaeozoic palaeogeography in the Grand Saint Bernard complex.* The External Crystalline Massifs and more generally the basement of the Alpine foreland belong to Hercynian internal zones. The basement of the Southern and Eastern Alps belongs partly to external Hercynian zones. In contrast, the terranes comprising the Grand Saint Bernard nappe probably do not record the Hercynian orogeny. The lines of opening for the Liguro-piemontais Ocean and Valais trough, which respectively delimit the 'SBR' block in pre-Alpine palaeogeography, appear therefore to have followed major ancient lineaments. Likewise, the lines of demarcation of Alpine thrust units appear to have also followed the structural trends controlling Carboniferous palaeogeography in the SBR block. Basement alone is annotated with conventional symbols. Cross lines marked 'Ligure' correspond to ophiolites derived from Tethyan oceanic crust. The Mesozoic cover is shown without ornament. SV-ML North, SM-N: Siviez-Mischabel North. Mont-Fort+SV-MV South, MF+SV-ML: Mont-Fort and Siviez-Mischabel South. *Section across the Alpine nappes simplified after Escher et al. (1987), and Lemoine Fig. 2.7, this volume. Reconstruction of the Saint Bernard Permian redrawn from Thelin et al. (1993), Fig. 8, p. 312. Main stratigraphical data from Neubauer and von Raumer (1993).*

the Hercynian metamorphic scheme. High–pressure and high–temperature metamorphic episodes are sometimes said to compare well with meta-morphic events of Ordovician to Latest Devonian age recorded in the

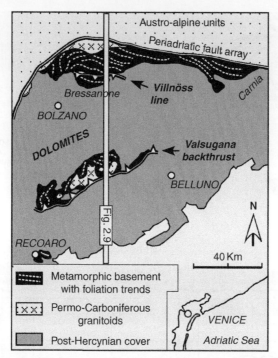

Figure 4.7 *Tectonic sketch map and main outcrops of the Hercynian basement in the Southern Alps.* The major structures of the Southern Alps result from south-verging Valsugana line of Alpine thrusts and north-verging Villnoss back-thrusting (see Figs. 2.9 and 2.17. The outcrops of basement in the Southern Alps are distributed between the Peri-adriatic fault array (Bressanone to the north) and the border of the Po Plain (Recoaro to the south). The outcrops consist of psammitic schists that have been subjected to low-pressure Hercynian metamorphism. A decreasing metamorphic gradient is observed from east to west. The anchizone is reached in Carnia: this shows the proximity of the Hercynian front in these regions. *Simplified from Sassi and Spiess (1993), Fig. 1, p. 600 and from Ring and Richter (1996).*

External Crystalline massifs. This idea is supported by 40Ar/39Ar ages of 340–360 Ma reported by Monie (1990) and cited by Thelin (1993; see Fig. 4.6) on micas from the pre–Permian basement of the Ambin Massif. In consequence, certain pre–Alpine metamorphic phases, albeit poorly dated, can be argued to belong to the Hercynian cycle.

In contrast, the Namurian and younger beds of the Zone Houillere (Fig. 4.6) have not been subjected to Hercynian metamorphism. It can also be noted that there are few absolute age measurements for the interval between the basal Devonian and the Permian. U/Pb dates cited by Thelin (1993) from zircons and pitchblendes and from whole rock Sm/Nd are either much older, 406 ± 50 Ma, or more recent, 275 ± 12 Ma. Therefore, Hercynian metamorphism may not have affected the basement of the Saint Bernard nappe and possibly that of the whole SBR block. If this is the case, the SBR block with absent Hercynian metamorphism can then be more consistently juxtaposed against the highly metamorphosed European Hercynian basement involved in the External Crystalline Massifs.

The main characteristics of the basement forming the framework of the Saint Bernard nappe (Fig. 2.7) compared to that of the Dauphinois–Helvetic domains are as follows:

(a) Non-deposition between the Ordovician (about 460 or 450 Ma) and/or the underlying Precambrian and the Namurian (about 325–320 Ma).

(b) Deposition of a very large coal basin, whose characteristics are those of the southern Hercynian foreland.

(c) The relatively local origin of clastic sediments, i.e., pebbles within the conglomerates indicate a Penninic provenance only and not a Dauphine or Helvetic source. In addition, the general current transport directions are northerly, i.e., European in the Alpine foreland, in contrast to southerly in the Pennine domain. The Late Palaeozoic basins of the external Alpine zones and Penninic domain were therefore respectively filled with sediments derived from different source areas.

(d) Probable absence of Hercynian metamorphism.

These different points constitute much of the argument for demonstrating the peculiarity of the Saint Bernard–Monte Rosa ('SBR') block *vis-à-vis* the External Crystalline Massifs and for considering it as relatively exotic. It was possibly accreted to southern Europe from the external Hercynian domain perhaps by major end Palaeozoic strike-slip movements after the acme of Hercynian metamorphism and before the Triassic.

Another point of view would compare the 'SBR' block and more generally the basement of the Brianconnais and Piemont zones with nuclei of Gondwanan origin (Fig. 4.1) such as 'Mancelia' (Armorican massif), 'Ebroia' (Iberia) or 'Moravia' (Bohemia), which more or less escaped Hercynian orogenesis after being detached from the fringe of North Gondwana (Fig. 4.3B). Mancelia and the SBR block were nuclei derived from Pan-African basement isolated within the Hercynian fold belt. Ebroia in contrast was close to the Hercynian foredeep and Moravia was attached to the North Hercynian foreland.

Basement nappes of the Penninic domain

Today, the SBR block, which consists of six stacked main nappe units, is separated by major Alpine sutures from the Helvetic and Dauphine domains to the north and from the Austro-alpine and South-alpine to the south (Fig. 4.6). These sutures coincide respectively with the lines of closure of the Valais basin and Ligurian Tethys on the Apulian side, which therefore correspond with two major palaeographic discontinuities extant since the Palaeozoic (Chapter 2, Middle Penninic Assemblage and Fig. 2.7).

Further discussion and analysis is limited as a consequence of the major gap in the stratigraphic record between Ordovician or older beds and those of Namurian or younger age in the Pontis and Siviez-Mischabel nappes. Moreover, basement is unknown at present beneath the plant-bearing beds of the *Zone Houillere* (= Coal zone) implying that an admissible reconstruction of the Pennine domain in the Late Palaeozoic must await new data.

Most of the outcrop area in the Penninic domain consists of pre-Mesozoic basement. Within the basement, a lower stratigraphic section constitutes the core of the overthrust nappes while upper stratigraphic units comprise the outer perimeter of the nappes. Within each individual nappe, the lower section has been polymetamorphosed by successive phases of pre-Alpine and Alpine metamorphism. The upper stratigraphic units are monometamorphic beds of Namurian and younger age affected by Alpine metamorphism only (Table 4.1).

The nomenclature, which is classical in Switzerland, was established following the paper by Escher et al. (1987; see this book, Figs. 2.7 and 4.6). The modes and causes of the relative coincidence between palaeogeographic Tethyan units and Alpine structural units are explained elsewhere in this book (Chapters 1, 12, 13, 15).

Table 4.1 Nomenclature and paleogeographic situation of structural units constitutive of the Saint-Bernard Monte-Rosa nappe complex

Suture of the Valais basin	Lower Pennine	Middle Pennine					Suture of the Ligurian Tethys
	Lower Zone Houillere	Central Alps					
		Saint Bernard Nappe Complex				Monte Rosa Nappe	
		Upper Zone Houillere	Pontis Nappe	Siviez-Mischabel Nappe	Montfort Nappe		
		Western Alps					
		Subbrianconnais	Brianconnais	Piemont zone			

The history of the polymetamorphic basement of the SBR block is closely analogous to that of basement fragments detached from North Gondwana during the Proterozoic and Early Palaeozoic to form the accreted Hercynian terranes comprising much of Europe. The polymetamorphic basement consisted originally of magmatic rocks of Proterozoic age and a thick series of metapelites with marine organic matter reportedly of Ordovician age. These rocks are cut by mafic, ultramafic and calc-alkaline granitoids. Several of the mafic rocks comprising the Siviez-Mischabel nappe, dated on average at 475 ± 50 Ma (Ordovician), have geochemical affinities with T-MORB suggesting a trend towards rifting and early spreading at this time.

Two phases of pre-Alpine metamorphism are known. The highest, eclogite grade (HP: high pressure), was succeeded by amphibolite grade metamorphism (HT: high temperature). However, their respective ages are unknown, as is the duration of the interval between the two events. While clearly pre-Namurian in age, any age estimates rely perforce on hypotheses based on comparison with metamorphic events affecting the basement of the external Alpine domain.

On the SBR block, extensional Permo-Carboniferous basins are underlain by pre-Namurian basement and infilled by volcaniclastics, clastics and coals according to locality. Basin margins were defined by extensional faults that controlled axes of subsidence and focussed the contemporaneous ascent of magmas. It now seems likely that the thrust planes separating the Alpine nappes of the SBR block follow the fault boundaries between the different Permo-Carboniferous basins (Fig. 4.6).

The Permo-Carboniferous basins are (1) the Mont-Fort and South Siviez-Mischabel basins, which form the monometamorphic part of the basement of the Mont-Fort nappe, (2) the North Siviez-Mischabel basin, which forms the monometamorphic part of the Siviez-Mischabel nappe s.l., (3) and finally, the basin of the Zone Houillere whose underlying basement is unknown as noted earlier.

The Zone Houillere is involved in two, superposed Alpine nappes. Though relatively narrow, the outcrop is elongate and can be followed in a semi-continuous manner for more than 100 km from Briancon to just beyond Switzerland. The sedimentary successions in these nappes are distinctly different, though both are of Carboniferous age.

The *upper nappe*, which constitutes the Zone Houillere and is referred to as the middle Pennine, is the most external of the Saint Bernard nappe complex. The succession comprises beds dated by plants as Namurian and

Westphalian A unconformably overlain by the Granon conglomerates of Stephanian age. The succession of the *lower nappe*, derived from the SE side of the small Valais basin (internal Valais or lower Penninic), comprises the undated Gres de la Praz below beds of Westphalian D age that are succeeded by the Early Stephanian and, in turn, Permian lava flows. The contrasting stratigraphy of the two nappe units indicates derivation from two separate Carboniferous basins or perhaps two sub-basins separated by syn-sedimentary faults. In contrast, the overlying Permian sediments are uniform across both basins and sub-basins.

 3. CONCLUSIONS

3.1 Some key points

The palaeogeographic boundaries of the Permo-Carboniferous basins comprising part of the basement of the Alpine nappes of the 'SBR' block (Monte Rosa, Mont-Fort, Siviez-Mischabel, and Pontis) coincide with the structural limits of these nappes. Moreover, thrusts between Alpine nappes often coincide with extensional faults defining Tethyan half-grabens. It is thus possible, if not probable, that Carboniferous extensional faults were reactivated during Jurassic Tethyan, rifting and, in turn, that the presence of these surfaces subsequently induced the location of thrust planes at the onset of Alpine shortening during Tertiary time.

The distinction between the two nappes of the Zone Houillere was first drawn from differences in their Mesozoic stratigraphy. However, the greater contrast in their Carboniferous stratigraphy shows clearly that the Valais (lower Zone Houillere–lower Penninic) and Brianconnais (Upper Zone Houillere–middle Penninic) units were already differentiated from Late Palaeozoic time.

The Zone Houillere, originating from the SE side of the Valais basin, is separated by the Valais (oceanic or intracontinental?) suture (trace of the Valais trough) from other basement nappes (e.g., the Simplon–Ticino nappes) derived from the NW side of the Valais basin (see Chapters 2 and 7). The history of the NW side of the Valais trough is consistent with that of the External Crystalline Massifs while, in contrast, that of the SE side has Penninic affinity.

The differences in the history of the Hercynian basement of the Alpine External Zones and that of the Penninic domain shows that the Valais rift and suture line follows and inherits a major basement discontinuity in the structure of Hercynian, pre-rift terranes. If the Carboniferous, and not Cretaceous, age of ophiolites previously reported to the Valais is confirmed, the possible existence of Palaeozoic oceanic crust in the Valais domain reinforces the importance of this discontinuity.

Given the relative consistency between the history of the basement of the lower Austro-alpine nappes and External Crystalline massifs, in contrast to the relative unique geology of the SBR block, the line of break-up of the Ligurian Tethys may coincide also with a Hercynian or older lineament.

3.2 Some unresolved problems

Nonetheless, many poorly resolved or unresolved problems remain. One concerns the initial setting of the Zone Houillere and, more generally, the SBR block in a Carboniferous palaeogeographic and structural context. The only certainty is that continuity with sub-aerial Gondwana is shown by the Gondwanan rather than Laurussian affinities of the flora, i.e., very clear differences from the flora of the coal basins of northern France and Belgium.

Another problem concerns the timing of the approach and accretion of the SBR block to form part of the European basement comprising the Tethyan pre-rift. Whether the block was accreted to the European domain as a result of late/end Hercynian, Late Carboniferous–Permian, strike-slip faulting, or prior to the Hercynian paroxysm like other nuclei of Pan-African basement derived from the North Gondwana margin remains unknown. It is known only that faulting had ceased by the start of the Triassic that is marked by the basal unconformity of the Werfenian (Scythian) sandstone. However, the kinematics remain uncertain and available published reconstructions differ greatly (see Chapter 7).

FURTHER READINGS

Matte (1991); Keppie (1994); Stampfli and Borel (2002).

PART TWO

The Tethys Phase

As classically described for present-day passive margins, the early stages of Tethyan passive margin evolution can be divided into three phases of which the first is the **pre-rift**. The phase of rifting associated with stretching and the thinning of the continental lithosphere pre-dating oceanic spreading is called **syn-rift**. The creation of rift systems together with the associated extensionally driven subsidence created the initial accommodation space for sediment deposition. The third **post-rift phase** is marked by the first appearance and spreading of oceanic crust in the axis of the rift, together with passive thermal subsidence and, theoretically, the end of extension in the newly formed, conjugate passive margins.

An essential part of the understanding of the geometry and evolution of structures of Mesozoic and Tertiary age in the Alps is lost if the Hercynian and older heritage is neglected in describing the pre-rift history (Chapter 4).

The history of the Palaeozoic ancestor of Tethys or Palaeotethys is characterized by the detachment of microcontinents from its southern margin, i.e. Gondwana. These microcontinents moved northward to collide or be subducted beneath the north margin, known as Laurussia (Fig. 3.2), resulting in a series of orogenies of which the Hercynian was the last in Western Europe before the Tethyan cycle.

The net result was the progressive construction of the southern part of Laurussia by the collage and incorporation of heterogeneous fragments. These include microcontinents of Gondwanan continental crust and fragments of oceanic crust (ophiolites) that represent the remnants of the small ocean basins originally separating these microcontinents.

The importance of the distribution of the rheological and structural discontinuities which separate these different Palaeozoic units lies in that they have largely conditioned or determined the geometry of the main branches of the Tethyan rift, that of the line of opening of the Tethys in the Jurassic and start of the Cretaceous and, later again, the main Alpine structural units from the Cretaceous.

The earliest phase of extension in the Western Alps is related to late Hercynian orogenic collapse. However, this phase must be clearly distinguished from later, more important, discrete extensional episodes. Each of these later episodes can be related to a step in continental break up and onset of the creation of Tethyan oceanic crust.

The Triassic extensional episode did not lead to oceanic spreading in Western Europe. However, this phase may be related to the development of the Tethys in the Eastern Mediterranean. In this respect, it may be linked to the pre-rift phase of the Jurassic Tethys also known as the Ligurian Tethys. However, it can also be considered as forming part of the history of Tethyan rifting in the Mesozoic. While the choice of the onset of rifting then becomes a question of personal preference, a key point to recognize is that each individual extensional episode records, at distance, the westward steps in the progression of rifting and oceanic spreading, during some 80 Ma of Mesozoic time (Figs. 3.2 and 3.4).

In the Ligurian segment of the Tethys, the onset of the main phase of **oceanic spreading**, that is the syn-rift/post-rift boundary, is dated as Middle Jurassic. In consequence the Ligurian-Tethys rift phase must be dated to part of the Middle Jurassic and earlier (Chapter 6).

During the Tethyan post-rift phase, the western European craton was again submitted locally to extension and rifting dated as Early Cretaceous in the Valais trough and Provence. This was followed by the creation of oceanic crust in the Bay of Biscay. Even the Ligurian-Tethyan oceanic crust itself was submitted to extension (Chapter 8, Section 2, p. 175).

The above discussion shows that, while convenient, the concept of the succession of pre-, syn- and post-rift phases cannot be applied in an overly simplistic or rigid manner.

CHAPTER FIVE

The Age of the Onset of Tethyan Rifting in Western Europe

Contents

Summary

The rift phase that marks the onset of the Alpine cycle and preceded spreading of the Liguro-piemontais Ocean was active during Liassic and part of Middle Jurassic times, i.e., from 205 Ma to the 165 Ma onset of spreading.

Prior to the Ligurian rift phase, Western Europe was, from Middle Carboniferous time (320–330 Ma), differentiated into areas undergoing regional subsidence by extension along normal and strike-slip faults. Towards the end of the Hercynian cycle in Permo-Carboniferous times, intra-continental sedimentary basin development took place.

Triassic times mark a turning point between the Hercynian and Alpine cycles. The extensional history of the Triassic was linked to, and contemporaneous with, that of the Neotethys both in the eastern Mediterranean and further east as well as the Central North Atlantic.

The Western Alps, from Rift to Passive Margin to Orogenic Belt, Volume 14
ISSN 0928-2025, DOI 10.1016/S0928-2025(11)14005-5

© 2011 Elsevier B.V.
All rights reserved.

Triassic rifting started as early as earliest Triassic times (Scythian) in the internal zones of the future Western Alps and Southern Alps. In contrast, it started in Late Triassic times (Carnian) in the graben of the Rhone valley where it is characterised by deposition of several hundred metres of evaporites.

The Triassic major peak transgression is dated as Early Ladinian (Fassanian). It is recorded by coeval marker beds in the Internal and External Alpine zones as well as in the Western European intra-cratonic basins as far North as the Barents Sea.

The first phases of Ligurian Tethys rifting date probably from the Late Triassic. However, the gross palaeogeography of the Triassic does not herald the future Jurassic Tethyan rift except locally.

 1. THE PRE-RIFT AND SYN-RIFT PHASES OF THE JURASSIC TETHYS

The construction of the Hercynian fold belt during Devonian and Carboniferous time created a single supercontinent, Pangaea. The Tethys Ocean then formed a re-entrant widening and opening towards the east (Fig. 3.2 and 3.4).

Like the preceding Hercynian cycle, the Alpine cycle consists of a succession of extensional phases that progressively fragmented Pangaea and were followed by a series of shortening phases due to the convergence of Africa with Europe that resulted ultimately in the formation of the Alpine fold belt.

Each oceanic basin formation event, including that of the Jurassic, was characterized in principle by three phases: pre-rift, syn-rift and post-rift (see Introduction to Part 2). However, given the prolonged history of extension of Pangaea and its terranes from the Namurian to the Early Cretaceous in Western Europe, the precise age of rift onset in the Liguro-piemontais Tethys, and in turn the onset of the Alpine cycle, is debatable to some extent though it is not earlier than the Late Permian.

During the Late Carboniferous and Early Permian, large subsiding coal basins, such as the St. Etienne basin, were controlled by a network of transtensional faults (Fig. 4.5) associated with the final, late orogenic collapse of the Hercynian fold belt. A close analogue is offered by the present day pattern of Alpine neotectonics (Chapter 15). The late Hercynian deformation was thus completely separate in space and time from the onset of the syn-rift phase in the Ligurian Tethys.

The Late Permian and part of the Triassic corresponds in time to the progressive east to west opening of the Neotethys in the Middle East (Fig. 3.2). In Western Europe, extension during this period created a complex network of faults and subsiding troughs but did not evolve to spreading even though contemporaneous with rifting and spreading in the Neotethys. This phase can be best described as aborted rifting. Thus, the Triassic of Western Europe represents a turning point between the complete and consolidated Hercynian cycle and the yet to commence Alpine cycle.

2. THE TRIASSIC TRANSGRESSIVE–REGRESSIVE CYCLE OF WESTERN EUROPE: A TURNING POINT IN THE REDEVELOPMENT OF TETHYS

2.1 Geographic overview; a subaerial continent to the west and a marine domain open to the east

Extension and subsidence, contemporaneous with the opening of the Neo-tethys, allowed marine transgression from the east or the southeast over part of Western Europe and North Africa (Fig. 5.1). To the west, deposition took place in continental, coastal or marginal marine environments with sands fringing the land areas. By contrast, in the Eastern Alps, the pelagic Late Triassic Hallstatt ammonoid-bearing limestones herald the oceanic domain, located further to the east where remnants are found in the Carpathians and Dinarides. Between these two extremes, the intermediate area, site of the future margins of the Liguro-piemontais Ocean, was dominated by deposition of platformal carbonates and evaporites, variable in time and space (Figs. 5.1 and 5.2).

2.2 The Triassic transgressive half-cycle

Its duration extends from the Early Triassic throughout half of the Middle Triassic (Anisian and lowest Ladinian). In Western Europe, the areal distribution of the typical quartz sandstones of the basal Triassic decreases with time due to westward migration of river drainage patterns. This change reflects the construction of very large carbonate platforms from Middle Triassic time especially on the locus of the future European margin of the Liguro-piemontais Ocean. In contrast, in the Southern Alps, differential subsidence linked to rifting created a more complex palaeogeography on the border of the future Apulia margin. In this region, carbonate platforms were separated by relatively deep troughs infilled by hemipelagic sediments, which are interbedded with tuffs and

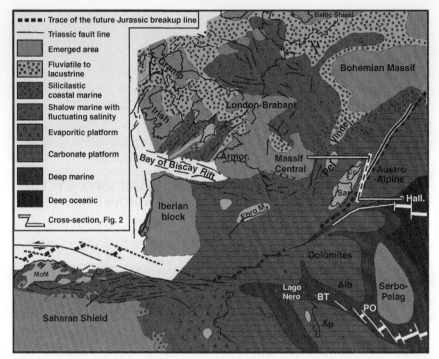

Figure 5.1 *Triassic palaeogeography and the Tethyan rupture between Europe and Africa.* The Neotethys, reincarnation of the Tethys from the Triassic consisted of many narrow gulfs and opened progressively from the end of the Permian but was closed towards the west (Fig. 3.2). The westernmost gulf was the Pindos-Olonos (Greece)–Budva (Yugoslavia)–Lago Negro (Italy) trough. It separated the platforms of the Southern Alps (Dolomites) and northern Albania from those of Tunisia, Sicily and the south of southern Apulia. However, the line of Jurassic break-up was not initiated from this trough. Rather, it developed westward from the Vardar branch of the Neotethys along its northern edge. Alb: Albanian Alps; Ap: Apulia; Armor; Brittany massif; BT: Budva Trough; Ebro M: Ebro massif; Gram: Grampian High; Hall: Hallstatt (see Fig. 8.2); Irish: Irish massif; MoM: Moroccan meseta; Pel: Pelvoux; PO: Pindos-Olonos Trough; Serbo-Pelag: Serbo-Pelagonian massif; Vindel: Vindelician. *Simplified after Dercourt et al. (2000).*

dated as Ladinian by ammonoids (Fig. 5.3). Hemipelagic sediments progressively covered the carbonate platforms as the area of carbonate deposition decreased with time.

The acme of the transgression, which was reached in the Early Ladinian, is contemporaneous with a period of accelerated extension in the Dolomites (C. Doglioni, pers. comm., 2002). It is also recorded by the maximum extent of ammonoid-bearing sediments in Western Europe including the Paris Basin

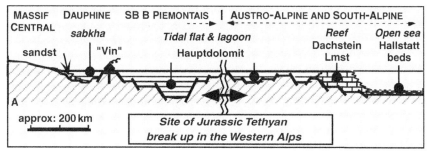

Figure 5.2 *Schematic section of Norian stratigraphy (Late Triassic) between the emergent Massif Central to the West and the deep Hallstatt basin to the East.* Variations in the observed thickness of Norian deposits require explanation by a system of normal faults bounding unequally subsiding blocks. Although these faults have not been reported from available field studies, their probable presence is a key sign of persistent extensional tectonism during the Norian; the continental volcanics of the Pelvoux massif provide similar clues. In addition, the Vindelician sill (Vin, Fig. 8.1) separating the Rhone valley trough from the subsiding Piemontais domain was not covered by sediment during the major but brief Carnian transgression. To the east, the Dachstein limestones in the Eastern Alps formed another sill, which separated the Hauptdolomit carbonate platform from the pelagic domain. The pelagic Hallstatt Beds, composed of marls and limestones with cephalopods, were deposited in a deep marine basin opening to the Neotethys (Location: see Fig. 5.1). *After Megard-Calli and Faure (1988). Modified from Lemoine et al. (2000), Fig. 7.2, p. 89.*

and Southern Alps (Dolomites) as well as a short-lived two-way migration of Tethyan and Germanic ammonoid faunas. In the Briancon area (Western Alps), it is recorded by crinoid and foram-bearing marker beds.

2.3 The start of the Middle to Late Triassic regressive half-cycle: the Carnian tectonic phase and sedimentary crisis

The short regressive interval that followed the peak of the Ladinian transgression was interrupted by a deformation event known as the **Carnian crisis**. The deformation is variously recorded by erosion, tilting, formation of giant breccias and the destruction of the carbonate platforms in both the future Alpine domain and adjacent the platformal areas. The crisis marked a new phase of extension that affected a large part of the West European craton and led to a resumption of subsidence. As a result, hitherto emergent areas were from then on drowned as in the lower Austro-alpine thrust sheets of the Grisons (Switzerland) and Corsica (Fig. 5.1). In areas of localized rapid sedimentation and subsidence, several hundred metres of anhydrite and salt were deposited in places as well as in the Aquitaine, Paris and German Basins, Rhone Valley and the future European margin of the Ligurian Tethys.

The transgressive–regressive cycle centred on the Carnian modulated temporarily, during an interval of 3–4 Ma (?), the major regressive phase (duration: 17–18 Ma) which encompassed part of the Middle Triassic and Late Triassic in the Southern and Eastern Alps. However, this complication has not been recognized in the Northern Calcareous Alps, German or Paris Basins.

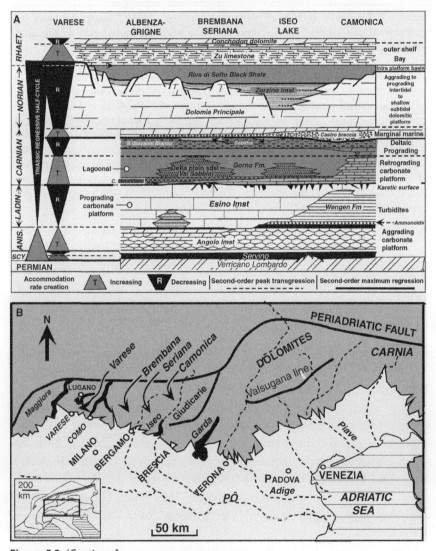

Figure 5.3 (*Continued*)

Figure 5.3 **The major transgressive–regressive facies cycle of the Triassic in Lombardy and the initiation of rifting in the Ligurian Tethys.** One of main points of this diagram is to illustrate the existence of a distension phase in the Late Norian followed by deposits of more and more open marine facies that are succeeded by those of the Liassic. This phase could then be proposed as marking the onset of rifting of the Ligurian Tethys. The same diagram also schematically depicts the major Triassic cycle which can be subdivided into four second-order transgressive–regressive cycles. These cycles were each controlled by discrete intervals of renewed distension and subsidence, which demonstrate at distance the successive phases in the development of the Neotethys further to the east. The first of these second-order cycles is mainly Scythian in age. Deposition of sands and conglomerates coincided with the end of the extensional–transtensional phase, which allowed emplacement of the underlying Verrucano red beds dated as Permian. The transgressive–regressive facies cycle of the end of the Scythian and Middle Triassic is characterized by at least two superimposed carbonate platforms: the first is chiefly aggradational and mainly Anisian in age (e.g. Angolo Limestone) while the second is mainly progadational and of Ladinian age (e.g. Esino Limestone). The troughs separating the carbonate platforms were infilled by turbidites and volcaniclastics (e.g. Wengen Fm.). The peak of the volcanism, maximum subsidence and the greatest extent of the hemipelagic domain date to the latest Anisian and earliest Ladinian. The peak of the transgression corresponds to the temporary drowning of the carbonate platforms and is defined at various places by an ammonoid horizon dated as Early Ladinian. The peak of this transgression is the most areally extensive of all the Triassic. It is known in other Alpine regions, in the German and Paris Basins and as far away as the Barents Sea. Infill of the basin was completed during the final part of the Ladinian. It is recorded by a karstified subaerial surface. Time equivalents of this surface are known throughout Western Europe including the Boreal regions. The succeeding transgressive–regressive facies cycle is centred on the Carnian. Though poorly expressed in the rest of Western Europe, it is characteristic of the Southern and Eastern Alps. It is characterized by mixed lithologies that include carbonates (Breno Fm.), volcaniclastics (Val Sabbia Fm.) and siliciclastics (San Giovani Bianco Fm.). The importance of erosion of the basement at this time is shown by the contribution of detrital quartz in the Late Carnian unknown in the Middle Triassic. Deposition of frankly prograding deltaic sands then sabkhas infilled the basin, which then became emergent. The following transgressive–regressive facies cycle is mainly Norian in age. The *Dolomia Principale (Hauptdolomit)*, which locally exceeds 2000 m in thickness, was deposited solely in upper tidal to shallow infratidal environments demonstrating that the production of carbonate was sufficient to fully compensate for subsidence at each point. The *Dolomia Principale* prograded over troughs characterized by hemipelagic sedimentation. These troughs would not be infilled again during later intervals. The end of the history of the very widespread carbonate platform corresponds to the maximum progradation, which marks the end of the major transgressive–regressive cycle of the Triassic. Prior to the end of the Norian, a new phase of dissection of the dolomite platform records a recurrence of distension and subsidence. From then on, more and more open marine sediments accumulated rhythmically until the start of the Toarcian, which was the time of the peak transgression of the following major cycle. This succeeding cycle resulted directly from the rifting of the Ligurian Tethys (Chapter 6). Approximate horizontal distance: 150 km. No vertical scale. *Simplified from Gaetani et al. (1998), Fig. 2, p. 794.*

2.4 The end of the Late Triassic half-cycle; the Norian overall regression

In the whole of the Southern Alps, the latest Carnian and/or the earliest Norian are marked by renewed extensional tectonism that induced new subsidence and transgression (see the Castro breccia, Fig. 5.3A). In the Western Alps area, discrete and localized extension took place again, notably on the Ardeche border of the Massif Central (Fig. 5.4) and near Briancon, where it may have triggered early salt movement (Fig. 5.5). However, slow subsidence compared to sedimentation rate prevented the establishment of a pelagic sedimentation regime.

The deposition of the aggrading to prograding *Dolomia principale* (=*Hauptdolomit*) over wide areas records the widespread regression of Norian age. Contemporaneous regressive siliciclastic formations are also developed in intra-cratonic European basins, such as the Paris Basin (*Marnes Irisees* and *Gres de Chaunoy* Formation) and the German Basin. The Norian overall regression is recorded as far north as the Barents Sea. It marks the end of the Triassic transgressive–regressive major cycle.

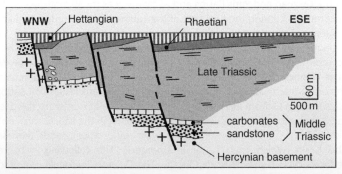

Figure 5.4 *Along the Eastern border of the Massif Central: Cevennes trend Liassic syn-rift faults active in the Late Triassic.* In the Ardeche, near Les Vans, certain syn-sedimentary NE–SW Late Triassic faults, which form part of the Cevennes fault system known in the Alps (Figs. 6.10, 6.17 and 6.21), remained active during the Liassic resulting in NW tilting of fault blocks. The sandstones and carbonates of the early-Middle Triassic show little variation in thickness across the faults. In contrast, the thickness of the multi-coloured argillaceous sandstones of the Late Triassic passes from 2.50 to 50 m across the first fault and to more than 200 m to the east across the second. This results from syn-sedimentary movement on three normal faults, visible in the field, in a region located outside that affected by Alpine deformation. *Simplified after Elmi (1984). Modified from Lemoine et al. (2000), Fig. 7.4, p. 91.*

Figure 5.5 *The southeast face of Mont Janus, south of Montgenevre, seen from the Chenaillet massif (Western Alps); an onlap of Norian age in the Piemontais domain.* In the part of the Piemontais zone situated in the Ubaye Valley and north of the Montgenevre pass as well as in the Brianconnais zone on the same transect, Norian dolomites are of *Hauptdolomit* facies as in the Austro-alpine and South Alpine units. These sediments were deposited in very shallow supratidal, intertidal and infratidal environments under the influence of tidal currents. In the Piemontais and Brianconnais, the *Hauptdolomit* forms two parallel tiers covered by an onlap surface. These beds were tilted several degrees with respect to the horizontal before deposition of bed 2 as shown on the photograph. However, no normal faults have been identified anywhere to date. Dolomitic breccias intercalated in the section may be due to collapse of the corresponding fault scarps, if present. It is possible that this style of early deformation marks the onset of *Raft tectonics* (see below, Fig. 6.10); following one of last episodes of rifting during the Late Triassic (Norian), which would have locally formed slopes. Blocks of *Hauptdolomit* might have become detached from each other to form rafts, which slid on a plastic bed formed by the underlying evaporites of Carnian age. This interpretation is not verifiable. This is because the structural units to which the *Hauptdolomit* belongs are included in the main thrust nappes where the relationship between the Hercynian basement and the Mesozoic cover has been lost by detachment. It is therefore unknown if the basement has, or has not, been subjected to vertical extension. *Photo Roberts. Modified from Lemoine et al. (2000), Fig. 7.3, p. 90.*

3. COMPARISON OF THE TRIASSIC MAJOR TRANSGRESSIVE–REGRESSIVE CYCLE FROM THE BRIANCONNAIS TO THE RHONE VALLEY

As shown for the Southern Alps (Fig. 5.3), the Triassic long term facies cycle can be subdivided into four second-order transgressive–regressive cycles. These four cycles are well recorded in the Brianconnais and Piemont zones, but not in the external zones (Fig. 5.6). Such data show that

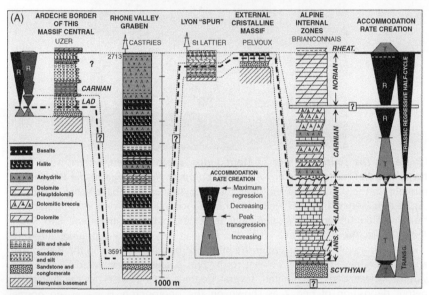

Figure 5.6 *(A) Comparison of Triassic transgressive–regressive facies cycles from the Brianconnais to the Rhone valley.* The internal Alpine Triassic, long-term, facies cycle can be subdivided into four second-order transgressive–regressive cycles. In the Alpine external zones, the Triassic major transgressive–regressive facies cycle is represented by its peak transgression and regressive half only. The sedimentary infilling of the subsiding Rhone graben is not older than late Middle (Ladinian) to Late Triassic. The acme of the Triassic transgression was caused by mild extensional phases that commenced probably by latest Anisian or earliest Ladinian times. But the thickness of Middle Triassic deposits is considerably smaller than the Late Triassic ones. The gypsum and salt layers mostly developed during the Carnian in the time of maximum subsidence. The latter suggests that the extension and subsidence became active in the Rhone graben not earlier than the Late Triassic. *Uzer: simplified from L. Courel et al. (1988). Castries: simplified from Baudrimont and Dubois (1977). Saint Lattier: completed after Courel et al. (1984). Brianconnais: after Megard-Calli and Faure (1988). (B) Location sketch map.* The outcrops and well-log sections of Fig. 5.6A are shown on this sketch map drawn for the Carnian (Late Triassic). Triassic volcanics are known around the Pelvoux massif and in Provence, north of the city of Toulon. They are well developed to the west in the North Pyrenean Triassic graben; 800 m is the isopach curve for Late Triassic beds. *Completed after Debrand-Passard et al. (1984), Fig. 11.5, p. 585.*

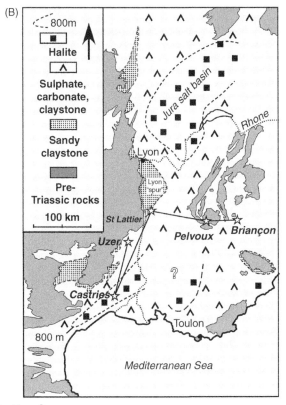

Figure 5.6 (*Continued*)

episodic Triassic rifting started as early as earliest Triassic times (Scythian) in the Internal Alpine Zones but only during the early Carnian in the External Zones, particularly in the Rhone Valley graben. This shows that the structural boundary between Internal and External Alpine Zones, that is the Penninic Frontal Thrust, is inherited from structures of Triassic age.

The **Lowest Triassic second-order facies cycle** is well developed in the Brianconnais zone only and is related to the Scythian (=Werfenian) quartzites. The upper part, composed of shales and evaporites, formed one of the decollement surfaces of the Alpine thrusts. It is most probably not recorded in the external zone.

In the Brianconnais, the **Middle Triassic** (Anisian–Ladinian) **second-order facies cycle** comprises several superimposed aggradational carbonate platforms. The peak transgression is marked by a few crinoid and foram-bearing marker beds. It is tentatively correlated with the Early Ladinian (Fassanian) major peak transgression known in both the Tethyan and the Western Europe areas.

In the Alpine external zones, the Ladinian peak transgression has been identified by a specific pelecypod-bearing marker bed in the cover of the Pelvoux External Crystalline Massif that allows long distance correlation from the internal Alpine zones to the foreland.

On the eastern border of the Massif Central that coincides with the western shoulder of the Triassic Rhone graben, the Middle Triassic second-order facies cycle is represented by alternating sands, silts and clays that grade basinward towards more evaporitic lithologies deposited in fluviatile to sabkha environments. The Early Ladinian (Fassanian) major peak transgression is dated by specific fish-scales, pelecypods, foraminifers, palynomorphs and conodonts. The corresponding layers are intercalated close to the Permian/Triassic contact. It forms a sharp 30-km-wide onlap on the crystalline basement of the Ardeche border of the Massif Central.

The base of **the Carnian second-order facies cycle** in the Brianconnais is marked by a strongly erosional surface that can be correlated with a sharp resumption of the extension and subsidence known on wide parts of Europe. Normal faults and slumped beds and gigantic breccias are indicative of a renewal of active syn-sedimentary extension characteristic of the Brianconnais zone area.

The Late Ladinian record is probably missing below this strongly erosional surface because the Early Ladinian marker bed that records the Triassic major peak transgression is situated close to the geometrical top of the Brianconnais Ladinian section.

The Brianconnais Carnian comprises a complex lithological association of sandstone, shale, dolomite and evaporites dated by terrestrial plants (specific *Equisetum*).

In the Rhone valley graben, the Carnian is characterized by the development of evaporites, noticeably by salt and anhydrite. The maximum thickness of salt may reach more than 500 m or even 800 m in places. Dating is provided by palynomorphs, foraminifers and pelecypods.

The Brianconnais and Piemont **Late Triassic second-order facies cycle** corresponds to the **Norian Hauptdolomit**. This comprises an aggrading, stable and slowly subsiding carbonate platform deposited under inter- to shallow-subtidal conditions. It is nearly entirely dolomitised. The deposition of the Hauptdolomit terminates the regressive Triassic major half-cycle with similar facies in many parts of the Apulia side of the Tethys margin and as far as Greece and Turkey.

In the Rhone valley graben, the Late Triassic succession is characterized by the presence of anhydrite in the more subsident part and a return to prograding conglomerate and sandstone at the top on the borders. The Norian stage has not been so clearly dated as the underlying Carnian and Ladinian. A tentative age is interpreted from lithological comparison with the Jura, Paris and Germany basins.

4. RELATIONSHIP OF THE MIDDLE TRIASSIC AND CARNIAN EXTENSIONAL EVENTS TO RIFTING OF THE LIGURIAN TETHYS

In the stable platforms weakly affected or even unaffected by Alpine deformation, extensional faults, ostensibly active from the Triassic to the Cretaceous, are well documented. One example, classically cited for the Dolomites (Southern Alps), is the system of faults and/or syn-sedimentary flexures, which form the N–S (N0° to N20°E) boundary between the subsiding Carnia-Belluno basin and the adjacent platform (Fig. 10.2, below). These faults were active from the Permian to the Jurassic; Permo-Triassic sediments reach 2 km thickness on one side of the fault and 4–5 km on the other.

In the sedimentary basin of SE France, it is often stated that the network of contemporaneous faults controlling Late Triassic sandstone distribution is not everywhere coincident with the Early Jurassic fault pattern. For example, in the Bourg d'Oisans half graben, a 1987 analysis of the population of local palaeofaults showed a N–S extension direction for the Middle and Late Triassic compared with N110°E for the Liassic. This conclusion led to a radical distinction between the strain fields of the Triassic and the Early Jurassic. Nonetheless, a recent (2002) re-evaluation has shown that the orientation of part of the Triassic palaeofault system has been rotated by Alpine deformation. Furthermore, reconstruction of the palaeofaults of Triassic and Hettangian age now demonstrates a single elongation direction-oriented N90°E to N110°E that is practically constant for the Triassic and Jurassic (see below, Fig. 6.16).

Another good example is provided by the long array of faults separating the Rhone graben from the Massif Central (Fig. 5.4, this Chapter and Fig. 6.16, below), which are hard linked but also detached in Triassic evaporites. Another is the Triassic rift of Aquitaine which may herald the line of opening of the Bay of Biscay during the Early Cretaceous (Fig. 5.1). In the Aquitaine, Paris and German basins, the distribution of the main Carnian evaporite depocentres is governed by networks of extensional faults, e.g., the NW–SE *Pays de Bray* fault in the Paris Basin.

While these relationships are well established in the extra-Alpine areas, an assessment of the structural control on evaporite distribution is more problematic in the Alpine domain. The main evaporite bodies are now completely detached from the underlying units and form major detachment zones at the base of overthrust nappes. If the evaporites were originally located in strongly subsiding half graben as elsewhere, it is possible that some Alpine thrust planes may have been induced by Triassic extensional faults. Indeed, the structure observed at Mont Janus near Briancon may have originated by foot wall diapiric uplift (Fig. 5.5) though no clear demonstration is however possible.

On the other hand, the major Triassic transgressive–regressive facies cycle is quite separate from the major transgressive–regressive cycle extending from the Late Norian through Aalenian (Early Middle Jurassic) times. It thus seems reasonable to conclude that the Triassic extension and the associated transgressive–regressive cycle is a record of distant events in the western arm of the Triassic Tethys and therefore quite distinct from that of the Jurassic Tethys. Thus, the Triassic forms part of the pre-rift history of the Ligurian Tethys.

5. TRIASSIC PALAEOGEOGRAPHY AND THE LINE OF OPENING OF THE LIGURIAN TETHYS IN THE JURASSIC

Part of the Triassic extensional fault system remained active during the Jurassic as shown earlier. Nonetheless, the line of Tethyan opening crosses the main Triassic palaeogeographic trends. The Jurassic break-up propagated from the northern Vardar trough of the Neotethys whose continuation is now found in the Hellenides. The break-up line crossed the immense Triassic carbonate platform reactivating in places Triassic faults (Fig. 5.1, 5.2).

6. END OF THE TRIASSIC (LATE NORIAN AND RHAETIAN) AND ONSET OF RIFTING IN THE LIGURIAN TETHYS

The wide carbonate-dominated platform astride the future European and Apulia margins of the Ligurian Tethys was faulted everywhere by the end of the Norian. In the Southern Alps, an intra-Norian discontinuity rests unconformably on the *Dolomia Principale (= Hauptdolomit).* The latter is cut by a network of extensional faults and is abruptly buried below black shales (Fig. 5.3). A comparable picture can be observed in Carnia (East of the Dolomites).

In the Subalpine zones and on the eastern border of the Massif Central (Ardeche: Fig. 5.4), dating of the Late Triassic is uncertain or absent. Nonetheless, Rhaetian beds with *Rhaetavicula* rest discordantly on underlying Triassic beds, which are there cut by a system of extensional faults. An analogous surface has been identified between the Norian and Rhaetian in the Brianconnais Peyre Haute nappe (Fig. 6.11).

Before the end of the Norian and from before the start of the Rhaetian subsidence resumed allowing the return of open marine conditions though the faunal assemblage lacks pelagic organisms such as ammonites. Platforms separated by hemipelagic basins were the probable main palaeogeographic elements. Some faults inherited from the Middle Triassic were reactivated while new fault arrays appeared for the first time. Both correspond to the network of Liassic extensional faults on the European craton. However, fault throws were minor compared to those of the Late Liassic.

Precise dates can be assigned to the rift onset in the Ligurian Tethys, either by convention or personal bias. Assignment to one phase or another of the Late Triassic tilted blocks in the Western Alps is ambiguous.

In conclusion one can propose that pre-Ligurian rifting began at the end of the Triassic and not at the start of the Jurassic in the strict sense.

FURTHER READINGS

Stampfli and Borel 2002 for the Alpine realm; Withjack, Schlische and Olsen 2002 and 2008 for comparison with the North America eastern coast.

CHAPTER SIX

The Pre-Ligurian Tethys Rift Phase on the European Margin

Contents

Summary

Extension first took place on the European margin of Tethys between the end of the Triassic (225 Ma) and start of the Liassic (205 Ma). Later, episodic but progressive extension during the Liassic and Middle Jurassic resulted in complete continental rupture and the onset of spreading at the end of the Middle Jurassic (165 Ma). The extensional structures typify those of rifts and include crustal scale tilted blocks or half-graben, transverse strike–slip or transfer faults and extensionally triggered diapirs.

To illustrate this chapter, type examples have been selected from the Central and Western Alps where many of the original extensional structures

The Western Alps, from Rift to Passive Margin to Orogenic Belt, Volume 14
ISSN 0928-2025, DOI 10.1016/S0928-2025(11)14006-7
© 2011 Elsevier B.V.
All rights reserved.

are exceptionally well preserved because of weak Alpine deformation. Thus, in the external zone of the Western Alps and on the border of the Massif Central, the NE–SW trend of the Cevennes family of faults corresponds to that of the rift phase faults of Jurassic age. Similarly, the Pelvoux–Argentera trend is one of the major transfer faults that partition the margin into different structural compartments.

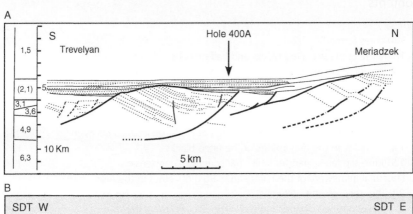

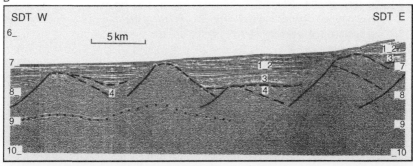

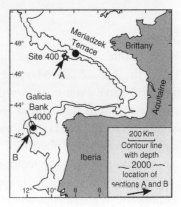

Figure 6.1 (*Continued*)

Though the Tethyan rift phase lasted for 60 million years, many discrete phases of active extension alternated with intervals of relative tectonic quiescence. Each phase of extension was thus followed by one of subsidence. The last phase of pulsed rifting led to complete continental rupture and the onset of spreading of the new Liguro-piemontais Ocean. Rupture was followed by passive thermal subsidence of the new European margin.

 1. STRUCTURES FORMED BY RIFTING

Rifting of the European margin of the Liguro-piemontais Ocean began in places in the Late Triassic (Chapter 5) and continued through the Early to the Middle Jurassic. Normal faults, tilted blocks and strike-slip or transfer faults characterize this rift phase. These structures have been well preserved in both nappes and the Alpine foreland. However, structures preserved within the nappes are difficult to restore to their original orientation because of possible rotation of the nappes during translation. The most significant structural observations have thus been made in areas subject to weak Alpine deformation such as the External Crystalline Massifs, the Sub-alpine Chains, the eastern border of the Massif Central (Cevennes and Ardeche), and more distantly in the Paris and Aquitaine and Biscaye (Fig. 6.1) Basins.

1.1 Rift phase faults, tilted blocks and half-grabens

The network of extensional faults defines a series of alternating half-graben corresponding to areas of relative uplift and subsidence. This pattern of tilted

Figure 6.1 *Examples of the geometry of tilted blocks beneath the Atlantic margins of Western Europe: points of comparison with the Tethyan margin.* The type of fault geometries bounding tilted blocks beneath the passive Atlantic continental margins of west Europe was illustrated for the first time in 1979: see Figures A and B.. These images inspired reconstructions of the syn-rift structure of the Tethyan margin involved in Alpine deformation. Refer to Fig. 11.8E. *(A) Geological cross section through Meriadzek Terrace (north of the Bay of Biscay) showing listric faults that bound the tilted blocks.* Near the base of listric faults is a horizontal reflector corresponding to the interface between two different layers defined by seismic refraction data. Moho is at 12 km. Horizontal and vertical scales are the same. *(B) Seismic reflection profile immediately west of Galicia Bank showing tilted blocks and listric faults and a horizontal reflector below.* The seismic section shows a sub-horizontal undulating reflector blow the tilted blocks at a depth corresponding to the brittle–ductile boundary in the continental crust. The idea has been largely confirmed by later observations. *Reproduced from Chenet et al. 1982, Fig. 2, p. 704.*

blocks and half-grabens closely mimics the classical style observed in many rift systems (see Introduction, Part II).

Rift phase normal faults are the expression of extension in the brittle, upper part of the crust. As noted earlier, these faults are well preserved in the external zones of the Western Alps and are known also, from observation or reconstruction, in the Helvetic nappes, the internal zones and nappes of the Pre-alps (Figs. 6.2 and 6.5–6.11).

Deep seismic profiles across many continental margins show that the normal or listric faults curve in depth to flatten at the brittle–ductile transition. On the Apulian margin, deep erosion has exposed the curvature of the major Lugano listric normal fault (Fig. 10.3). On the European margin, geometrical constraints imply that the main extensional faults flatten at depth like those on the Apulia margin (Figs. 6.5 and, below, 10.3). However, erosion has not been deep enough to expose the curvature of the fault planes at outcrop.

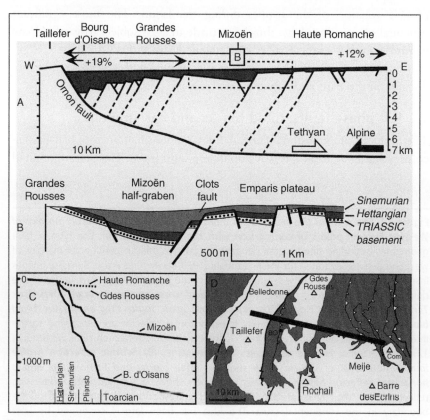

Figure 6.2 (*Continued*)

1.2 Stratigraphic and facies architecture of syn-rift sequence

The overall geometry of the syn–rift sequence in half-grabens exhibits thickening towards the footwall of the adjacent tilted block. The wedge shaped or triangular cross-sectional form of the sequence reflects contemporaneous movement of the bounding fault during infill of the half–graben (Figs. 6.1 and 6.3). However, the stratal architecture of the infill does not consist of strata disposed regularly in a simple fan shape. Instead, the typical succession is broken by unconformities and/or onlap surfaces that result from the episodicity of rift driven subsidence and may also be linked to sea level change. Comparable stratal architectures are well known from the Late Jurassic syn-rift sequences of the North Sea (Fig. 6.3).

The wedge shaped syn–rift sequence thins towards the raised edge of the adjacent tilted block where the age equivalent section is condensed and or eroded locally. The raised edges of the tilted fault blocks have undergone

---

Figure 6.2 *Development of the Bourg d'Oisans half-graben during the early Jurassic. (A) Reconstruction of the whole Bourg d'Oisans tilted block during the Early Jurassic.* The proposed profile of the listric fault responsible for the formation of the Bourg d'Oisans half-graben consists of series of ramp and flats. In effect, the quasi-continuous creation of accommodation space in the main half-graben situated at the foot of the Ornon fault cannot be explained by a model composed only of horizontal flats. The main basin corresponds to the upper ramp. The rollover anticline corresponding to the upper flat is represented by the Grandes Rousses–Rochail block. The basin corresponding to the lower ramp is the Mizoen antithetic half-graben. Finally, the Emparis Plateau–Haute Romanche area corresponds to the lower flat. The minor creation of accommodation space in the latter area requires a horizontal deep flat. Stretching of the whole with respect to the pre-rift state is estimated at 12% and 9% for Bourg d'Oisans half-graben alone. *(B) Detail: The Mizoen half-graben.* The half-graben was first subjected to diffuse extension with faults of minor throw sealed by the Sinemurian. The Mizoen half-graben was later formed by displacement on the Clots fault antithetic to the Ornon master fault and separating it from the Emparis Plateau to the east. The resulting syn-rift sediments pass laterally and progressively into the condensed section of the Grande Rousses. *(C) Relative subsidence curves for each of the parts of the Bourg d'Oisans half-graben.* The sector of the half-graben situated at the foot of the Ornon fault exhibits the strongest subsidence. The Mizoen half-graben shows significant but much less subsidence. The intermediate sector (Grande Rousses and adjacent units) as well as the Emparis Plateau and units of the upper Romanche valley are characterized by condensed sections and small amounts of subsidence. For the thicker sediments, breaks in the subsidence curves record discontinuities in extensional fault activity during rifting. *(D) Location diagram.* Simplified after Chevalier 2002; Chevalier et al. 2003.

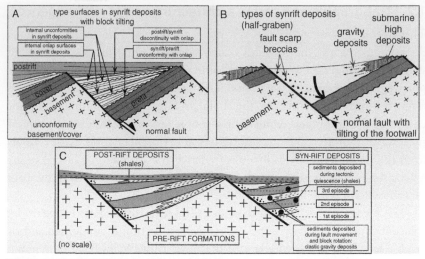

Figure 6.3 *A, B* and *C Sedimentary signatures and record of extensional tectonism and episodic rifting from Late Jurassic half-graben in the North Sea.* These figures of tilted blocks and half-grabens from the North Sea provide points of comparison with those of the Tethyan margin. In the case of the Tethyan margin, the marine syn-rift sediments were deposited from the onset of rifting. Examples of syn-rift stratal and facies architecture are found in the Jurassic of the Bourg d'Oisans 'syncline' (Fig. 6.5). The tilted blocks have widths of 10–20 km and are bounded by faults with throws of 1–5 km. During each episode of extension, displacement along the normal faults resulted in rotation of the hanging wall block as well as the creation or rejuvenation of submarine topography. Submarine erosion resulted in interbeds of clastic sediments in the infill of the half-graben. *Simplified from Ravnas and Steel 1997. Modified from Lemoine et al. 2000, Fig. 2.4, p. 30 and Fig. 2.6, p. 33.*

less subsidence than the half-graben and also possibly footwall uplift. As a result, sediments were deposited in shallower water depths, are less thick and may have been subject locally to sub-aerial erosion (Fig. 6.1).

The lateral transition between beds deposited on the foot wall and those in the half-graben is marked by progressive thickening of the sequence and deepening of depositional environments. Shallow water sediments deposited on the high block pass laterally into deeper water shales wherein discontinuities are commonly marked by turbidites and debris flows (Fig. 6.3–6.5).

1.3 A type example: the Bourg d'Oisans half-graben

The recognition of closely analogous syn-sedimentary structures of Jurassic age in the Bourg d'Oisans half-graben (Figs. 6.2 and 6.5) to those observed on seismic profiles across present day passive continental margins has largely

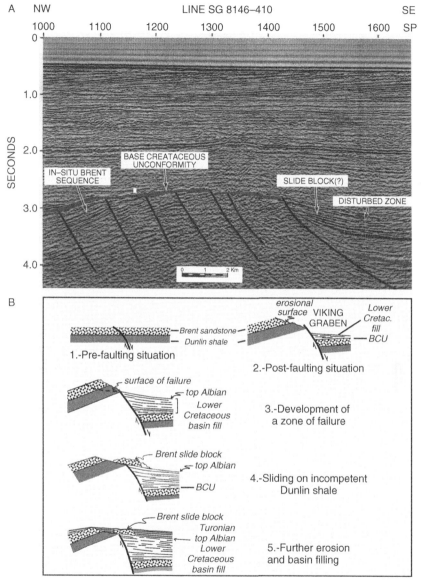

Figure 6.4 *Example of a slumped block emplaced during the syn-rift phase on the flank of the Viking graben (North Sea). (A)* Seismic section, and, **B**, interpretation. (B) no scale on this cartoon. Refer to Fig. 6.6. *Simplified from Alhilali and Damuth 1987, Fig. 9.*

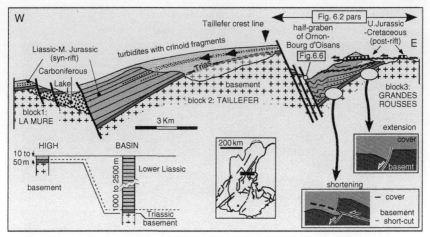

Figure 6.5 *Schematic cross section of three large tilted blocks east of Grenoble.* The post-rift deposits are here separated from the original substrate by a thrust. As a result, the original thickness of the section above the basement cannot be determined. *Modified from Lemoine et al. 2000, Fig. 8.3, p. 96.*

motivated the application of passive margin structure to pre–Alpine tectonics (Figs. 6.1, 6.3 and 6.4). Obviously, the reconstruction of the geometric characteristics of the Bourg d'Oisans graben prior to Alpine deformation was necessary to support such comparisons. Several models have been proposed of which the most recent relies on the application of biostratigraphic and stratigraphic analysis at the scale of ammonite sub-zones to the modes of syn-sedimentary extension as well as the analysis of palaeofault populations.

1.3.1 Reconstruction of the geometry of the Bourg d'Oisans half-graben in the syn-rift phase

Restoration of the syn-rift geometry utilizes the basement/cover interface as a key marker to which zero altitude is assigned given the sandy and evaporitic nature of the Triassic sediments resting on basement. The restoration shows three compartments that are, successively, from west to east (Fig. 6.2):

- The main half-graben with a thick sedimentary infill limited to the west by the Ornon master fault (Fig. 6.6) whose estimated throw at the end of the Liassic was about 1300 m. The infill corresponds to a line of basin depocentres aligned parallel to the foot of this fault and about 3 km in longitudinal extent. It passes progressively eastward into a sector showing progressively less subsidence as far as the eastern limit of the Grandes Rousses massif.

Figure 6.6 *View of the breccia on the Ornon fault (near the village of Chalp de Cantaloupe).* The photograph faces south and the cross section below is E–W. Exotic blocks, up to tens of metres in size, are composed of Triassic dolomites and basalts (210–220 Ma) or Liassic limestones (195–205 Ma) and are enclosed by highly cleaved marlstones of Late Liassic (185 Ma) age dated by ammonites. The in situ Triassic beds from which these blocks were derived lie 1700 m above this outcrop. The present day total throw on this Liassic fault is 3500–4000 m. For location see Fig. 6.2 (See colour plate 7). *Photographs by courtesy of Thierry Dumont.*

- The next compartment to the east is the half-graben of Mizoen. It is bounded to the east by the main Clots fault which is antithetic to the Ornon fault although with a smaller throw (Fig. 6.2B).
- The easternmost part consists of the Emparis plateau followed by other units, which are cut by the Romanche valley, and disappear beneath the

Pennine frontal thrust. This part of the Bourg d'Oisans half-graben is characterized by condensed sediments whose subsidence has been modest.

The difference in behaviour of the three compartments through time can be visualized by their relative subsidence curves (Fig. 6.2C).

If the Ornon fault is considered as the emergent planar upper part of a listric normal fault, geometrical rules can be applied to reconstruct the profile of the fault in depth using surface observations: the most appropriate profile is that formed of several ramps and flats (Fig. 6.2A).

1.3.2 Modes of crustal extension: diffuse extension then focussed extension

Two modes of extension of the Bourg d'Oisans half-graben can be distinguished for the start of the syn–rift phase (Fig. 6.7).

- From the Triassic to the Early or Middle Hettangian, diffuse extension was distributed across the basin. Crustal stretching was accommodated by minor throw on many, small and relatively closely spaced faults (Figs. 6.6–6.9). The numerous small syn-sedimentary faults dispersed throughout the area of the basin (Fig. 6.7) have lengths averaging 1 km and not more than 1.5 km. Throws are everywhere less than 150 m and on average 50 m. Spacing is less than 1 km with an average of 500 m. These faults are typically sealed by sediments of Late Hettangian age (Angulata zone).

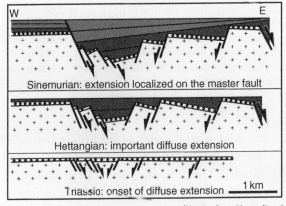

Figure 6.7 Diffuse then focussed extension. Simplified after Chevalier 2002.

Figure 6.8 *Aerial view of tilted blocks of probable Hettangian age in the Besson Lakes area near Alpe d'Huez.* View looking south with west to the right in contrast to the orientation of Fig. 6.6. The Grande Rousses External Massif dipping to the right is faulted into small third-order size tilted blocks. These belong to the same group shown in Fig. 6.6 and are thus of probable Hettangian age (c. 200 Ma). The Liassic sedimentary cover has been stripped by recent erosion. The remaining cover consists of a few metres of Triassic dolomites resting on gneissic basement and preserved in the axes of the half-grabens (See colour plate 8) *Photograph courtesy of Cl. Kerchove.*

- From the middle or Late Hettangian (Liasicus or Angulata zones) to the Late Liassic, extension was focussed along the Ornon master fault and some antithetic faults (Figs. 6.2A, 6.2B, 6.5 and 6.6).

The length of the Liassic Ornon fault is about 35 km. As noted earlier the throw is estimated at 1300 m. It does not mark a single surface but rather a corridor of faults with a sinusoidal envelope, several kilometres wide in places resulting from the connection of many segments. Other faults, of lesser importance and antithetic to the master fault, include the Clots fault which delimits the Mizoen half-graben to the east (Fig. 6.2B). The neighbouring Petites Rousses fault was active only during the Sinemurian and Early Pliensbachian. Lengths in map view are of the order of 5–10 km and throws are less than 500 m. The stretching direction for the assemblage of small and large fault lies between N90°E and N110°E.

1.4 Hierarchy in the dimensions of the tilted blocks

The tilted blocks can be grouped into three orders of size depending on width and also the magnitude of throw along the bounding faults.

The smallest, third-order blocks have widths of a few hundred metres and are bounded by normal faults with throws ranging from a few to several tens of metres. Well-documented examples are known on the border of the

Figure 6.9 (*Continued*)

Massif Central (Fig. 6.10), in the cover of the External Crystalline Massifs (Figs. 6.8 and 6.9) as well as in the Brianconnais (Fig. 6.11) and Piemontais (Fig. 6.12) domains. These third-order faults date to the earliest, latest Triassic or Early Liassic phase of rifting. While clearly seen at outcrop, they are typically sub-seismic in scale because of their small throws.

The second-order tilted blocks have widths of between 5 and 20 km (Figs. 6.2 and 6.5) and encompass the small, third-order faults. The second-order faults often utilize faults of the preceding generation but have throws between 1 and 3 km. These second-order blocks are seismic-scale structures and are closely comparable in all respects to tilted blocks imaged beneath passive margins and present day rifts (Figs. 6.1 and 6.3). Though clearly visible in the field, these blocks were traditionally interpreted as Alpine

---

Figure 6.9 *The west flank of the Rochail tilted block (southern extension of the Grandes Rousses Massif): a small half-graben of Hettangian age.* The western flank of the Rochail basement massif corresponds to the gentle slope of the Bourg d'Oisans major half-graben (Fig. 6.2). The associated Ornon master fault is illustrated on Fig. 6.6. The small-scale fault of Hettangian age on views B and C of this figure is shown on the model, Fig. 6.7. Triassic basalts (210 Ma) and shallow marine Hettangian limestones (206–202 Ma) have been tilted by displacement along normal fault F on the left or west of the photograph (C). The fault throw is about 10 m and the width of the small half-graben is a few hundred metres. Triassic basalts outcrop to the left of the fault trace. The later sediments have been dated as Late Hettanginian (203–202 Ma) by ammonites. Here they are sub-horizontal and onlap the underlying basalts (photograph **D**). These beds were deposited in deeper marine environments than the underlying strata due to initial subsidence following rifting. The time interval between the rift event (recorded by fault movement and bed tilting) and the initial subsidence is less than the total duration of the Hettangian (c. 5 Ma). Blocks of comparable dimension and significance, but without sedimentary cover, are shown on the photograph of Fig. 6.8 (Alpe d'Huez). **(A)** Overview: From left to right, i.e., from east to west: firstly outcrops of the pre-rift including the pre-Triassic crystalline basement above Lake Vallon; yellowish Triassic sands and dolomites a few tens of metres in thickness; overlying flows of Triassic basalts (dark cliff). Secondly, outcrops of the syn-rift: limestones and clays of Liassic and Middle Jurassic age. Finally, outcrops of the post-rift represented by a thin resistant layer of the Late Jurassic and the Early Cretaceous. Below and to the left (small rectangular frame), a small syn-rift fault displaces the basalts shown on Figs. 6.9C and 6.9D. **(B)** Close view on the small-scale half-graben. The upper frame shows slumped beds of Late Hettangian age. **(C)** Close-up view on the fault that bounds the half-graben of Hettangian age. **(D)** Early Hettangian beds onlapping the Triassic basalts. **Tr:** Triassic basalts. **Hett:** Early Hettangian platform carbonates. **Sin:** Late Hettangian to Early Sinemurian ammonite-bearing pelagic marlstones (See colour plate 9) *Photographs courtesy of Thierry Dumont. A and C: modified from Lemoine et al. 2000, Figs. 8.5 and 8.6, Table VIII.*

structures until deep seismic data across present day passive margins showed this interpretation was incorrect.

The largest first-order blocks are several hundred kilometres wide and include several second-order blocks. The Brianconnais is one major uplifted first-order block. It can be schematically subdivided into two first-order blocks between 30 and 50 km wide which are in turn composed of 5–8 km wide blocks (Figs. 6.13 and 6.14). In contrast the 150–200 km wide Dauphinois basin, which has a syn-rift section between 2 and 3 km in thickness, is an example of a complex, first-order half-graben (Fig. 6.13).

Finally, a comparison can be usefully made with the present day passive margin west of Iberia. This margin is structured into half-graben and major blocks, such as Vigo and Galicia Banks, of the same order of size as the Brianconnais platform and Dauphinois basin (see below, Fig. 12.2).

1.5 Strike-slip faults, transfer faults, transpression, transtension

The extensional fault blocks are limited in their along strike length by faults which transfer the extension from one block to another. Where seen in the field, these faults are strike-slip in character and typically trend

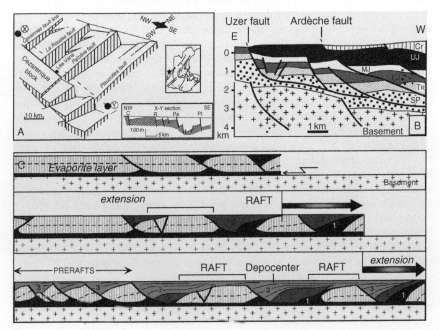

Figure 6.10 (Continued)

Figure 6.10 *Extensional structures on the eastern border of the Massif Central (proximal part of the Tethyan margin). (A) Schematic block-diagram of the structure of the Les Vans area of the Ardeche border of the Massif Central.* Tilted blocks and syn-rift faults which trend NW–SE (Cevennes) or NW–SE (Pelvoux–Argentera) define an extensional fault fabric unaffected by Alpine deformation. Faults: **C**, Cevennes; **R**, Rousse; **Pa**, Paiolive; **Pl**, Plauzolles. *Simplified from the field work data by Elmi 1984. Modified from Lemoine et al. 2000, Fig. 8.2, p. 94. (B) Deep Structure.* Interpretation of seismic profiles in the same area show that the major faults visible at outcrop are the surface expression of a common detachment surface that links these faults at depth. This surface utilizes incompetent horizons in the Stephanian or Permian. The fault was subject to only modest inversion during Alpine deformation in contrast to comparable surfaces located, however, in distal margin positions (Fig. 6.5 and 13.3, below). Note also the small extensional faults, which displace the Hercynian basement surface. The common detachment surface separates the structure into two parts in a vertical sense. In the Early panel, the Hercynian basement is cut by a few normal faults of modest throw probably initiated in Stephanian time. In the upper panel, the faults which separate tilted blocks in the Mesozoic cover do not extend down into the basement. Their number and attitude suggest that the rate of extension was appreciably more important in the cover than in the basement with the caveat that no rigorous analysis of this problem is available. **Cr**: Early Cretaceous. **LJ**: Early Jurassic. **MJ**: Middle Jurassic. **SP**: Stephanian and Permian. **TR**: Triassic. **UJ**: Late Jurassic. *Simplified after Bonijoly et al. 1996, Fig. 11, p. 621 and Razin et al. 1996, Fig. 3, p. 627. (C) Processes of raft tectonics during thin-skinned extension.* The type of structure interpreted on the eastern margin of the Massif Central can be compared with that formed by *raft tectonics* which is an extreme form of thin skin extension. Extension in the cover above undeformed, brittle basement is enabled by an intervening weak ductile, decollement layer consisting of evaporites or shales. Along the border of the Massif Central, the decollement horizon is the shales of the Permo-Triassic succession. For the Norian *Hauptdolomit* blocks in the internal Alpine units, the detachment would occur at the level of the underlying gypsum beds if this model is applicable there also. In the case of the figure shown here above (Fig. 6.10B), the individual blocks dissected in the cover by listric faults blocks rest in contact with each other and are termed *pre-rafts*. When the blocks are separated by a distance sufficient to remove contact they are termed *rafts sensu stricto*. The separation favours the formation of trough-like depocentres in which sediments accumulate contemporaneously or during successive phases of block translation. This type of geometry has been clearly identified on the east and west margins of the south Atlantic, most notably on the margin off Angola. The mechanism for the translation of the rafts is assigned to gravity sliding on the slope of the margin. The width of the Angola margin where these phenomena have been observed is of the order of 440 km. This width is sufficient to allow formation of individual rafts, 10–20 km wide, separated by depocentres 5–10 km wide. Although the same type of mechanism might be applied to the eastern (Cevennes) border of the Massif Central during the Mesozoic, the width of the slope to the axis of the Rhone Trough is only a few tens of kilometres (cf. Angola) and too small for the development of the raft stage. *Simplified after Duval et al. 1992, Fig. 1, p. 388.*

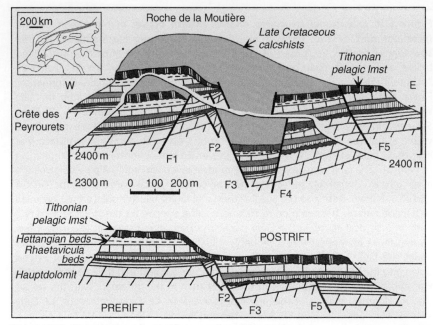

Figure 6.11 *Superimposed syn-rift unconformities in the Brianconnais Peyre Haute Nappe (South of Briancon).* In the Peyre Haute Nappe, the Norian Hauptdolomit is the latest pre-rift formation. Several erosional unconformities dated as latest Triassic (Rhaetian: *Rhaetavicula*-bearing beds) were induced here by small-scale syn-sedimentary extensional faults. These are closely comparable to similar structures in the external unit in the Bourg d'Oisans area (Figs. 6.2 and 6.7) that resulted from the initial, diffuse extension. The major surface at the base of the Tithonian pelagic limestone corresponds to the Jurassic break up unconformity. In the other Brianconnais nappes, neither Rhaetian nor Liassic beds are preserved below this sub-aerial erosional surface. At other nearby localities, this surface is overlain by Late Bathonian platform carbonates that correspond to the earliest post-rift deposits. *Adapted from P. Tricart 1988, unpublished.*

NW–SE. In principle, they prefigure the transform faults of the future oceanic crust. Examples of clearly exposed but relatively small transfer faults have been described from the Ardeche border of the Massif Central (Fig. 6.10). When oriented obliquely to the extension direction the displacement determines the development of transtensional structures, e.g., the small, NW–SE elongated half-graben at the Col des Marmes on the Pelvoux massif (Fig 6.15), or transpressional structures, e.g., the inversion structures in the Saint Laurent area of the Median Pre-alps of Chablais (Fig. 6.14A).

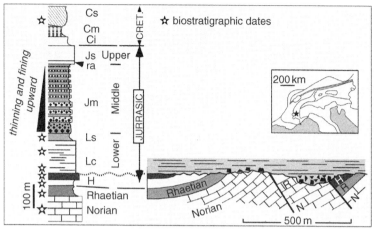

Figure 6.12 *Syn-rift sedimentation in the Piemontais unit of Rochebrune (East of Briancon).* Two successive phases of rifting are shown on this section. The Early Liassic phase is expressed by an unconformity on small tilted blocks and by the deposition of clastic sediments. The second, Late Liassic phase, is marked by deposition of turbidites. The clastic sediments were derived by erosion of the sub-aerial Brianconnais platform. The succession shows from base to top: (a) upward thinning of the beds, (b) upward fining of grain size from decimetres to millimetres and (c) a progressive change in the nature of the clastic debris. At the base, the debris consists mainly of dolomites originating from erosion of Triassic carbonates; increasing quartz and mica towards the top reflects erosion of the basement after the Triassic cover was removed by mechanical and karstic erosion. **N**: Norian. **R**: Rhaetic. **H**: Hettangian. **Lc**: Sinemurian and Pliensbachian. **Ls**: Toarcian. **Jm**: Middle Jurassic. **Ra**: radiolarian-bearing cherts. **Js**: Late Jurassic. **Ci, Cm, Cs**: Early, 'Middle' and Late Cretaceous, respectively. *Modified and supplemented after Dumont 1988. Adapted from Lemoine et al. 2000, Fig. 8.11, p. 99.*

The most important of the transfer faults is the Pelvoux–Argentera line (Figs. 6.16 and 6.17), which divides the Tethyan margin into two distinct structural segments.

 ## 2. STRUCTURAL FRAMEWORK OF THE EUROPEAN MARGIN OF TETHYS IN THE ALPS

The Alpine foreland is comprised of a well-preserved and relatively complex network of rift phase faults grouped into several main trends (Fig. 6.16).

– A NE–SE family of mainly extensional faults called the Belledonne or Cevennes trend.

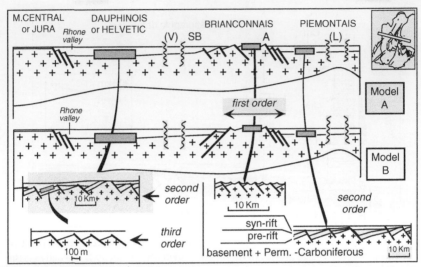

Figure 6.13 *Schematic structure of the European Tethyan margin along the Briancon transect during Jurassic rifting.* Rank in the order of size of the tilted blocks is shown in italics. **SB**: Subbrianconnais; **V**: Valais domain in the Central Alps: this part of the transect marks a complex suture during the Alpine collision along which one or more terranes of unknown width would be eroded or subducted. Terranes and consequently field data are missing along the Brianconnais/Subbrianconnais nappe boundary due to Alpine thrusting. Therefore, two models are presented here. **Model A**: almost all the faults dip towards the axis of rift and site of the future Ligurian Ocean: in this model, the Brianconnais comprises as a whole a first-order block. **Model B**: the faults bounding the Brianconnais dip symmetrically in opposite directions: in this case, the Brianconnais is a very large horst. **A**: Acceglio zone, located in the internal Brianconnais, was deeply eroded during rifting. **L**: axis of the rift, which would be ruptured at the end of the Middle Jurassic to initiate the Ligurian Ocean. **V**: axis of the rift, which would be ruptured by Late Jurassic or Early Cretaceous times to initiate the Valais Ocean: the age of the onset of the Valais Ocean, if it exited, is debated (see Chapters 7 and 11). *Adapted from Lemoine et al. 2000 Fig. 8.13, p. 103.*

- A NW–SE family of strike-slip or transfer faults known as the Pelvoux–Argentera trend.
- A family of N–S to N10 faults characterized by oblique slip.

2.1 The role of the different rift phase fault families: NE–SW (Cevennes-Belledonne) and NW–SE (Pelvoux–Argentera)

The NE–SW Cevennes trend of the major tilted blocks is apparently parallel to the axis of the Liguro-piemontais Ocean. In contrast, the orthogonal Pelvoux–Argentera trend arguably represents that of transfer

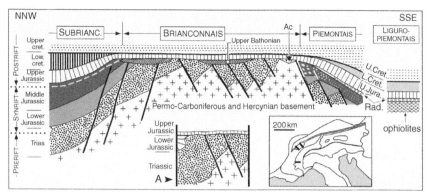

Figure 6.14 *Restored cross section of the SBR block before Alpine collision showing the positions of the Subbrianconnais, Brianconnais and Piemontais domains.* This section integrates studies made across Briancon (Western Alps) and the Pre-alps (Central Alps). The succession of units in the Pre-alps (Fig. 2.8) comprises, from NW to SE, the Mediane Plastique nappes (equivalent to the Subbrianconnais), the Mediane Rigide nappes (equivalent to the Brianconnais) and the Breche nappe (equivalent to the Piemontais). The width of the Brianconnais zone before folding was between some tens of kilometres and a hundred kilometres depending on the cross section. The Brianconnais area was sub-aerial and partly eroded during the syn-rift phase. The first post-rift marine sediments on the Brianconnais platform are platformal limestones of Late Bathonian age. **AC**: Acceglio zone where syn-rift erosion and tectonic denudation has exposed the Permian and crystalline basement. Rad: radiolarites. **A**: In the Chablais Pre-alps, the compressional structure is probably due to local transpression. Section is schematic and not to scale. *Simplified after Baud and Septfontaine 1980; Amaudic du Chauffaut et al. 1986; Septfontaine 1995. Adapted from Lemoine et al. 2000, Fig. 8.10, p. 98.*

faults in the evolving rift. Two lines of evidence corroborate this hypothesis.

Statistical analysis of fault populations in the area east of Grenoble allows reconstruction of the approximate syn-rift fault geometry and deformation. Microtectonic studies of the orientations of slickensides also provide a detailed insight into extension directions. For the Late Liassic rift phase, the extension direction was NW–SE, perpendicular to the tilted blocks and faults of the Cevennes–Belledonne trend and parallel to the Pelvoux–Argentera trend.

In addition, NW–SE small circles defining the rotation of Eurasia with respect to Africa during Middle to Late Jurassic spreading give an indication of the extension direction during the immediately prior rift phase (Fig. 6.16). The NW–SE Pelvoux–Argentera line is closely tangential to these small circles whose low curvature reflects the distant location of the Euler pole near the Gulf of Bothnia.

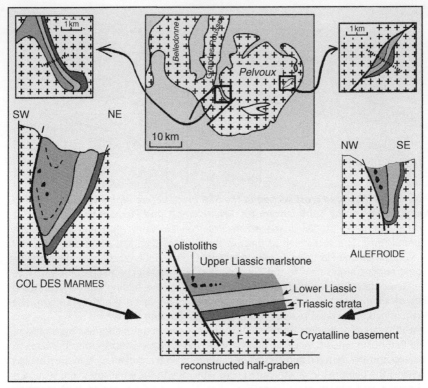

Figure 6.15 *Jurassic reconstruction of two synforms of Triassic and Liassic sediments in the Pelvoux External Crystalline Massif.* To the left, map and section of the synform or 'pinched syncline' of Desert in Valjouffrey Col des Marmes., to the right, the 'pinched syncline' of Ailefroide. These synforms are asymmetric since the Triassic is absent above the Hercynian basement on one flank. These synforms can be interpreted as syn-rift half-grabens later tightened during the Alpine collision. The Late Liassic beds include breccias derived from fault scarps and comparable to those at the foot of the Ornon fault (Figs. 6.4 and 6.5). While the reconstruction of the half-grabens of the Marmes Col and Ailefroide is straightforward, the grabens do not have the same strike. The half-graben of the Marmes col is defined by a NW–SE trending palaeofault nearly perpendicular to the regional extension direction but parallel to strike of the major transfer faults (Fig. 6.15). This half-graben has probably formed by transtension. In contrast, the fault bounding the Ailefroide half-graben trends NE–SW and parallel to the strike of the major tilted blocks. *Modified from Lemoine et al. 2000, Fig. 8.8, p. 97.*

2.2 Segmentation of the margin by major transfer faults: the NW–SE Pelvoux–Argentera line

This line, which marks the southern limit of the External Crystalline Massifs of the Pelvoux and Argentera (Fig. 6.16), is a particularly clear example of a transfer zone or fault. It separates two segments characterized by distinctly

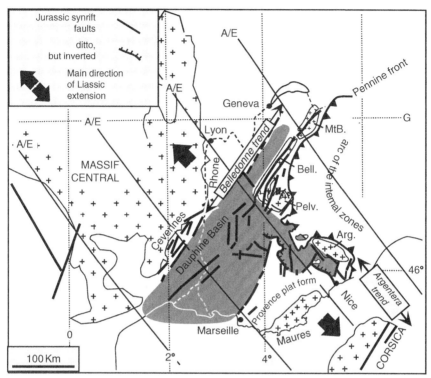

Figure 6.16 *Syn-rift faults, extension directions and small circles of rotation of Africa with respect to Europe.* **A/E**: Small circles of rotation of Africa with respect to Europe from the start of spreading. External Crystalline Massifs: **MtB**: Mont Blanc; **Bell**: Belledonne; **Pel**: Pelvoux; **Arg**: Argentera. Grey: Dauphinois Basin with thick hemipelagic Jurassic sediments. *Modified from Lemoine et al. 1989 and Lemoine et al. 2000, Fig. 13.2, p. 144.*

different fault fabrics and moreover Alpine structures (Fig. 6.17). The segment located SW of the Pelvoux–Argentera line comprises the 150–200 km wide Dauphinois Basin situated between the Massif Central to the NW and the Provence platform to the SE. The Dauphinois Basin is characterized by up to 1000 m of Triassic evaporites mobilized into diapirs during rifting. The total thickness of the Mesozoic infill is between 10 and 15 km in the Languedoc including 2000 m of Bathonian-Oxfordian black shales alone. There are no outcrops of the External Crystalline Massifs. The same structural segment includes Corsica when the latter is restored to its original position (see below, Fig. 9.1). The available reconstructions for Alpine Corsica shows that the Tethyan margin, on this transect, is much narrower than that in the Briancon area (Fig. 9.2).

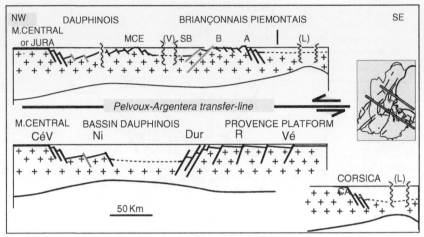

Figure 6.17 *Jurassic structural style of the Tethyan continental margin on either side of the Pelvoux–Argentera transfer.* The segment situated SW of the transfer consists of the following elements: the single, 150 km wide, subsiding Dauphinois basin, the Provencal carbonate platform, the Hercynian basement of the formerly contiguous Maures-Esterel-Corsica block and the narrow adjacent Piemontais zone adjacent to the Ligurian oceanic crust (L) of Corsica. The segment located NE of the transfer consists of two large high areas separating three subsiding basins. These are, from NW to SE: the narrow, 50 km wide northward continuation of the Dauphinois basin in the Western Alps and further north as the Helvetic Basin of the Central Alps; the high is represented by the trend of the External Crystalline Massifs (MCE): the subsiding basin on the site of the future Valais ocean (V); the Brianconnais platform of the Western Alps (SB, B, A; about 100 km wide); the Piemontais basin adjacent to Ligurian ocean crust (L). **A**: Acceglio zone of the internal Brianconnais deeply eroded during rifting. **B**: Brianconnais. **Cev**: Cevennes faults. **D**: Durance fault. L: axis of the Jurassic rift, which ruptured to initiate spreading in the Ligurian Ocean. **MCE**: External Crystalline Massifs. **N**; Nimes fault. **R**: Rouaine fault. **SB**: Subbrianconnais. **V**: Future Valais Ocean break-up line or through line: **Ve**: Vesubie fault. *Modified from Lemoine et al. 2000, Fig. 8.13, p. 103.*

In contrast, the structural segment situated north of the Pelvoux–Argentera transfer is composed of a small branch of the Dauphinois basin. This branch is only 50–60 km wide in comparison to the 200 km wide basin to the south and has a much smaller infill of only 5 km of Mesozoic sediments at Grenoble. The basement of the adjacent External Crystalline Massifs forming the blocks of La Mure, Taillefer and Grandes Rousses (Figs. 6.5) was relatively close to outcrop during the syn-rift phase. Today, the basement of these blocks forms one of the main lines of peaks in the Alps. Along these crests, the thickness of the Mesozoic cover is reduced to a few tens of metres in contrast to the thick section found in the intervening half-grabens, such as the Bourg d'Oisans 'syncline'. Beyond the Pennine Front to the SE, the stack of the Brianconnais and Piemontais nappes was derived from the distal part of the Tethyan margin, here several hundred kilometres wide but absent in Corsica.

Syn-rift diapirism is largely confined to the Dauphinois Basin to the SW because of the presence of thick evaporites there compared to their thinness or absence in the segment immediately to the NE. However, Triassic evaporites also occur on the distal part of the Tethyan margin near Briancon and Val d'Isere and may also have been mobilized as diapirs during rifting.

It is quite probable that other transfer faults of Jurassic age existed to the northeast and east as in the Grisons (=Graubünden) area (see below, Figs. 12.8 and 12.10). However, palinspastic reconstruction is more problematic than in the Western Alps because of the intensity of later Alpine deformation.

3. EVIDENCE OF EXTENSION AND PULSED RIFTING DURING THE LATE TRIASSIC AND JURASSIC

The displacement history of normal faults controlling the tilting of crustal blocks can be shown at outcrop from several characteristic sedimentary structures. These include timing of the cessation of fault movement from the age of the overlying section unconformities or onlaps which indicate tilting even if the fault is not exposed (e.g. see Fig. 6.9). Intercalated slumped beds, olistoliths, syn-sedimentary breccias and turbidites collectively indicate differentiation of strong submarine topography (Figs. 6.6, 6.12, 6.18).

On the European margin of Tethys of the Western Alps, these different pieces of evidence can be linked in space and time. Correlation shows that the rifting was pulsed and not continuous (Figs. 6.20) and furthermore that each pulse of rifting was immediately followed by sharp, initial subsidence as might be expected.

3.1 The timing of extension rates during the Liassic

Detailed correlation shows two (or three) distinct phases of rifting during Triassic and three during Liassic to middle Jurassic times (Fig. 6.19). These phases are separated by time intervals of relative tectonic quiescence. Nonetheless, extensional movements along faults did not cease between the main episodes. The motion along a single fault was discontinuous. Figure 6.20 shows that the duration of the peaks in displacement rates were brief and of the order of the duration of an ammonite sub-zone, i.e. less than 500,000 years or 300,000 years. Extension was diffuse during the

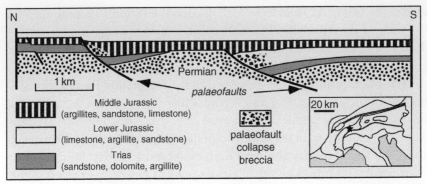

Figure 6.18 *Tilted blocks of Early and Middle Jurassic age in the Glarus Alps (Central Switzerland).* The Glarus Alps belong to the Helvetic nappes which represent the continuation of the Dauphinois domain in Central Switzerland. Because of the thin Triassic section, exposure of the fault scarps has allowed erosion of Permian sandstones which has supplied siliciclastic debris of variable sizes to accumulate at the foot of the scarps. These palaeofaults are today oriented NW–SE or E–W. However, these trends may not represent the original strike because of possible rotation during thrusting. *After Trümpy, in Lemoine and Trümpy 1987. Modified from Lemoine et al. 2000, Fig. 8.9, p. 97.*

first phases, with displacement velocities less than 0.5 mm per year. The maximum rate, focussed along the Ornon fault, reached more than 1.5 mm per year locally. It is dated as late Sinemurian (Turneri zone). Peaks of fault activity were therefore not simultaneous from one part of the profile to another.

A convenient way of understanding the evolution of rifting is to use relative subsidence curves for half-grabens defined by rift phase faults on the one hand (Fig. 6.2C: subsidence) and the rates of movement of the main faults determined from ammonite zones or sub-zones (Fig. 6.20) on the other. The rate of fault movement is calculated from the biostratigraphic record of two well-correlated sections situated on one part or another of the fault. The resolution is of the order of 100,000–400,000 years for the Hettangian–Early Pliensbachian interval. The relative error in the estimation of rate can be important because of errors in the available absolute ages and the relative imprecision in dating by ammonites. Nonetheless, the systematic variation in errors gives a semi-quantitative value to the results. In addition, it appears that the movement of the main faults during the Liassic was discontinuous with periods of quiescence separating those of active movement, each of different rates. Periods of stasis may be expressed in the observed onlap patterns in half-graben.

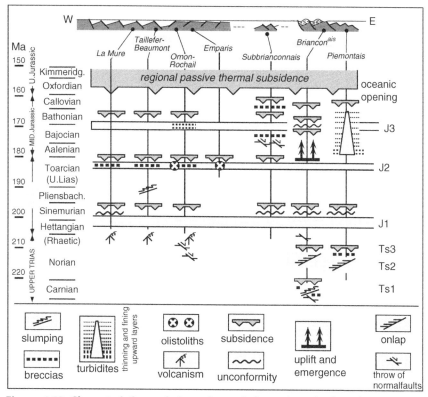

Figure 6.19 *Characteristics and chronology of the main episodes of Triassic and Jurassic Tethyan rifting on the Grenoble transect.* Ts1, Ts2, Ts3, J1, J2, J3: successive episodes of rifting on the Grenoble transect. Two other major episodes of rifting induced widespread unconformities and palaeogeographic renewals in the Subalpine zone, such as in the city of Digne area. One is dated at the Early/Late Pliensbachian boundary and the other to the Pliensbachian/Toarcian boundary (Figs. 6.20 and 16.4). However, they have not been classically described on the Grenoble transect. *Modified from Lemoine et al. 2000, Fig. 8.12, p. 101.*

This method of high resolution dating shows that the peaks in activity were not all simultaneous along the margin (Fig. 6.20). The faults in a middle position were the most active and had faster rates of movement than faults in a proximal position.

3.2 Episodic syn-rift diapirism

Over a large part of southwest Europe, the post-Hercynian platform is covered by evaporites deposited during the transgressive phases of the

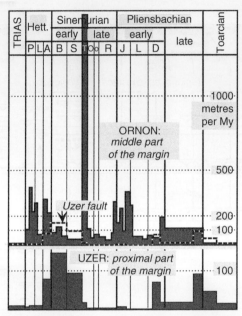

Figure 6.20 ***Comparison of rates of movement of two syn-rift faults at the start of the rift phase.*** The first phases of movement of the Ornon fault (location: middle part of the Tethyan margin; Bourg d'Oisans area) are dated as Early and middle Hettangian (Planorbis and Liasicus zones). These phases are not recorded on the border fault of the Massif Central (Uzer) in a proximal position where, in contrast, the phases of activity dated to the Late Hettangian (Angulata zone) and the Early Sinemurian (Bucklandi zone). The maximum rate dated to the Late Sinemurian (Turneri zone) is contemporaneous with the peak transgression in the wider Sub-alpine domain. Rates of displacement along faults range from 0 to 1000 metres per million years. The two figures are at different scales because of the more important throws and rapid rate variations along the Ornon fault. **Hett.**: Hettangian. Ammonite zones: **P**, Planorbis; **L**, Liasicus, **A**, Angulata; **B**, Bucklandi; **S**, Semicostatum; **T**, Turneri; **O**, Obtusum; **o**, Oxynotum; **R**, Raricostatum; **J**, Jamesoni; **I**, Ibex; **D**, Davoei. *Simplified after Chevalier 2002 unpublished, and Chevalier et al. 2003.*

major Triassic cycle (Chapter 5). In the lower Rhone valley, about 1000 m of evaporites underlie younger sediments up to 10 km in thickness (Fig. 5.6). While gypsum is common at outcrop, salt, anhydrite and gypsum are all present in the sub-surface. From the start of the Alpine cycle, salt diapirs, rising from Triassic evaporites have, in places, deformed beds of Jurassic and Cretaceous age on the European margin of the Tethys.

The structural characteristics of the syn-rift diapirs are as follows (Figs. 6.21–6.23):

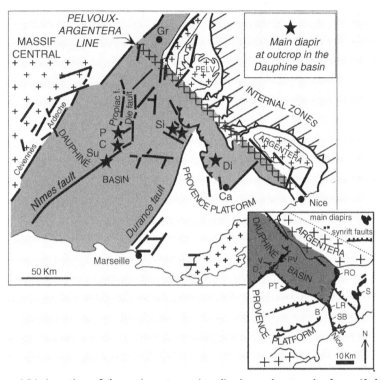

Figure 6.21 *Location of the main outcropping diapirs and network of syn-rift faults in the Dauphinois domain and Maritime Alps.* **B**: Bonson; **C**: Condorcet; **D**: Dalluis; **Di**: Digne area; **LP**: La Roquette sur Var; **M**: Montaulieu; **PT**: Puget-Theniers; **PV**: Peone-Valberg; **RO**: Roquebilliere; **S**: Sospel; **SB**: Saint Blaise; **Si**: North of Sisteron area (Rocher de Hongrie; Mont Rond); **Su**: Suzette; **T**: Toudon; **V**: Villetalle; **Ca**: Castellane; **Gr**: Grenoble. *Simplified after Dardeau and Graciansky 1990, Fig. 1, p. 145.*

- common occurrence of diapirs on rift phase faults or at the nodes of the extensional fault network (Fig. 6.21);
- progressive thinning of beds towards the salt diapir demonstrating contemporaneous diapir rise due to downbuilding of sediment (Fig. 6.22);
- interbedded syn-sedimentary breccias and slumped beds whose restriction to the immediate area of the diapir indicates syn-sedimentary slopes;
- erosion surfaces within the beds surrounding and covering the diapir;
- common development of radial and concentric syn-sedimentary faults above the crest of the diapir.

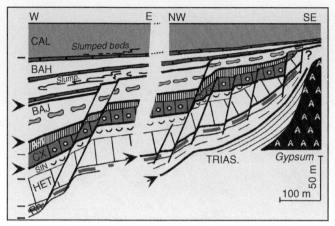

Figure 6.22 *Restored section across the Villetalle diapir (Maritime Alps).* The datum is the top Middle Jurassic (top Callovian). Each rift episode coincides in time with a phase of diapiric growth. Both are recorded by syn-sedimentary faults, local unconformities accompanied by hiatuses, by breccias or slumped beds. The small arrows to the left show each episode. The net result is thinning of the beds towards the crest of the diapir. Movement did not stop throughout the entire period of rifting from the Triassic to the Bathonian. Movement ceased from the onset of deposition of the Terre Noires in Callovian time, i.e., the start of passive thermal subsidence of the whole margin. **A-T**: Aalenian-Toarcian; **BAH**: Bathonian clays and limestones; **BAJ**: Bajocian cherty limestones; **CAL**: Callovian Terres Noires; **CX**: Pliensbachian cherty limestones; **Rh**: Avicula beds; **SIN**: Sinemurian nodular limestones with *Gryphea*; TRIAS: varicoloured Triassic shales overlying the evaporites. Location on Fig. 6.20. *Simplified after Dardeau and Graciansky 1990, Fig. 10, p. 445.*

The syn-sedimentary discontinuities noted above demonstrate episodic diapirism. Periods of acceleration in the rise of the diapirs can be dated from the age of eroded sediments and internal erosion surfaces. In terms of the history of the margin, a particularly significant point is the coincidence of the timing of episodes of diapir ascent with those of rifting (Figs. 6.19, 6.22, 6.23) suggesting diapir rise was triggered by extension.

4. MODES OF SUBSIDENCE DURING RIFTING

Each of the discrete phases of rifting was accompanied by sharp subsidence shown by rapid deepening of the depositional environment.

Rifting in the Dauphinois area, dated to the Hettangian/Sinemurian (200 Ma) boundary illustrates this phenomenon (Fig. 6.9). In many places, oyster-bearing or coarse-grained limestones were deposited in shallow

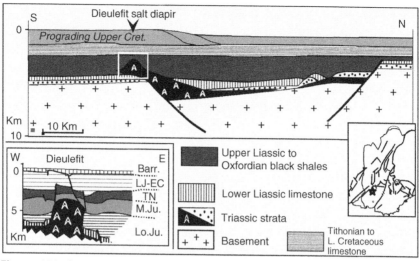

Figure 6.23 *Cretaceous palinspastic reconstruction of the future rim-syncline and Jurassic diapir of Dieulefit.* The two sections have been interpreted from seismic profiles as the diapir does not outcrop. The longer section shows the diapir has grown adjacent to an extensional fault active during the Tethyan rift phase. The detailed section shows faults displacing Liassic limestones are sealed by the overlying Late Liassic sediments just as the faults displacing the Late Jurassic are sealed by Early Cretaceous beds. Synformal beds much closer to the diapir are younger and probably represent rim-synclines. Location: 20 km NW of Condorcet (C, Fig. 6.21). **Barr**: Barremian limestones; **LJ-Ec**: Late Jurassic and Early Cretaceous limestones and shales; **TN**: Middle Jurassic and Oxfordian Terres Noires; **M.Ju**: Middle Jurassic limestones and shales; **Lo.Ju**: Liassic beds, limestones below, shales above. *Simplified after Roure and Coletta 1996 Fig. 13b, p. 202 and after Dardeau and Graciansky 1990.*

water depths under wave influence. These limestones are overlain by fine-grained limestones or marls containing a pelagic fauna of ammonites deposited in much greater water depths than the underlying sediments. It can also be shown that a surface of non-deposition is onlapped by the overlying beds. These relationships clearly show that tilting of the limestones following extension was followed by deepening of the seafloor and deposition of pelagic sediments. Sediments below and above this surface are dated as Hettangian.

Another, more subtle, example is found on the distal part of the margin (Fig. 6.12). Here fine-grained ammonite-bearing limestones deposited in relatively deep water were succeeded by red clays deposited below the carbonate compensation depth (CCD: see Chapter 1) after a phase of rifting dated to end Middle Jurassic time (Callovian).

4.1 Correlation of uplift history to rifting

Two processes cause uplift of the crests of the tilted blocks. The first, a well-known consequence of listric normal faulting, results in the return of the crest of the downfaulted block to a common regional elevation producing uplift. The second is elastic rebound due to removal of the load on the footwall by downfaulting. Depending on initial water depth, sub-aerial exposure may result.

The most spectacular example is the Brianconnais platform. Extending from the Mediterranean to the Grisons area of eastern Switzerland, the platform was about 100 km wide and several hundred in length. Sub-aerial exposure began during the second phase of Jurassic rifting (J2; 185–190 Ma; Figs. 6.19 and 6.20) and lasted for 10 million years. During this period, the thick pre-rift Triassic carbonates were subjected to karstification. The contemporaneous products of mechanical erosion are found interbedded in the syn-rift sequences of the adjoining deepwater Piemontais domain.

4.2 Thermal subsidence

Passive thermal subsidence began at the end of rifting and the onset of seafloor spreading. In the Western and Central Alps, the subsidence is marked by deposition of deepwater sediments across the European margin as well as the submergence of previously sub-aerial areas such as the Brianconnais platform.

5. SUMMARY: THE FUTURE EUROPEAN CONTINENTAL MARGIN IN THE WESTERN AND CENTRAL ALPS DURING THE RIFT PHASE

The most typical structures resulting from stretching of the lithosphere and continental crust of Pangaea in the Liassic are second-order tilted blocks, some 5–20 km wide, separated by half-graben infilled with sediments. Our type example here is the Bourg d'Oisans half-graben (Fig. 6.2). Another type is the Rhone valley graben that contains a more than 10-km-thick sequence of Mesozoic and Cenozoic sediments. A master fault array separates the emergent Massif Central at the west from the Rhone valley half-graben at the east (Figs. 6.13, 6.17). Rifting created a differentiated topography of high blocks and relatively deep subsiding troughs or half-grabens. In terms of scale, the largest, first-order elevated blocks or islands are about 100 km wide at least, as are the main subsiding basins (Fig. 6.12).

Seismic sections (see below, Fig. 10.3) combined with field observations (Fig. 6.2) show that the faults bounding tilted blocks link to deep detachment surfaces. The majority of these fault planes formed zones of weakness, which would be later utilized in compression from the onset of Alpine deformation (Chapter 13).

In the proximal, western part of the margin, these surfaces affected mainly the ductile horizons of the cover most importantly the Triassic evaporites and also the argillaceous beds of the Stephanian and Permian (Fig. 6.10), the marls of the Liassic, Late Jurassic and Early Cretaceous (see below, Fig. 13.2).

In the middle part of the margin occupied by the External Crystalline Massifs, deep decollement surfaces cut the basement tending to flatten towards the boundary between the brittle upper crust and ductile lower crust. This extension led to significant thinning of the continental crust probably on the scale of that observed in the Bay of Biscay and west of Iberia (Figs. 1.8 and 6.1).

With regard to the internal zones belonging to the section of European margin in the Alps, the intensity of Cenozoic Alpine deformation does not allow clear identification of the structural style linked to rifting. In contrast large amounts of extension represented by low angle block tilting are preserved in parts of the nappes derived from the distal Apulia passive margin (Figs. 10.8 and 12.6, below).

The rift system was divided into two structural segments by the NW–SE Pelvoux–Argentera transfer fault (Figs. 6.16 and 6.17). Within the two segments on either side of the transfer fault, the disposition and number of first-order subsiding basins and blocks are quite different. The SW segment was subject to much greater subsidence than the NE segment indicating a change in the polarity of normal faulting across the transfer zone.

Rift related structures formed in a discontinuous or episodic manner. Successive phases of extension were separated by periods of relative quiescence whose alternation determined the cyclicity of transgression and regression (Figs. 6.19, 6.20 and 16.5, below).

Faulting migrated from a proximal to an axial position during rifting. One of the most highly extended rift basins in an axial position was ruptured completely during the Jurassic (165 Ma), thus initiating spreading in the Liguro-piemontais Ocean and resulting in the structural fabrics of the conjugate margins of Europe and deformation (Chapter 10 and 12). However, during the post-rift phase, a later extensional episode was superimposed on the European margin including the Brianconnais. This episode leaded to the widening of the small Valais basin between the External Crystalline Massifs and the Brianconnais (Chapter 7).

The rift related structures have considerable importance because many were reactivated during Alpine compression as inversion structures, thrusts and nappes (Chapters 13 and 14).

FURTHER READINGS

Stampfli and Borel 2002 for the Alpine realm; Withjack, Schlische and Olsen 2002 and 2008, for comparison with the North America east Atlantic margin.

CHAPTER SEVEN

Late Jurassic and Early Cretaceous Development of the European Margin Spreading of the Liguro-Piemontais Ocean

Contents

Summary

Stretching of the originally thick continental crust led to seafloor spreading and the cessation of extensional tectonism by the end of the Middle Jurassic on the European margin. During spreading of the Tethys Ocean, the post–rift sediments were deposited in progressively increasing water depths because of passive thermal subsidence of the continental margin and the adjacent oceanic crust.

The Western Alps, from Rift to Passive Margin to Orogenic Belt, Volume 14
ISSN 0928-2025, DOI 10.1016/S0928-2025(11)14007-9
© 2011 Elsevier B.V.
All rights reserved.

However, a second phase of extensional tectonism took place during the Late Jurassic and Early Cretaceous. Part of the earlier network of Liassic to Middle Jurassic syn-rift faults was reactivated and a new network of faults developed during the Late Jurassic. Like the Tethyan rift phase, this later extension affected much of the European margin including the future Alps, the Paris Basin and the Aquitaine Basin including that part which would become the future Pyrenees. This new phase of rifting heralded the opening of the Bay of Biscay and adjoining west margin of Iberia. In the Alps, this phase of rifting has been classically considered to have resulted in the formation of the Valais Trough, a small oceanic basin. However, this view is now a matter of contention.

 ## 1. THE AGE OF THE ONSET OF THE POST-RIFT PHASE AND SPREADING OF THE LIGURO-PIEMONTAIS OCEAN

The Liguro-piemontais and Central Atlantic Oceans opened contemporaneously (Fig. 3.4) resulting in the separation of the Eurasia–North America block from the South America–Africa block to which Apulia was attached.

In the Alps, Corsica and the Apennines, the oldest sediments deposited on the early oceanic crust represented by the Liguro-piemontais ophiolites are dated by radiolaria to the upper part of the Middle Jurassic (see below, Chapter 11; Bathonian to Callovian, about 160–170 Ma; Fig. 11.1). Results from DSDP drill sites have provided comparable ages for the oldest sediments that rest directly on the oldest oceanic crust of the contemporaneous Central Atlantic Ocean.

 ## 2. ORGANIZATION OF THE EUROPEAN MARGIN OF THE TETHYS IN THE LATE JURASSIC AND EARLY CRETACEOUS

The submarine palaeogeography was characterized by large-scale domains inherited from the pattern of syn-rift faults, syn-rift subsidence and stretching of the crust. As a result of the opening of the Liguro-piemontais Ocean, the post-Middle Jurassic palaeogeography comprised the following domains from the proximal to the distal margin (Fig. 3.6):

(A) **The relatively static and unstretched European platform** on which 1–2-km thick platform of carbonates were deposited in the Jurassic to form the two main carbonate platforms of the Causses and Jura in the Alpine foreland.

(B) **The stretched European continental margin**, on which pelagic and hemipelagic sediments were deposited. This area comprised the large subsiding **Dauphinois–Helvetic** basin (Figs. 2.4, 6.16 and 6.17). There, the tens of kilometres wide arrays of half-graben and intervening high blocks formed during the Early and Middle Jurassic rift subsided as a whole due to post-rift subsidence and were progressively buried by an increasing thickness of sediments.

(C) The **Valais zone or trough**, a subsiding area that bounded the Saint Bernard–Monte Rosa (SBR) elevated block to the west and northwest (Figs. 2.7 and 12.7). The Valais zone was classically considered to have formed during a Late Jurassic–Early Cretaceous extensional event that led to oceanic spreading in the Valais trough. The Liguro-piemontais and Valais domains merge together in the Eastern Alps as a single basin because of the eastward termination of the SBR block (see below, Figs. 12.8 and 12.10). Further to the east, the distinction between the Valais basin and the Liguro-piemontais Ocean thus has little meaning. On the other side that is to the west and SW, Valais sedimentary units are not known further south than the Tarentaise (= Isere) valley.

(D) A relatively high area, several hundred kilometres wide known as the SBR–Brianconnais block was located to the northeast of the present-day Pelvoux–Argentera line (Figs. 6.16 and 6.17). The SBR–Brianconnais area was emergent during the Early and Middle Jurassic rift phase (Late Sinemurian to Early Bathonian) but subsided as a whole from Callovian times onward due to post-rift subsidence and starved of sediments. On the Corsica transect to the south, the **Provencal platform** represents this high (Fig. 6.17).

(E) The actively spreading **Liguro-piemontais Ocean**, site of the major part of the future ophiolite and Upper Penninic Alpine nappes (Figs. 2.4 and 2.5).

(F) The Apulia continental margin now represented within the Southern Alps and Austro-alpine nappes which are equivalent in palaeogeographic terms (Figs. 2.4, 2.5 and 10.2).

3. INTERPRETATION OF THE VALAIS ZONE: AN OCEANIC BASIN OF EARLY CRETACEOUS AGE OR A FAILED RIFT SUPERIMPOSED ON THE TETHYAN MARGIN?

3.1 Present-day structure of the Valais Zone

The Valais zone extends from the Hohe-Tauern window in the Eastern Alps as far southwest as the Versoyen region of the Western Alps on the Franco-Italian border (Fig. 7.1). It lies tectonically between the overlying internal Brianconnais domain and the external, underlying European plate (Fig. 7.2). Because of the structural position of the Valais zone, its rocks have been interpreted as being derived from a former basin, the Valais Trough, located between the European plate and the SBR block (see below, Figs. 12.8 and 12.10).

The Valais Zone forms part of a broad zone of Late Jurassic–Early Cretaceous extension that can be followed from the Bay of Biscay through the Aquitaine Basin and Provence into the Western Alps (Figs. 12.8 and 12.10, below). In the Bay of Biscay, this extension heralded the onset of spreading in post-Aptian times. Within the Valais Zone, mafic magmatism, stratigraphically inferred to be of Early Cretaceous age, has led to the

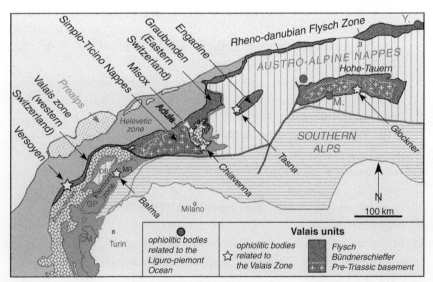

Figure 7.1 *Location of Valais units in the Western, Central and Eastern Alps.* Refer to Figs. 2.2, 2.4 and 4.3. DB: Dent Blanche; DM: Dora Maira; MR: Mont Rose; Tam: Brianconnais Tambo nappe. Misox: see to Fig. 7.2. *Completed from Beltrando et al. 2007, Fig. 2 p. 87.*

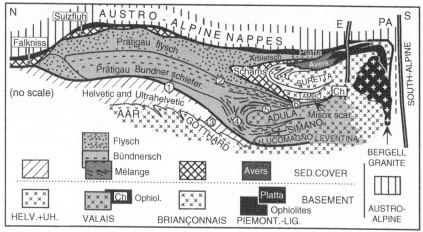

Figure 7.2 *Schematic N–S structural cross section of the eastern margin of the Central Alps crossing the Grisons.* The assemblage of Valais units (= lower Pennine or north Pennine) is sandwiched between the Helvetic and Ultrahelvetic nappes below and SBR and Piemont-Liguro-piemontais nappes has to be (upper Pennine or south Pennine) above. The Austro-alpine nappes cover the whole. The *Bündnerschiefer*, dated as Liassic by *Gryphaea* at the northern front of the Adula nappe and by palynomorphs in the Misox Scar, are included in a 'tectonic melange' (*Steinman 1994*). These sediments are interpreted as syn-rift deposits with respect to the development of the Valais basin. **Ch:** Chiavenna ophiolites. **Platta**: ophiolites and sedimentary cover belonging to the Platta nappe. **E:** Engadine fault. **PA:** Peri-adriatic fault. Available dates in the Valais domain – **1:** Klus series, Early Cretaceous palynomorphs; **2:** Via Mala, Cenomanian palynomorphs; **3:** Bedretto, Middle to Late Jurassic; cover of the Gotthard massif; **4:** Terri, Liassic Lithyotis bearing limestone; **5:** Vals, Early Liassic limestone with *Gryphaea*; **6:** San Bernardino, Middle Jurassic palynomorphs. *Lemoine 2003, unpublished.*

hypothesis that the Valais Trough is underlain by Early Cretaceous oceanic crust. However, there are obvious kinematic problems in reconciling the presence of oceanic crust in the Bay of Biscay and the Valais Zone with its absence in the intervening area of Provence.

The structural organization of the Tethyan margin in the Late Jurassic summarized above has been classically agreed in the Alpine geological community. Nonetheless, this view has become open to debate following new results which suggest quite different views of the Tethyan margin and its evolution. Either the Valais Trough is underlain by oceanic crust as in the traditionally agreed interpretation since 1960 or the age of the oceanic crust in the Valais Zone (Trough) is Middle Jurassic and not Early Cretaceous as in the latter traditional view. Or it is composed of thinned continental crust as suggested by new observations made since 2000.

3.2 Lithostratigraphy of the Valais units; their European affinities

The Valais stratigraphic succession is now involved in the Lower Penninic nappe system between the external Helvetic nappes below and the internal Brianconnais Zone Houillere above (Fig. 7.1). The Brianconnais Zone Houillere is the lower element of the Middle Penninic (Table 4.1 and Fig. 2.7).

The Continental basement: The gneissic cores of the Simplon–Ticino nappes in the Central Alps are variously named as the Lucomagno, Simano, Adula (Fig. 7.2) and Monte Leone, Lebendum, Antigorio, Simiano, Ticino and Adula nappes (Fig. 4.3). Equivalent units in the Eastern Alps constitute the *Zentralgneiss* of the Hohe-Tauern window (Fig. 2.4). All originate from the northern, i.e., the external, European side of the Valais basin. These units do not have equivalents in the Western Alps.

The Mesozoic and Palaeogene sedimentary cover: The sedimentary section consists of two main formations (Fig. 7.3). The lower, the Valais *Schistes Lustrés* or *Bündnerschiefer* is of Valanginian to Albian age and outcrops as the cover of the Simplon–Ticino nappes of Central Switzerland, and in the Grisons (= Graubünden). The *Bündnerschiefer,* lying above or against the Adula gneiss, and the Pratigau *Bündnerschiefer* also form part of this group (Fig. 7.2).

The upper formation, of Cenomanian to Early Eocene age, is comprised of the flysch of the Isere valley (Tarentaise) in Savoy, the flysch of the Niesen nappe of the Prealps (Fig. 2.8) and that of Pratigau and Arblatsch in the Grisons (= Graubünden; Figs. 7.1 and 7.2).

Comparison of the Mesozoic successions of the Valais and Liguro-piemontais domains (Fig. 7.3) shows the European affinity of the Valais succession. Notably, resedimented limestones in the Early Cretaceous of the Valais are derived from a carbonate platform unknown at outcrop. These limestones are absent in the equivalent age section of the Liguro-piemontais where the Early Cretaceous is composed of alternating black shales and micritic pelagic limestones (*Argile a Palombini* Formation).

3.3 The classical stratigraphic interpretation of the Valais succession: the existence of the Early Cretaceous Valais ocean (Antoine 1968–1992)

The main structural unit derived from the Valais palaeogeographic domain is the Sion–Val Ferret–Courmayeur–Tarentaise nappe (Fig. 7.4). The nearest Valais continental basement outcrops about 100 km to the northeast in the core of the Simplon–Ticino (= Lower Penninic) nappes (Fig. 4.3).

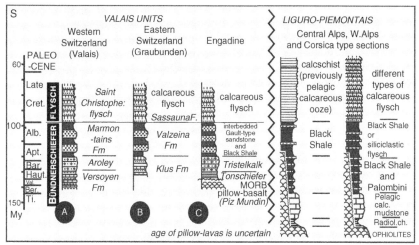

Figure 7.3 *Comparison of the Valais and Liguro-piemontais stratigraphic successions.* Location of sections: see Fig. 7.1. Compare with Figs. 11.12 and 11.13. In western Switzerland (Valais), the Versoyen limestones rest on either gneiss, Triassic quartzites or dolomites. In eastern Switzerland (Graubünden = Grisons), the Klus formation is detached from unknown basement. The synthetic sections of the Liguro-piemontais succession are representative of the Western, Central Alps and Corsica. As a result of continued spreading, different sedimentary formations of the Late Jurassic and Early Cretaceous are found resting on ophiolitic basement. The lower part of the *Bündnerschiefer* of the Valais comprises the Aroley and Tristelkalk limestones redeposited from a carbonate platform unknown at outcrop. These limestones, Early Cretaceous in age, are found only in the Valais domain. In contrast, the coeval South Pennine Liguro-piemontais succession consists of alternating black shales and micritic pelagic limestones known as the *Argile a Palombini* Formation. The upper part of the Valais *Bündnerschiefer* comprises intercalated black shales and stilpnomelane bearing sandstones. These slightly metamorphosed rocks are derived from glauconitic clastics of the same facies (Gault) and age, i.e., Late Aptian to Albian. They are very similar to those found in the epicontinental basins of Western Europe. In contrast, outcrops of Liguro-piemontais *Schistes Lustres* of the same age consist of black shales interbedded in places with a clay-quartz flysch unknown in the Valais domain. *Lemoine 2003, unpublished.*

Three post-Jurassic lithostratigraphic subdivisions characterize this nappe. These are the *Aroley* limestone, the *Marmontains* quartzites and black pelites and the *Saint Christophe* 'flysch' (Fig. 7.3A). The existence of this 'Valais trilogy' has justified stratigraphic correlations over a distance of several hundred kilometres between the Tarentaise of the French Western Alps and the Grisons (Graubünden) of Switzerland.

Another key unit considered as part of the Valais is the Versoyen ophiolite (Fig. 7.4) which comprises serpentinites, gabbros, dolerites,

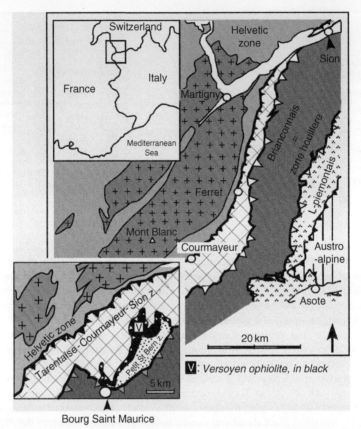

Figure 7.4 **Structural sketch map of the Valais zone and surrounding units on the Versoyen transect.** The Valais zone includes the Tarentaise–Sion–Courmayeur, Versoyen and Petit Saint Bernard units which all constitute the Lower Penninic. The Brianconnais Zone houillere belongs to the 'SBR' block that is Middle Penninic. The Liguro-piemontais zone is Upper Penninic. Redrawn and simplified from Masson et al. 2008, Fig. 1.

pillows basalts and associated black schists. It rests above one of the typical units of the post-Jurassic Valais succession (Aroley limestone; Figs. 7.3A, 7.4 and 7.5) around the Petit Saint Bernard pass on the France--Italy border. Within the Valais zone, only the Versoyen unit exhibits the characteristics of oceanic crust and overlying sediments.

The classical interpretation of the Versoyen ophiolite is that it is derived from oceanic crust of Early Cretaceous age. This interpretation rest on the following assumptions (Fig 7.5A):

– The Versoyen ophiolite represents the basement of the sedimentary succession of the Sion–Courmayeur–Tarentaise nappe.

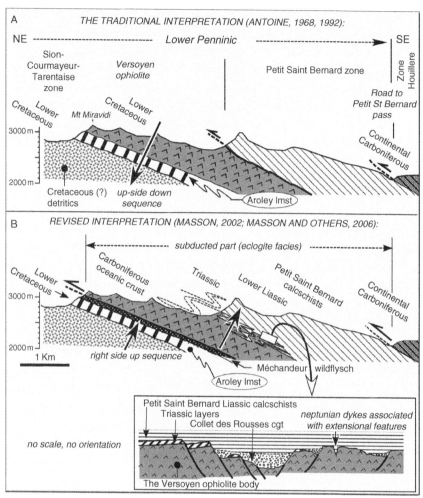

Figure 7.5 *Two structural interpretations of the Versoyen-Valais ophiolite. A:* In the traditional interpretation, the Versoyen ophiolite is derived from Tethyan Early Cretaceous oceanic crust that formed part of the basement of the Valais zone. **B**: According to the revised interpretation, the Versoyen ophiolite is a slab of Early Carboniferous oceanic crust incorporated in the European craton much earlier during the Variscan orogeny. These renewed views are still in debate among Alpine geologists. *Redrawn and interpreted from Masson et al. 2008, Figs 2 and 3.*

– The contact between the Versoyen ophiolite and the Aroley limestone (= base of the sedimentary succession of the Sion–Courmayeur–Tarentaise nappe) below is upside down.

– Since the Aroley limestone has been dated to the upper part of the Early Cretaceous, it follows that the original oceanic crust from which the ophiolite was derived must then be of Early Cretaceous age also.

Leading to the same conclusions is a comparison between the stratigraphy of the Valais *Bündnerschiefer* and that of the Liguro-piemontais *Schistes Lustrés*. This suggests an analogy between the two basins, oceanic basement and sedimentary cover, including close similarities in age (Fig. 7.3).

Starting from these ideas, the notion of the Valais Ocean was established in the Versoyen and then generally applied to the whole of the fragmentary ophiolites located within the Valais Zone in the Central and in the Eastern Alps, as far as the Hohe-Tauern window. In Eastern Switzerland, these are the Chiavenna unit (Fig. 7.1) between the gneissic cores of the Tambo (SBR = Brianconnais-type unit; Fig. 12.1) and Adula (Valais) nappes, respectively (Fig. 7.2). Mafic magmatism dated at ca. 93 Ma has been recently reported from the Chiavenna ophiolites (Liati et al. 2003 and 2005). In the Eastern Alps (Fig. 7.1), various ophiolites outcrop beneath the Austro-alpine nappes in the Rechnitz, Hohe-Tauern (Glockner nappe s.s) and Lower Engadine (Tasna nappe) windows (Figs. 2.4 and 7.6).

3.4 An alternative: the Tethys Ocean and its Valais branch were contemporaneous

The idea that the onset of spreading in the Tethys and the Valais trough might be contemporaneous and of Middle Jurassic age first resulted from numerical modelling (Manatschal et al. 2001; Lavier and Manatschal 2006 and Fig. 12.9, below).

This hypothesis has been reinforced by combined field observations and radiometric age dates (Manatschal et al. 2006) from the Tasna nappe (Fig. 12.7 below) situated in the Engadine window of SE Switzerland (Fig. 7.6).

In the Engadine window, the succession comprises from bottom to top:
– The Valais *Bündnerschiefer* (Fig. 7.3 C) is an assemblage of highly deformed calcschists containing a few metabasaltic bodies. It is surmounted by a melange (*Roz-Champatsch/ Stammerspitz* zone) which includes a flysch containing blocks of continental and oceanic basement. This succession typifies the infill of the Valais trough.
– The south Penninic Arosa zone which is strongly tectonized and reputedly derived from the ocean–continent transition (OCT) south east of the Liguro-piemontais Ocean.
– The *Tasna-Ramosch zone*. The Tasna nappe is composed of serpentinites, granite-gneiss basement and sedimentary cover.

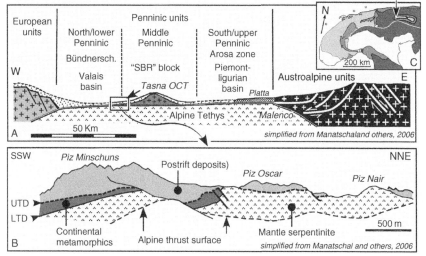

Figure 7.6 _Reconstruction of the Alpine Tethys on the Tasna nappe transect at the NE extremity of the SBR block._ This reconstruction shows the existence of a Middle Jurassic to Early Cretaceous oceanic basin beneath the Valais Trough between the European basement and the 'SBR' block conforming to the model agreed in Alpine geology since Antoine (1968). It accepts that the Tasna nappe is derived from the prolongation to the NE of the SBR block (see Fig. 12.7), according to the interpretation of Trümpy (1972). The Brianconnais–SBR block is interpreted as representing a continental allochton block separating the Valais from the Piemont-ligurian basins. Malenco, Platta: see Fig. 10.4, below. **UTD** is Upper Tasna Detachment. **LTD** is Lower Tasna Detachment. _Adapted from Manatschal, Engström et al. 2006, Fig. 2 p. 1852 and Fig. 11, p. 1864._

– Austro-alpine units, derived from the Apulia-Adriatic block, i.e. from the SE continental margin of the Liguro-piemontais Ocean.

The main part of the Tasna nappe consists of continental basement and its sedimentary cover. Following Trumpy (1972), it has been considered to represent the north eastward prolongation of the 'SBR' block (Briancon-nais/Saint Bernard–Monte Rosa; Fig. 12.10 including the Sulzfluh and Falkniss nappes; Fig 7.2). The argument is that the stratigraphic succession is comparable in these different units.

A small part of the Tasna nappe _sensu stricto_, called the Tasna OCT (Manatschal et al. 2006) consists of continental basement, serpentinites and overlying sediments. It may legitimately represent the OCT between the SBR block and Valais ocean crust, if the latter exists, and if the correlation of the Tasna OCT with the NE border of the SBR block is truly demonstrable. Study of the Tasna OCT has shown the following:

1. The age of denudation of the Valais mantle at the north-eastern extremity of the SBR block is middle Jurassic in age as for the Liguro-piemontais. The appearance of Valais ocean crust does not then date to the early Cretaceous as previously supposed but is older.

   In effect, Ar/Ar age determinations made on phlogopites contained within spinels and websterites yield dates of 169–171 Ma, including errors. The rocks of mantle origin are mainly lherzolites with fertile spinels and numerous levels of websterites with spinels. Palaeotemperature measurements, on pyroxenes preserved despite the degree of serpentinization, yield temperatures of $900° \pm 50°C$. These low temperatures are in agreement with the temperatures of exhumed sub-continental peridotites that were within the uppermost mantle before tectonic exhumation during Jurassic rifting.

   In addition, U/Pb dates from a garnet–plagioclase–quartz pegmatite and a metagabbro yield ages of 310 and 221 interpreted as indicating an earlier Variscan event (Cf. Versoyen).

2. The first sediments resting on oceanic basement are thin undated argillite beds (*Tonschieffer*: Fig. 7.3C). The oldest dated beds towards the base of the *Bündnerschiefer* comprise the *Tristel Formation*, of Late Barremian age (Fig. 7.3C).

   The start of post-rift sedimentation would then largely post-date the submarine denudation of the mantle as has been observed on the abyssal plain west of Iberia (Chapter 12).

Discussion

The Tasna OCT units have yielded particularly useful results applicable to the practical details of stretching of the continental crust especially in its final stages. The related structural interpretations rest on tectonic models, drawn also from west of Iberia, which are tenuous because of the structural complexity as noted by Manatschal et al. (2006). There may be an element of circular reasoning in the sense that the existence of an oceanic Valais arm is assumed a priori without discussion. Such points of discussion show that the existence of oceanic crust beneath the Valais trough remains uncertain at present as will be seen below.

3.5 Revision of the traditional interpretation: a Palaeozoic age for the Versoyen ophiolite and not Cretaceous (Masson et al. 2008; Beltrando et al. 2007)

Since 2000, new work, structural and stratigraphic as well as radiometric age dating, has shown that the Versoyen ophiolite, classically considered as typical Valais, is of pre-Triassic age, more precisely Early Carboniferous (Visean) to Permian (Fig. 7.5B). The observations presented by Masson (2007) are both stratigraphic and radiometric. These are as follows:

– The Versoyen ophiolitic body is not the basement of the Sion–Courmayeur–Tarentaise sedimentary unit. On the contrary it is thrust above as is particularly shown by the inverse jump in degree of metamorphism on either side of the contact between the two terrains. In effect, the underlying Sion–Courmayeur–Tarentaise nappe was not subjected to a pressure of more than 10 Kb. In contrast, the ophiolite above was subjected to eclogite grade metamorphism (T approx. 350°C and $P = 15$–17 Kb) and later retrograde metamorphism to blueschist or green schist facies.

– The Aroley limestone, reputedly representing the base of the Sion–Courmayeur–Tarentaise succession, is cut by a ravinement surface, overlain in stratigraphic contact by Wildflysch (called the *Mechandeur* Formation), containing blocks of all sizes closely similar in nature to the Versoyen ophiolite. The Aroley limestone is therefore not upside down. Thrusting of the Versoyen ophiolite thus took place in submarine conditions.

– The Versoyen ophiolite body, as it outcrops today, is cut again at its summit therefore to the southeast (Fig 7.5) by a subaerial erosion surface overlain by discontinuous Triassic beds. Ravines cut the Triassic beds and are infilled by a breccia (the 'Collet des Rousses' conglomerate) overlain by belemnite bearing limestones of Liassic age. The Triassic and Liassic beds comprise the calcschists of the Petit Saint Bernard. They have the same degree of metamorphism as the Versoyen ophiolite. The consequences are (a) that the ophiolite is in place and (b) that the oceanic crust from which the ophiolite originated is pre-Triassic in age. The Petit Saint Bernard unit has not therefore been thrust above the Versoyen ophiolite contrary to the previous classical interpretation.

– Geochemical analyses show that the ophiolitic rocks were derived from tholeiitic magmas of intermediate to transitional affinities (I-MORB to T-MORB).

- Earlier obtained radiometric dates of Early Carboniferous (Visean; 337 ± 4.1 Ma) were measured from the hearts of zircons found in a layered gabbroic intrusion. Recent dating by Beltrando, Rubato et al. (2007) has obtained Permian ages (267 ± 1 Ma and 272 ± 2 Ma) from zircons in a layered meta-leucogabbro and in a metagranite, respectively. The associated intrusive rocks forming the *layered rock complexes* of the Versoyen unit range from olivine gabbros to more acid rocks which may demonstrate fractional crystallization from the same magma. The more mafic types and the more acid types might therefore have formed during successive magmatic episodes of different ages. Since the serpentinites in the Versoyen massif are intruded by gabbros dated to the Permian, they were therefore already incorporated in basement older in age than Late Palaeozoic. The available results plus the state of deformation do not allow these conclusions to be extended further.

Consequences
Hercynian palaeogeography
It is difficult to conceive that the mafic rocks of Versoyen could be derived from Late Palaeozoic oceanic crust taking into account the available reconstructions for the Carboniferous and especially the Permian. According to the ideas of Beltrando et al. (2007), the magmatic rocks of the Versoyen would have intruded the basement during Permian times probably in an intra-plate extensional setting resulting from the extensional collapse of the Hercynian chain. The situation may be comparable to that proposed for the Permian gabbros of Malenco (Location of Malenco shown in Figs. 10.4 and 10.8, below).

Mesozoic palaeogeography
The newly available results do not permit acceptance of the view, held since Antoine et al. (1968 to 1992), that the Valais basin in the Western Alps was floored by Cretaceous oceanic crust. The results may also exclude the presence of Jurassic oceanic crust as proposed by Manaischal et al., (1999–2004); Lavier and Manatschal, (2006).

Consequently, this opens to the question of the correlation between 'ophiolites' reported to the Valais Zone in the Versoyen (Valais *stricto sensu*) and others located in the Central and Eastern Alps (Fig. 7.1).

On the northern transverse of the Western Alps, the Valais **Balma unit** (Fig. 7.1) is composed of serpentinites, eclogites and amphibolites. It lies in a complex structure composed of the Monte Rosa nappe (Piemontais basement) below, and a discontinuous gneissic unit above which is reported to the Brianconnais, in turn surmounted by typical Liguro-piemontais ophiolites.

On the transverse of the Central Alps, the **Chiavenna ophiolite** (Fig. 7.1) consists of serpentinites, carbonates and metabasalts. It forms a sandwich between the Tambo nappe (Brianconnais-type unit; Figs. 12.8 and 12.10) and Adula (Valais) nappes, respectively (Fig. 7.2).

Beltrando et al. (2007) note that if the two units of Balma and Chiavenna truly record the existence of magmatism and perhaps the formation of oceanic crust of Cretaceous age as proposed by Liati et al. (2003 and 2005), no direct and easy structural correlation can be made with the Versoyen ophiolite. The latter is also in a much more external position than the Balma and Chiavenna units. In consequence the palaeogeography of the Valais basin may have been more complicated in detail than a simple gulf opening to the northeast of the present-day position. This observation would seriously restrict or even result in the total disappearance of the notion of the Valais Ocean in Cretaceous reconstructions of the Tethyan margin.

A possible scenario that reconciles the new observations noted above is as follows. Two rifts developed contemporaneously during the rift phase that led to the formation of the Liguro-piemontais Ocean. One of these was the Valais trough located west of the SBR block. This rift widened from the SW to the NE, where stretching was greatest as shown by the Tasna nappe. Here crustal attenuation was greatest but did not reach the point of complete rupture and the onset of oceanic spreading. Crustal attenuation exhumed the lower continental crust and mantle, dated as Variscan in age both in the Tasna nappe and in the Versoyen 'ophiolite'. The second rift, located east of the SBR block, proceeded to complete rupture and the formation of oceanic crust in the Liguro-piemontais Ocean. The final stages of spreading in the Liguro-piemontais Ocean may be recorded by the Balma and Chiavenna ophiolites.

A possible analogue to the Valais trough may be the Porcupine Basin west of Ireland. Thick continental crust borders the north, west and east sides of the basin. However, within the basin, the magnitude of east–west stretching increases from north to south where extreme attenuation has been imaged on seismic profiles (Reston et al. 2001). The western bounding ridge is underlain by thick continental crust and separates the Porcupine basin from oceanic crust formed in response to contemporaneous rifting. This ridge may be comparable with the SBR block.

It should be stressed that work on this subject is not sufficiently advanced for an unequivocal demonstration and solution to the problem of the Valais trough. In all cases, they do not exclude the role of the Valais zone as a subsiding Early Jurassic extensional domain, later subjected to Late Jurassic– early Cretaceous extension, situated in an internal position in terms of Alpine polarity with respect to the Helvetic domain. However, the age – Early Carboniferous–Permian (Variscan cycle), Late Jurassic or Cretaceous (Tethyan cycle) – of the Valais ophiolites remains in debate.

In the Eastern Alps (Fig. 7.1), with regard to the various outcrops beneath the Austro-alpine nappes in the Rechnitz, Hohe-Tauern (Glock-ner nappe s.s.) and Lower Engadine windows (Figs. 2.4 and 7.6), the problem still remains an open question.

Structural inheritance in the Valais zone

In the classical interpretation (Antoine et al. 1968–1992), the Schistes Lustrés of the Valais or *Bündnerschiefer* rest on the continental basement of the external zone in the region of the Simplo-tessinois nappes and on oceanic crust internally. A major suture therefore divides the Valais sedi-mentary succession into two parts. This suture would be reused as an intra-oceanic subduction surface during the Alpine orogeny.

In the modified interpretation, the Valais *Bündnerschiefer* rests as a whole only on European continental basement. The major suture which separates the entire Valais domain (lower Penninic) from the SBR block (middle Penninic) is outlined by units derived from Hercynian basement (Chapter 4, Table 1). These were incorporated in European continental crust during the Variscan cycle and then later overlain by Triassic and Jurassic sediments.

During Alpine subduction, eclogite grade metamorphism affected both the Versoyen 'ophiolites' and the overlying Triassic and Jurassic sedimentary cover that form the Petit Saint Bernard zone. The subduction surface was intra-continental and controlled by a major discontinuity in the Hercynian basement. This is a nice example of the structural inheritance of a major Hercynian discontinuity which can be added to those discussed above in Chapter 4.

4. THE BEHAVIOUR OF THE SAINT BERNARD–MONTE ROSA ('SBR') BLOCK DURING THE LATE JURASSIC– EARLY CRETACEOUS RIFT PHASE

Role as a barrier between the Valais and Liguro-piemontais basins: On the SBR block, thin Late Jurassic and Early Cretaceous sedi-ments, ranging in thickness from only a few to, at most, tens of metres

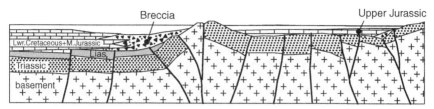

Figure 7.7 *Interpretation of the post-rift palaeostructure of the Schams nappe (Grisons, eastern Switzerland) in terms of local transpressive structures.* Simplified after Schmid et al. 1990. Adapted from Lemoine 2000, Fig. 9.3, p. 108.

(Fig. 6.14), reflect severe condensation or sediment starvation expressed by siliceous and/or phosphatic crusts separated by long hiatuses. The SBR block was a bathymetric high that acted as a barrier between the two basins on either side (Figs. 6.14 and 6.17). The existence of this barrier partly explains the differences between the sedimentary successions of the Valais trough and Liguro-piemontais basin.

Late Jurassic–Early Cretaceous extension on the SBR block: The SBR block was equally affected by the Late Jurassic–Early Cretaceous phase of rifting, which resulted in a system of submarine tilted blocks whose relief generated breccias. In the mountains of the SBR domain *s.l.*, such fault blocks are recognized in the Schams nappe (Fig. 7.7), and in the area around Briancon on the southern prolongation of the SBR block (Fig. 7.8). In the Breche nappe of the Piemont domain of the Prealps, the 'Upper Breccia' (Late Jurassic) is a classical example (Fig. 7.9).

5. RIFTING OF LATE JURASSIC–EARLY CRETACEOUS AGE IN THE WESTERN ALPS

The Vocontian basin and the Urgonian platform: One of the most significant rift structures of the Alpine Early Cretaceous is the Vocontian basin (Fig. 7.10). The basin was an eastward opening horse-shoe-shaped gulf characterized by pelagic deposition in the centre and bounded by the system of Urgonian carbonate platforms that are the southward continuation of those of the Jura. The palaeogeography of this domain is partly inherited from the previous Liguro-piemontais rift phase given that the majority of the faults defining the shape of the gulf were previously active during Liassic times (see Fig. 7.10B compared with Fig. 6.16).

Late Jurassic–Early Cretaceous rifting in the pyrenean-provencal domain

The whole extensional domain of Late Jurassic–Cretaceous age comprised a long, subsiding transtensional corridor characterized by a system of tilted blocks, half-graben and small pull-apart basins infilled by syn-rift sediments of mainly Cretaceous age. A network of sinistral strike-slip faults that bound the Massif Central to the SE cuts this corridor. Although this part of the rift system never developed into a spreading centre, greater extension in the Bay of Biscay resulted in the initiation of spreading probably in Aptian time.

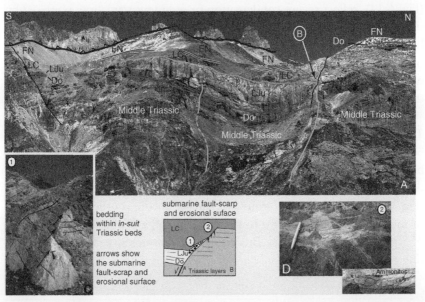

Figure 7.8 *Faults of the Laugier valley: an example of E–W normal faults and tilted block of Late Jurassic and Cretaceous age in the Brianconnais zone (Western Alps).* **A**: Twenty kilometres south of the Briancon transect, two E–W oriented normal faults (shown in light blue) have been transported without major deformation in one of the Brianconnais nappes. Fault **F2** was active only in the Late Jurassic. Fault **F1** was active during Late Jurassic as well as Late Cretaceous times. **B**: Diagram showing the submarine escarpment of fault F1 for location of (C) and (D). **C**: Submarine fault-scarp breccias and submarine erosional surface. **D**: Sedimentary dykes, dissolution cavities infilled by clays associated with phosphatic and manganese crusts. Some of the crusts have been imprecisely dated to an Albian to Senonian age range by planktonic foraminifera. **E**: Simplified scenario of the synsedimentary deformation. **Do**: Middle Jurassic, mainly late Bathonian, platform carbonates; FN: 'Flysch Noir', i.e., Palaeogene flysch; **LC**: Late cretaceous calcschists; **LJu**: Late Jurassic pelagic mudstone. (See colour plate 10) **A** to **D**: Photographs courtesy of T. Dumont. **E**: interpretation after Claudel et al. 1997. Adapted from M. Lemoine 2000, Fig. 9.2, p. 108.

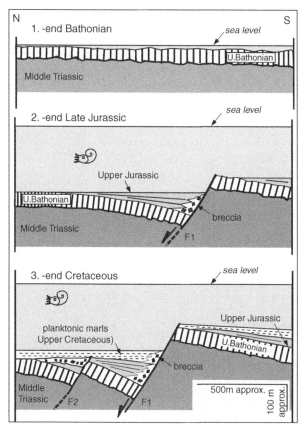

Figure 7.8 *(Continued)*

The whole extensional domain of Late Jurassic–Cretaceous age was later thrust northward by the Pyrenean orogeny, mainly Eocene in age. The elongate Pyrenean-provencal domain today consists of an array of different structures, which extend eastward from the Bay of Biscay to the Maures massif (Fig. 7.10). Major structural elements include the **North Pyrenean fault system, the nappe of the Eastern Corbieres** of the Lower Languedoc and, east of the alluvial fill of the lower Rhone valley, the **thrusts of Lower Provence** (e.g., Etoile and Sainte Baume), which die out in favour of the internal detachments of the Triassic overlying the Hercynian Maures massif. In Provence, the southern limit of this domain is the old established **Bausset thrust** and the Lower Provence basin which are considered to be the easternmost prolongation of Pyrenean structures (Floquet et al. 2003–2005).

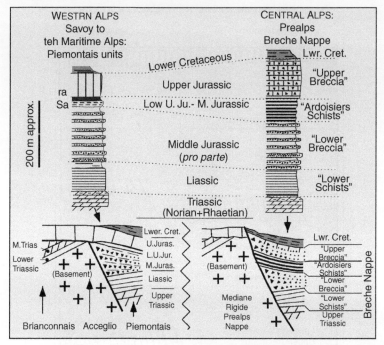

Figure 7.9 *Analogies and differences between the Piemontais Jurassic successions in the Western Alps along the Briancon transect and in the Central Alps along the Prealps transect (Breche nappe).* The submarine topography that generated the 'Upper Breccia' is interpreted to have formed by Late Jurassic reactivations of Liassic faults on the outer margin of the Brianconnais platform. Sa: clay-schists. Ra: radiolarian cherts. *Adapted from M. Lemoine 2000, Fig. 9.4, p. 109.*

6. SUMMARY AND CONCLUSIONS: DEFORMATION OF THE EUROPEAN MARGIN DURING SPREADING OF THE TETHYS

The extensional system that affected much of Western Europe is best developed in the Pyrenees, Provence and the southern Subalpine area (Vocontian basin: Figs. 7.10A and 7.11). This phase of rifting preceded the Middle Cretaceous (110 Ma) opening of the Bay of Biscay to the west.

In the Central Alps, the Valais units consist mainly of Hercynian continental basement (Simplon–Ticino nappes) and sedimentary Meso-zoic to Eocene cover. Rare ophiolites involved with the Valais units testify the presence of oceanic crust in the area. However, the age – Early

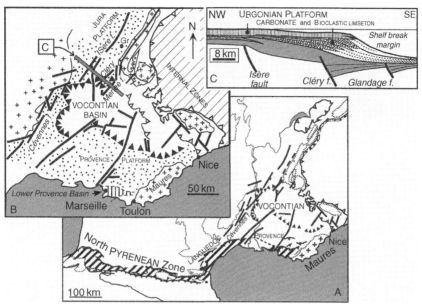

Figure 7.10 *Pattern of Late Jurassic–Early Cretaceous rifting in SW Europe.* **A:** distribution of the rift system in SW Europe. In this model, the rifting led to the formation of oceanic crust in the Valais domain, according to the classical reconstruction. However, the reconstruction of the original rift fabric is difficult because of Alpine deformation. Syn-sedimentary faults and the associated tilted fault blocks are well preserved in the Subalpine chains (see **B** and **C**, this figure and Fig. 7.11) because of modest Alpine deformation. However, oceanic crust is absent in this area as well as in the Pyrenees–Provence rift. The latter has been thrust northward. **B:** Dotted area, Urgonian platform in the Jura, Cevennes, and Provence. Vocontian basin and network of rift-related faults. **C:** Section crossing the boundary between the Urgonian platform and Vocontian basin (Location, see B, this figure). The basin boundaries were mainly controlled by faults, most of which were active from the start of the Jurassic. In the platformal domain, the subsidence rate was lower than in the basin and was matched by carbonate deposition. In shallow water areas, local tectonics and long-term changes in subsidence rate induced temporary flooding or emergence of the entire platform and an overall progradation of platformal deposits into the basin. *Adapted and completed from Jacquin et al. 1991, Fig. 4, p. 125.*

Carboniferous–Permian (Variscan cycle), Late Jurassic or Cretaceous (Tethyan cycle) – of the Valais ophiolites is now in debate.

In summary, one of the key events in the 'post-rift' history of the Liguro-piemontais margin, first rifted in Early to Middle Jurassic times, was the development of a successor rift system of Late Jurassic–Early Cretaceous age.

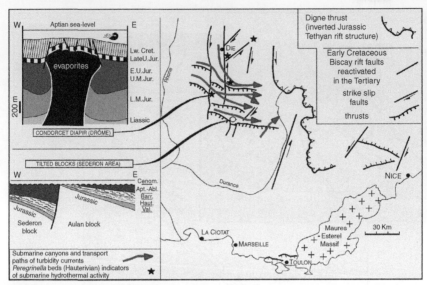

Figure 7.11 *Expression of Early Cretaceous rifting in the southern Subalpine chains.* The uplifted edges of blocks tilted during rifting formed submarine highs; the intervening lows acted as transport pathways for clastic sediments. Simultaneously, diapir rise, first triggered during Liassic rifting, was rejuvenated or also possibly initiated in the Early Cretaceous. The interpretation proposed for the outcropping Condorcet diapir (this figure) and its neighbour, the subsurface Dieulefit diapir (Fig. 6.23), is based on modelling that shows the two diapirs are located close to extensional faults affecting the basement. Deformation of successive beds by the Condorcet diapir demonstrates continuous halokinesis from the start of the Jurassic until the Early Cretaceous. Synforms corresponding to rim-synclines migrated towards the axis of the diapir between the Liassic and Early Cretaceous. Syn-sedimentary faults displacing the Tithonian limestone marker horizon (top of the Late Jurassic) record the main phase of rifting linked to the opening of the Bay of Biscay. Rim synclines and faults do not affect the Albian around the Condorcet diapir. **Val**: Valanginian; **Haut**: Hauterivian; **Barr**: Barremian; **Apt-Alb**: Aptian-Albian; **Cenom**: Cenomanian. *Adapted after Graciansky et al. 1987 and from Lemoine et al. 2000, Fig. 9.1, p. 107.*

However, the evolution of this system from a rift to oceanic spreading in the Late Jurassic or Early Cretaceous in the future Central and Eastern Alps must now be in question notwithstanding the remaining uncertainties.

FURTHER READING

Stampfli and Borel (2002).

CHAPTER EIGHT

The Late Cretaceous Phase and the Onset of the Alpine Shortening

Contents

Summary

The Late Cretaceous marks the onset of Alpine compression, dated in the Eastern Alps, as middle Cretaceous. In the Western Alps, it is best expressed in the external zones (Devoluy and Pelvoux areas) as well as in the internal Liguro-piemontais and part of Austroalpine zones. However, southern Provence remained in extension until the end of the Santonian as did the Brianconnais on the Briancon transect until the beginning of the Tertiary. These differences must be attributed to the nature and difference in ages of the relative movements between the European, Iberian and Apulia plates.

1. EVOLUTION OF THE EUROPEAN MARGIN OF THE TETHYS IN THE LATE CRETACEOUS: EARLY ALPINE DEFORMATION

During the Late Cretaceous (100–65 Ma), the area of the future Alps was subjected at various times and in different areas to compression or to extension reflecting the onset of subduction and, within the European plate, the effects of the movement of the Iberian plate.

The Western Alps, from Rift to Passive Margin to Orogenic Belt, Volume 14
ISSN 0928-2025, DOI 10.1016/S0928-2025(11)14008-0
© 2011 Elsevier B.V.
All rights reserved.

1.1 The expression of Cretaceous compression in the internal zones of the Western and Central Alps: flysch deposition and onset of subduction

Initiation of deposition of the first Helminthoid flysch is dated as Senonian. It rests generally on black-shales of tentative Cenomanian–Turonian age (Fig. 7.3). The flysch consists of mixed terrigenous siliciclastic and calcareous sediments deposited as deepwater turbidites in a proto-trench along the foot of the Apulia margin. The clastics were derived by erosion of sub-aerial topography created by the initial phase of orogenesis. The deposition of the flysch records the initiation of convergence along both the continent–ocean subduction zone of the Liguro-piemontais Ocean and the continent–continent subduction zone of the Valais basin (Fig. 8.1). The latter subduction surface was controlled by a major discontinuity in the Hercynian basement (Chapter 4). Subduction may have begun at the continent–ocean boundary according to the traditional view (Antoine et al. 1968–1973) or even at the continent–continent boundary following the later work of Masson et al. (1999–2008 and Chapter 7). However, subduction along both zones led to an approximately contemporaneous progressive closure of the Liguro-piemontais Ocean and Valais basin, respectively.

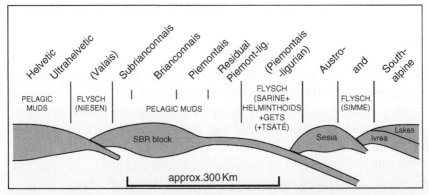

Figure 8.1 *Sketch showing the distribution of the principal Late Cretaceous lithotypes in the different zones along a Central Alps transect.* The palaeogeographic domains shown in parentheses were undergoing subduction or had already been subducted. Helminthoid flysch deposited in the area of trench slope break was derived from uplift of the overriding plate. Globigerinid pelagic limestones were deposited everywhere except for sites of flysch deposition. This very simple sketch applies to the Central Alps but not to the Western Alps south of Savoy where there are no equivalents to the Valais domain or the Niesen flysch. Ivrea: lower crust of the Ivrea zone. Lakes: zone of Lakes, comprising upper crust superimposed on the Ivrea zone. *Adapted from Lemoine et al. 2000, Fig. 9.5, p. 110.*

1.2 The expression of Cretaceous compression in the Austroalpine of the Grisons: flysch deposition and early thrusting

East of the Central Alps and within the confines of the Eastern Alps, the Austroalpine domain in the Grisons was subjected to a phase of compression that resulted in the westward displacement of nappes followed by a phase of extension. The existence and the age of such compressional events can be demonstrated from three types of observations:

(i) Detrital minerals characteristic of high-pressure–low-temperature (HP–LT) metamorphism are present in the Cretaceous flysch. For example, at the front of the Eastern Alps, a slice of flysch of Late Turonian to basal Coniacian age contains detrital minerals such as glaucophane and lawsonite that are typical of blue schists. This flysch was deposited in a temporary trench over which the Austroalpine nappes were progressively overthrust. In effect, it is surmounted by the main frontal thrust of the Austroalpine nappes. It rests on the Rheno-danubian flysch nappe which is derived from the Penninic domain (Winkler and Bernoulli 1986).

(ii) The sealing of lower-Late Cretaceous thrusts within Upper Austroalpine nappe pile in the Eastern Alps by the Gosau Formation dated as Late Cretaceous to Middle Eocene in age (Wiesinger et al. 2006).

(iii) Recent reliable radiometric age dates confirm the existence of HP–LT metamorphism in the Eastern Alps during the Cretaceous (Thöni 2006).

The compression resulted from the change in motion of the Apulia-African plate with respect to Europe (Fig. 3.5) that marked the onset of the Alpine collision.

1.3 Folding as expression of Cretaceous compression in the external zones of the Western Alps

Broad areas of the West European craton including the future Western Alps were affected by deformation of Late Cretaceous age. The Devoluy mountain, part of the Diois, as well as the Pelvoux crystalline massif, is composed of E–W-oriented folds and thrusts (Figs. 8.2–8.6) marking a compressive phase that commenced mainly during the Turonian (90 Ma) and continued episodically during the Late Cretaceous (Fig. 8.2). The folds of the Devoluy die out northward and their northern limit lies close to the

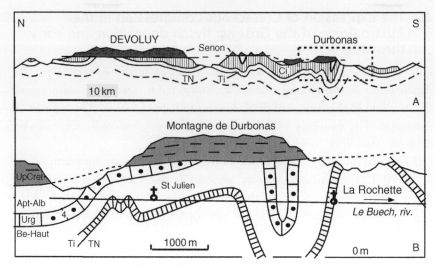

Figure 8.2 *East–West folds and the Late Cretaceous unconformity in the Devoluy (Western Alps).* (A) Overall north–south section across the Devoluy folds and Late Cretaceous unconformity. (B) Detail of the northern part of the Devoluy mountains (Western Alps). Be-Haut: Berriasian to Hauterivian strata; Urg: Urgonian (Berriasian to Early Aptian) platform carbonates; Apt-Alb: Late Aptian to Albian black-shales; Ci: Early Cretaceous; Senon: Senonian pelagic limestone. Location: see Fig. 13.5. *Modified from Agard and Lemoine 2003, Fig. 26b, p. 23.*

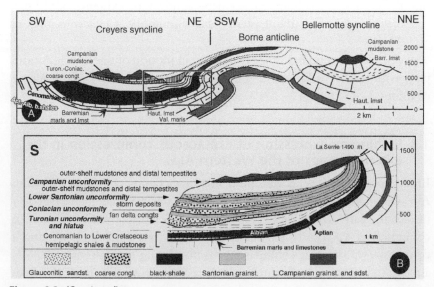

Figure 8.3 *(Continued)*

salt basin edge in the Dauphinois basin suggesting that the folds may detach at depth in Triassic evaporites. The underlying basement of the adjacent Pelvoux massif was involved in the E–W-oriented folding as well as the overlying Devoluy and Diois areas. The southern boundary of the Pelvoux massif (Fig. 8.6C and D) is cut by south–verging reverse faults and small thrusts. This deformation is dated by Late Eocene Nummulite-bearing limestones and the overlying Oligocene Grès du Champsaur. In a similar manner, the northern boundary of the Pelvoux massif is a set of north-verging reverse faults such as the Meije and Muzelle thrusts (Fig. 8.6A and B).

Figure 8.3 *East–West folds and superimposed Late Cretaceous erosional unconformities in the Creyers syncline (Eastern Diois area).* The characteristics of the unconformities shown here are as follows. The Mid-Cenomanian unconformity is defined by the angular relationship between Albian black-shales and alternating Late Cenomanian mudstone/marlstone. The Mid-Turonian unconformity corresponds to the angular relationship and sharp lithological change between alternating Late Cenomanian mudstone/marlstone below and Late Turonian fan-delta conglomerates above. The Coniacian, Early Santonian and Campanian unconformity surfaces are clearly erosional at the northern edge of the syncline. They display a fan-shape, southward diverging, attitude towards the subsiding part of the area located around the axis of the Creyers syncline to the South. All these unconformities are folded. The amount of shortening decreases from the bottom to the top illustrating the successive fold phases. Sub-aerial and subsequent submarine erosion of the adjacent Borne growth anticline to the North supplied the clastic material towards the subsiding area in the South. Three principal events are recorded here by the subsidence history: The first took place between the deposition of the ammonite-bearing Middle Cenomanian marlstones and the Late Turonian fan-delta conglomerates. The event corresponds to a regional uplift with associated sub-aerial exposure and erosion. The uplift was followed by a phase of rapid subsidence, allowing deposition of about 500 m of conglomerates. The duration of the uplift and subsidence event is at most 4 my which is the time-span of the entire Late Turonian and Coniacian. The second event is recorded by the Early Santonian angular unconformity and by the sharp gradation between proximal storm deposits below and outer-shelf tempestites above. This event is also followed by a phase of rapid subsidence of Late Santonian age. During end Late Cretaceous times, the basin filled up to become emergent around the Cretaceous/Tertiary boundary. Dating of the unconformities in the Late Cretaceous shows that the folding of Devoluy and eastern Diois must not be confused with the more recent Pyrenean-Provencal folding which began at the very end of the Cretaceous and continued until Eocene times. Apt.-Albian b.shales: Aptian-Albian black-shales; Barr. Lmst: Barremian limestone; Cenoman: Cenomanian; congt: conglomerate; Coniac: Coniacian; Haut. Lmst: Hauterivian limestone; L. Campanian grainst. and sdst.: Late Campanian grainstone and sandstone; sandst: sandstone; Turon: Turonian; Val: Valanginian. Location: see Fig. 13.5.

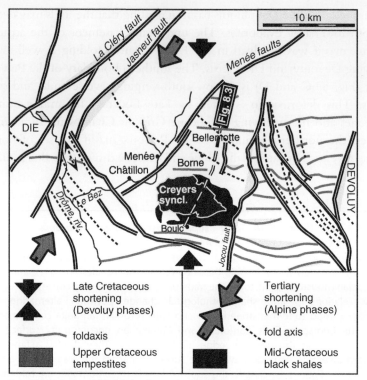

Figure 8.4 *E–W system of folds in the upper Drome valley (eastern Diois and west of Devoluy).* The E–W folds of the upper Drome valley are the westward continuation of those of the Devoluy, shown in Fig. 8.2. The north limb of the Creyers syncline shows superposed unconformities whose ages range from the middle of the Cenomanian to the start of the Campanian (Fig. 8.3B). The Late Cretaceous folds are deformed by later folds and strike slip faults of Alpine *s.s.,* i.e. Neogene age. Location: see Fig. 13.5. *Redrawn and simplified after Arnaud, 1974.*

In the north, the reverse faults apparently intersect Palaeogene strata. However, this anomaly might also be an effect of late Alpine disharmony. In consequence, the pre- or late-Palaeogene age of the north Pelvoux thrusting is not yet fully clear.

In Provence, i.e. in the region of the cities of Marseille, Nice and Digne (Barreme; Fig. 13.5), E–W folding, mainly Eocene in age, did not commence until the end of the Cretaceous (Fig. 8.5) as in part of the Eastern Pyrenees.

In summary, the N–S Pyrenean-Provencal shortening was succeeded at different dates from one region to another from the start of the Late Cretaceous until the end of the Eocene.

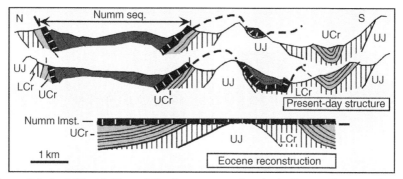

Figure 8.5 *East–West folds and the Eocene unconformity in Eastern Provence (NW of the City of Nice; south of the Western Alps).* The section shows a few beds of shallow marine Nummulitic limestone dated as Middle to Late Eocene depending on locality succeeded by thick Globigerinid-bearing silty marls mainly of Early Oligocene age. The latter rests unconformably on a sub-aerial erosional surface that cuts previously folded Mesozoic strata. Santonian and locally Campanian strata sub-crop the unconformity showing that the E–W folding is not older than Late Cretaceous as in Lower Provence (Toulon area; see below) and the Pyrenees. In consequence the Pyrenees–Provence and Devoluy folds belong to two different phases of Alpine deformation. UJ: Late Jurassic limestones; LCr: Early Cretaceous marls and shales; UCr: Late Cretaceous limestones; Numm lmst: unconformable, Nummulite-bearing limestone; Numm seq: Middle Eocene to Early Oligocene sequence. *Source: Ginsgurg and Montenat (1966). Modified from Agard and Lemoine 2003, Fig. 26d, p. 23.*

 ## 2. THE PELAGIC 'GLOBIGERINID-BEARING MUDS' OF THE WESTERN AND CENTRAL ALPS AND ASSOCIATED CLASTIC SEDIMENTS: DEMONSTRATION OF EXTENSIONAL TECTONICS

During the Late Cretaceous, calcareous pelagic sediments comparable to the Globigerina muds of the present-day oceans were deposited throughout the Helvetic, Ultrahelvetic, Subbrianconnais, Brianconnais and Piemont domains.

In the *Schistes Lustrés*, which draped the yet to be subducted Liguro-piemontais Ocean crust, the original lithologies have been largely obliterated by Alpine metamorphism. Nonetheless, in the Late Cretaceous section, two lithotypes can be distinguished (see Figs. 11.21, 11.22 and 11.23).

– First, Globigerina-bearing marls which are represented by the calcareous shales of Queyras and the *Serie Rousse* of the internal zones between Saint Bernard and Simplon. Clastics are of both oceanic and continental origin, depending on the locality.

– Second, distal flysch represented by fine-grained turbidites deposited far from the original sources of the material. This facies is exemplified by the

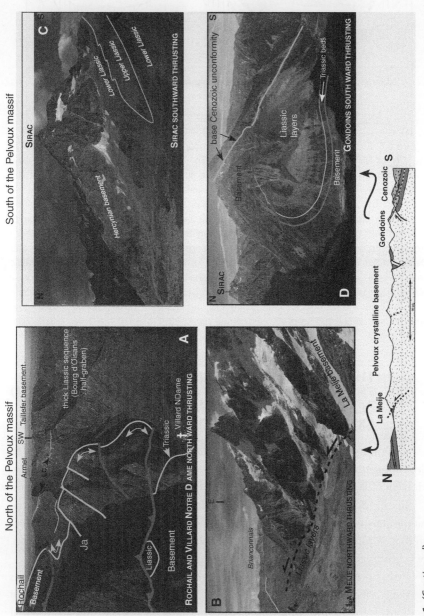

Figure 8.6 *(Continued)*

Valais *Bündnerschiefer* and the *Serie Grise* of the internal zones between Saint Bernard and Simplon (Pennine Alps).

The pelagic muds of the Late Cretaceous and Palaeocene to Early Eocene (100–50 Ma) in the Brianconnais are interbedded in places with breccias ranging in thickness from tens to several hundred metres. The clasts, whose size ranges from millimetres to many metres, were derived mainly by erosion of the Triassic carbonate platform and more rarely of Jurassic limestones, or, in the more internal zones, from Triassic and Permian quartzites as well as metamorphic rocks of the pre-Carboniferous basement.

Such breccias are well developed in the Vanoise mountains (Brianconnais zone of Savoy: 'SBR' block) and in the Briancon area. In contrast, they are absent or of minor importance in the Brianconnais of the Prealps (Central Alps) with the exception of the Gummfluh breccia of Palaeocene age.

In the case of the Melezein breccias near the city of Briancon and those found in the Guil and Escreins nappe windows to the south, there is no visible or exposed connection with palaeofaults. Nonetheless, submarine erosion surfaces intersect Middle or Late Triassic strata. The distribution of the overlying breccias shows that syn-sedimentary scarps were also present in the internal zones.

In some cases, field studies have shown that the breccias result from erosion along normal faults preserved despite Alpine deformation (Figs. 7.8 and 8.7). The fault surfaces have been preserved by manganese or iron crusts

---

Figure 8.6 *Late Cretaceous to Early Cenozoic folding and thrusting on both northern and southern sides of the Pelvoux crystalline core.* A north–south section across the Pelvoux massif shows it as a gigantic 'pop-up' structure that has been maintained permanently in elevated position during multiphase Late Cretaceous to Early Cenozoic deformation. (A) Superimposed Tethyan and Alpine deformation on the gentle slope of the Rochail massif. The top surface of the Hercynian basement and Triassic layers (in yellow) are affected by northward-verging wide folds and reverse faults. Axis of folds is E–W oriented. Preserved small-scale faults (Ja, in blue) of earliest Sinemurian age belong to an early Tethyan extensional phase (Fig. 6.13). For structural location, see Figs. 6.2D and 6.5. The Rochail massif corresponds to the southern prolongation of the Grandes Rousses. In the background, the Taillefer constitutes the footwall of the Bourg d'Oisans half-graben. LC: Early Cretaceous marker bed. (B) La Meije thrusting, seen from the orientation table of Le Chazelet, above the village of La Grave. The vergence of the La Meije basement is northward. Medium- and small-scale folds in the Triassic and Liassic layers in the footwall are E–W oriented. (C) The Sirac thrusting is associated with E–W-oriented reverse folds. These are intersected in the same area by a sub-aerial erosional surface that is overlain by conglomerates and Nummulite-bearing limestones. (D) The south-verging thrusting of the Gondoins basement and the pre-Cenozoic unconformity. The thrust basement is upside down. Axis of the overturned Liassic fold is again E–W oriented. Location: see Fig. 13.5. (See colour plate 11) *Unpublished data and photographs courtesy of Thierry Dumont 2006.*

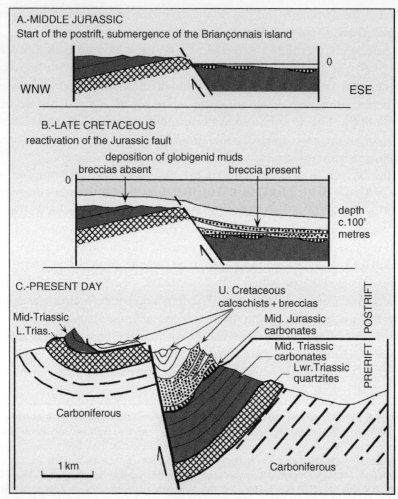

A.-MIDDLE JURASSIC
Start of the postrift, submergence of the Briançonnais island

WNW

ESE

0

B.-LATE CRETACEOUS
reactivation of the Jurassic fault

deposition of globigenid muds
breccias absent   breccia present

0

depth
c.100'
metres

C.-PRESENT DAY

U. Cretaceous
calcschists + breccias

Mid-Triassic
L.Trias.

Mid. Jurassic
carbonates

Mid. Triassic
carbonates
Lwr.Triassic
quartzites

POSTRIFT

PRERIFT

Carboniferous

1 km

Carboniferous

Figure 8.7 *Cerces fault (Brianconnais zone, Western Alps). Example of the syn-rift fault that remained active until the Late Cretaceous.* The Cerces massif, located NE of the city of Briancon, consists of two synclines separated today by a sub-vertical, N–S fault. The fault acted as a normal fault during Jurassic rifting so that the E compartment was downthrown and the W compartment uplifted and eroded. Further, comparison of the thickness of Triassic and Jurassic beds as well as the distribution of the Late Cretaceous breccias on various parts of the fault has shown that the throw had the same sense in the Jurassic and Cretaceous. Iron and manganese crusts have preserved the shape of the fault. The angle of dip remains that of a normal fault despite Alpine deformation. *Adapted from Lemoine et al. 2000, Fig. 9.6, p. 112.*

or beneath sediment wedges along palaeo-escarpments, which show that the faults were exposed at the seabed.

3. EVOLUTION OF THE DEFORMATION FIELD IN THE CENTRAL AND WESTERN ALPS DURING THE LATE CRETACEOUS

From the start of the Late Cretaceous (100 Ma), the initiation of Alpine folding strongly modified the deformation regime. Subduction began in the internal zones and the Valais (Fig. 8.1). Parts of the external zones of the Western Alps were also subjected to compression (the 'Devoluy phase', Figs. 8.2–8.4). In contrast, at least part of the Brianconnais domain was again subjected to extension (Figs. 7.7–7.9) just like southern Provence (see below; Floquet et al. 2003, 2005).

The evolution of the Tethys during the Early to Middle Jurassic syn-rift phase is closely comparable on both the Prealps and Briancon transects (Chapter 6). During the post-rift Tethyan spreading phase, however, their geological history was quite different. In the Prealps (Central Alps), the succession of the Breche nappe includes voluminous Late Jurassic breccias distributed over at least 100 km in an east–west sense between the Chablais and Swiss Prealps, perhaps related to the Late Jurassic–Early Cretaceous, Valais extensional event (Fig. 7.9). Equivalents are absent in the Vanoise and Briancon areas. In the Late Cretaceous, the reverse is true and the succession in the Vanoise and Briancon mountains includes important breccias while such breccias are minor in the Median Prealps.

4. COMPARISON WITH THE PYRENEES AND PROVENCE DOMAINS

Late Cretaceous deformation and correlative unconformities are widespread in the Alpine foreland, including the Rhone Valley and Paris basins. Many pre-1970 publications have recognized that the E–W Pyrenees–Provence folding (Fig. 8.5) began during the Late Cretaceous. By extension of this observation, it has been traditional to assign all E–W folds of the Sub-alpine domain to the Pyrenees–Provence folding phase, e.g. those of the Devoluy phase (Figs. 8.3–8.5). These fold phases were approximately correlated in time with folds and thrusts known in the Eastern Alps.

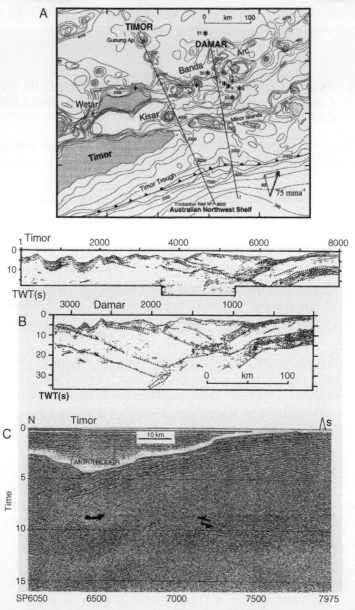

Figure 8.8 *(A) Bathymetric map and location of the BANDA deep seismic reflection survey lines DAMAR and TIMOR.* Shot point numbers are in italics. Earthquake epicentres are shown as stars. Numbers relate to McCaffrey (1988), Table 1; letters a, b and c to events of 30 May 1988, 8 August 1988 and 21 May 1991 (Harvard centroid-moment tensor solutions). Barbed line represents the plate boundary with barbs in the direction of subduction. Arrows indicate relative plate motion. Bathymetry in metres.

(Continued)

Sinistral strike-slip movement of the Iberia block resulted in pull-apart basins, delimited by tilted blocks that opened during the Late Cretaceous in the Pyrenean and Provence domains (Leleu et al. 2005). In lower Provence, in the area of the cities of Marseille, La Ciotat and Toulon, a characteristic example is the Late Cretaceous Lower Provence (= Bausset) basin (for location, see Figs. 6.16, 7.10 and 13.5). This narrow and elongated transtensional basin is dated as Cenomanian to early Santonian. The pebbles of crystalline basement in the Ciotat conglomerates of Turonian age demonstrate the existence of contemporaneous strong sub-aerial relief that formed a probable western extension of the Maures massif now submerged beneath the Mediterranean (Floquet et al. 2003, 2005). The first E–W striking folds appeared in the lower Provence area at the end of the Santonian. In the eastern North Pyrenean zone, the first E–W folds only appeared at the Cretaceous–Tertiary boundary (M. Bilotte, oral communication, 2000; Bilotte 2007). These movements continued during the Eocene.

Progressive understanding has therefore shown that, if the Diois and Devoluy in the Sub-Alps were truly folded from the start of the Late Cretaceous (Figs. 8.2 and 8.3), folding only began at the end of the Cretaceous in the Eastern Pyrenees and Southern (= Lower) Provence domains. The Sub-alpine and lower Provence domains thus belong to quite different structural units in this respect. Recent publications by Floquet et al. (2003–2005) have shown the role of structural inheritance of Alpine faults from both the Early Jurassic and Late Jurassic–Early Cretaceous rift phases.

From Richardson and Blundell, 1996, Fig. 1. (B) **Line drawing interpretations of the BANDA seismic sections: (a) Timor; (b) Damar.** Major reflectors and zones of high reflectivity are highlighted with dotted lines. *From Richardson and Blundell 1996, Fig. 3.* (C) **Migrated seismic section from the TIMOR profile in the vicinity of the Timor trough.** Post-Permian shelf sediments that can be tied to the Troubadour well (top right) are clearly seen beneath the sea floor on the inner trough wall, more recent undeformed sediments lie on the bottom of the trough. One arcuate reflector is indicated (arrow at SP 6500) in uppermost basement beneath the shelf sediments. It may represent an unrecognized Devonian–Carboniferous basin, or side-swipe from a fault plane parallel to the seismic profile and bounding such a basin. Moho lies at about 15 s beneath the trough. Numerous lower crustal reflectors occurring between 8 and 12 s on the right of the figure appear offset with a thrust sense (half-arrows) along south-dipping thrust shear zones near SP 7250. *Reproduced from Snyder et al. 1996, Fig. 3, p. 65.*

5. KINEMATICS AND UPLIFT DURING THE LATE CRETACEOUS

The development of the Western Alps during Cretaceous and Eocene times is largely a consequence of the interactions resulting from the relative movements of the European, Iberian and Apulia plates as well as the timing of these displacements.

If the kinematics of Iberia plate are relatively well constrained thanks to the pattern of well-dated oceanic magnetic anomalies in the Bay of Biscay and North Atlantic, those of the Apulia plate remain much less known. In addition, the timing of the Late Cretaceous extensional and compressional phases in the internal zones of the Central and Western Alps is not documented in sufficient detail to allow a useful comparison between the Sub-alpine, Provence and Pyrenean domains. As a result an integrated view of the deformation of SW Europe during the Late Cretaceous remains to be fully understood.

Nonetheless, some important observations can be made that will need to be addressed in developing future kinematic models. The conventional view has been to examine timing of deformation rather than the consequences as shown by uplift. It is important to stress that the Devoluy deformation phase (Figs. 8.2–8.4) resulted in massive regional uplift elevating hitherto deep basinal areas above sea level to be later eroded sub-aerially. The regional extent of this uplift is shown by the widespread deposition of shallow water sediments, mainly platform carbonates dated by Nummulites and other benthic foraminifers. These deposits covered hitherto basinal areas throughout much of the Dauphinois Basin and also beyond the Pelvoux–Argentera line into the Brianconnais. Clearly the morphology of the passive margin was profoundly modified by this event.

The second issue concerns the location of compressional and transtensional tectonism. First, the deformation observed within the Dauphinois Basin lies wholly within the European plate and may be linked to the movement of the Iberian plate. It thus seems unlikely to be related to the onset of subduction on the African plate revealed by deposition of the Helminthoid flysch. Second, the extension observed in the Brianconnais area might reasonably be caused by flexure of the European plate in response to the initial phase of obduction of Apulia–Africa. A present-day analogue is offered by extension of the margin off NW Australia in response to loading of the plate margin by obduction of Timor (Fig. 8.8). More detailed study is required to unravel the complexities of deformation linked to Iberia from obduction of Apulia–Africa.

CHAPTER NINE

The Tethyan Margin in Corsica

Contents

Summary

The eastern part of Corsica corresponds to a southern extension of the Alps although the structural zones are not exactly the same as those defined in the Western Alps. The palaeostructure of the reconstructed passive margin also differs in several important respects for which an explanation will be proposed later in Chapter 12. In terms of structure, Corsica can be divided into a western 'granitic', 'crystalline' or 'Hercynian' part and an eastern 'Alpine' part. The latter is characterized by a stacked pile of thrust slices and nappes, and by a west to east increase in Alpine metamorphism which is particularly intense in northern Corsica in the Castagniccia and Cape Corse areas. The eastern part of Alpine Corsica includes ophiolite nappes comparable to those of the Alps. A series of tectonic slivers found between these nappes and Hercynian Corsica was derived from the European continental margin which was much narrower than the reconstructed margin of the Western Alps.

1. HERCYNIAN AND ALPINE CORSICA

1.1 Granitic or Hercynian Corsica

In map view, Hercynian basement comprises most of the western half of Corsica (Fig. 9.1). It consists throughout of various granites, of Hercynian to late Hercynian age (Permian), with associated gneisses and micaschists.

© 2011 Elsevier B.V.
All rights reserved.

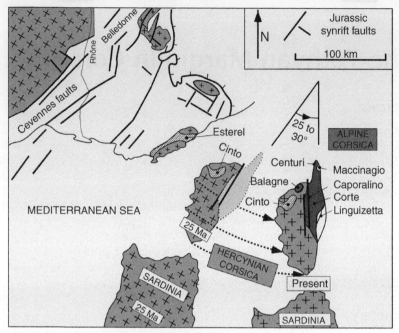

Figure 9.1 *Pre-Miocene reconstruction of the position of Corsica*. *In part after Gauthier and Rehault 1988. Adapted from Lemoine et al. 2000, Fig. 10.1, p. 115.*

Sedimentary and volcanic rocks of Carboniferous and Permian age also rest locally on the basement. In particular, the Permian volcano-sedimentary complex of Monte Cinto is comparable to that of Esterel in eastern Provence. To the north of Balagne as well as in southeast Corsica, the Mesozoic and Tertiary sedimentary cover rests unconformably on the Hercynian basement.

1.2 Alpine Corsica

Alpine Corsica comprises a series of nappes and slices thrust from east to west onto the Hercynian basement. An east–west section (Fig. 9.2) across Corsica shows three superposed assemblages below the thin Mesozoic cover of the internal border of the Hercynian basement. These are from west to east:

– Slices of sedimentary rocks such as the Corte and Caporalino units derived from the European continental margin.
– Nappes consisting of ophiolites and overlying Jurassic–Cretaceous sediments now represented as *Schistes Lustrés*.

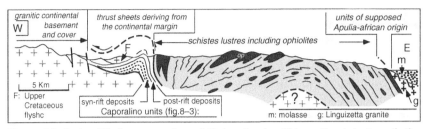

Figure 9.2 *A simplified cross section of Alpine Corsica*. This section is just north the city of Corte (as also Fig. 9.3). F: Late Cretaceous flysch of Tralonca-Santa Lucia probably of internal origin (?). *Modified from Caron and Bonnin 1980. Adapted from Lemoine et al. 2000, Fig. 10.2, p. 116.*

– Some slivers all originating from an internal position with respect to the remnants of oceanic crust. These are, for example, the Centuri gneiss and sedimentary slices of Maccinaggio situated respectively west and east of the northern tip of Corsica, and the Linguizetta granite situated at the internal margin of the *Schistes Lustrés* complex in eastern Corsica. Because of their geometric position, these internal units could well have been derived from the conjugate Apulo-African margin unless major back thrusting is postulated. Alpine metamorphism increases in intensity from west to east and has affected the internal border of Hercynian Corsica, the slivers and ophiolite nappes. The metamorphism includes an initial blue schist and eclogite phase due to subduction followed by much later, retrograde green schist phase.

2. THE INITIAL POSITION OF CORSICA DURING THE MESOZOIC

The past position of Corsica during the pre- and syn-rift phases, and later spreading (Triassic to Cretaceous; 230–65 Ma) has great relevance to reconstruction of the Ligurian Tethys. Palaeomagnetic studies of the Permian volcanic rocks of Corsica and the Tertiary Volcanics of Sardinia show that the Corsica–Sardinia block has undergone a post-Early Miocene anticlockwise rotation of 20–30 degrees. This rotation opened the oceanic Liguro-Provencal (Balearic) basin of the Western Mediterranean. Placing Corsica in its initial position thus requires closure of the Liguro-Provencal basin to the 2000 m bathymetric contours of Provence and Corsica using a rotation of 30 degrees. However, the exact position of the pole of rotation is uncertain and reconstructions therefore differ. In Fig. 9.1, a reconstruction has been chosen that places the Permian volcano-sedimentary complexes of Monte Cinto and Esterel in a conjugate position. Restoration of Corsica to its initial position suggests that the boundary between Hercynian and Alpine Corsica is approximately parallel to the NE–SW strike of syn-rift Jurassic faults such as Cevenole and Belledonne, i.e. the syn-rift faults which define the tilted blocks of the Ligurian Tethys.

3. THE TRANSITION FROM THE CORSICA–PROVENCE PLATFORM TO THE LIGURO-PIEMONTAIS OCEAN

3.1 The Corsica–Provence Platform

Restoration of western Hercynian Corsica to its Jurassic position juxtaposes its granitic-gneissic basement and Permian volcanics against the Maures and Esterel massifs. The remnants of the Jurassic cover of Hercynian Corsica visible on the internal eastern border correspond to a carbonate platform which would have been originally contiguous with that of Provence. The Ligurian Tethyan continental margin was situated to the southeast of this carbonate platform which was therefore constructed on the originally contiguous Corsica–Provence block.

3.2 The Jurassic passive margin in Corsica

Units derived from the continental margin are represented by an assemblage of overthrust slivers and small nappes such as Corte or the Caporalino units. In these continental margin units, there are few pre-rift sediments and the Triassic is commonly thin or absent. Syn-rift sediments are, except in certain units (Fig. 9.3B), rich in sandstones derived from the Hercynian basement because of rapid Jurassic erosion of the thin Triassic cover. The onset of post-rift sedimentation is marked by deposition of radiolarites or limestone of Middle to Late Jurassic age. In the Caporalino units (Fig. 9.3), the post-rift sequence consists of Belemnite-bearing fine-grained limestones that are interbedded with calcareous turbidites containing algae and corals. This association suggests a relatively deep marine environment in which pelagic sediments were intercalated with flows of calcareous debris derived from the adjacent platform. Similar facies types have been found within the sedimentary cover of the ophiolites. In the Balagne area, near the pass of San Colombano, radiolarites, overlain by late Jurassic pelagic limestones, rest on oceanic basalts. However, these radiolarites and pelagic limestones are interbedded with platformal limestones (notably algal limestones) indicative of very shallow water depths. This

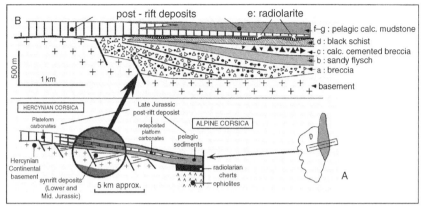

Figure 9.3 Sections across the Tethyan margin and ocean of Corsica during late Jurassic time. (A) Schematic section of the margin during the Late Jurassic. No scale. (B) Late Jurassic reconstruction of the Caporalino unit. Syn-rift sediments (Early and Middle Jurassic). a: breccias composed of basement and sandstone; b: sandy flysch; c: carbonate cemented breccias; d: black schists. Post-rift sediments (Late Middle Jurassic and Late Jurassic); e: radiolarites; f: carbonate cemented conglomerates with basement pebbles; g: limestone, interbedded pelagic sediments and resedimented beds derived from very shallow water carbonate platform. *Modified after Amaudric Du Chauffaut 1982 and adapted from Lemoine et al. 2000, Fig. 10.3, p. 118.*

unusual succession and association provoked a celebrated controversy on the depth of deposition of radiolarites during the first part of the 20th century. The association of radiolarites with green algae-bearing limestones including *Dasycladales*, indicative of a maximum water depth of 10–20 m (i.e. the photic zone) can be readily explained by reworking and redeposition of calcareous muds derived from the adjacent platform. This model is compatible with the deposition of radiolarites below the carbonate compensation depth (CCD) which was probably located between 1000 and 3000 m water depths during Middle Jurassic time. In addition to the redeposited limestone beds, relatively coarse siliciclastic layers are intercalated within the radiolarites. These could originate either from granitic Corsica or from the conjugate Apulia margin. However, there is no way to choose between either provenance.

3.3 Lack of continuity with the structural zones and specific palaeogeographic domains of the Western and Central Alps

Direct equivalents of the western Alpine structural units constituting the Tethyan margin cannot be defined in Eastern Corsica. The difference may reflect segmentation of the margin by a transform fault separating the originally contiguous Corsica–Sardinia block from the future Alps *stricto sensu*. Only the Corsican Liguro-piemontais units that comprise the ophiolites and their pelagic sedimentary cover are comparable to the Liguropiemontais units of the Alps. The radiolarites, pelagic limestones and *Schistes Lustrés* share the same characteristics as those found in the Early and Late Cretaceous of the Alps.

 ## 4. SUMMARY: CHARACTERISTICS OF THE TETHYAN MARGIN IN CORSICA

The passive margin in Corsica retains many of its original features. Pre-rift or early syn-rift beds of Triassic carbonates are thin or absent. In consequence, the syn-rift sequence is rich in siliciclastics eroded from the Pre-Triassic crystalline basement. The post-rift pelagic sediments comprise radiolarites and limestones of Middle to Late Jurassic age deposited on both the continental margin and oceanic crust. Also present are interbeds of sediments derived from the continent. These sandstones contain pebbles of basement and platform carbonates deposited in very shallow water depth. Such observations imply that the first oceanic crust was probably located close to the platform. The Tethyan continental margin of Corsica was therefore much narrower than that of the Western and Central Alps. The difference may reflect segmentation of the margin by a transform fault separating the originally contiguous Corsica–Sardinia block from the future Central Alps (Chapter 12).

FURTHER READING

Caron (1980).

CHAPTER TEN

The Apulia-African Margin of the Liguro-piemontais Ocean: The Transition from Continent to Ocean

Contents

Summary

Before examining the development of the Liguro-piemontais Ocean, the evolution of the Apulia-African margin is briefly reviewed using examples from the Southern Alps of northern Italy and southern Switzerland, and from the Austroalpine nappes of southeastern Switzerland. Since the greater part of the Apulia margin escaped subduction, Tethyan (pre-Alpine) structures are there particularly well preserved, especially those of the former continent–ocean boundary in the Graubünden region of the Central Alps. In these areas, the age of pre-, syn–and post-rift sediments are comparable to those of the European margin.

The Western Alps, from Rift to Passive Margin to Orogenic Belt, Volume 14
ISSN 0928-2025, DOI 10.1016/S0928-2025(11)14010-9
© 2011 Elsevier B.V.
All rights reserved.

One of the most remarkable structures in the Southern Alps is the Lugano fault. This fault is one of the first to have been described and compared with similar structures observed on seismic profiles across present-day margins. The fault bounds a major half-graben infilled with syn-rift sediments. It can be followed laterally for 30 km. Towards its base, the fault penetrates the basement as far as 15 km palaeodepth where it flattens progressively. Its trace is marked by the presence of mylonites and ultra-mylonites developed in green schist metamorphic facies.

In the proximal parts of the Apulia-african margin in the Graubünden area, the structures responsible for thinning the brittle upper crust are high-angle listric faults that define tilted blocks and typify rifting. The thinning of the lower crust results from ductile stretching that follows the model described for the Iberian margin and the exhumed margin in the Southern Alps.

In contrast, thinning of the crust on the distal part of the margin resulted in a different structural style. A system of low-angle detachment faults that cuts the crust and mantle dismembered the upper crust and exhumed the sub-continental mantle but without the formation of the classical listric faults of rifts. For this reason, the emplacement of the initial axis of the rift defining the morphology of the top of the pre-rift and that of the line of continental break-up are shifted in space. Inheritance from earlier structures may be the reason for the shift.

The characteristics of the Apulia-african margin will be presented below by means of two sub-parallel transects: one across the Southern Alps between northern Italy and southern Switzerland (Fig.10.1) and a second in eastern Switzerland in the Graubünden region (Fig.10.2). Some comparative points will be made with the Western Alps transect, particularly the evidence of episodic rifting, whose phases are broadly contemporaneous with those known from the European margin.

1. SOUTHERN ALPS TRANSECT: NORTHERN ITALY TO SOUTHERN SWITZERLAND

The Southern Alps lie between the Peri-adriatic fault array to the north and the Tertiary foreland basin of the Po plain to the south (Fig. 10.1: Insubric line; Fig. 10.2A; Fig. 2.4: PC, PG, PT, S, E, M lines). In this area, mild, southerly verging Alpine deformation has had relatively little effect on the rift fabric formed in Liassic to Middle Jurassic times. The main faults and tilted blocks strike nearly north–south, i.e. perpendicular to Alpine structures. In this region, reconstruction of the Tethyan margin involved in

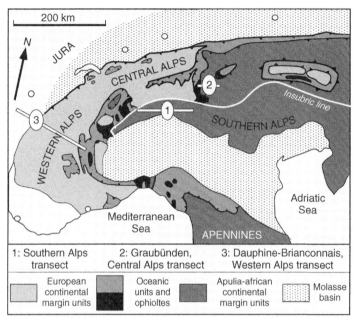

Figure 10.1 *Location of cross sections in the Southern Alps, Graubünden (= Grisons) and the Western Alps in the context of the main Alpine structures.* Refer to Fig. 2.2. *Modified from Lemoine et al. 2000, Fig. 5.2, p. 64.*

Alpine folding was first made from a comparison with the present-day western margin of the Central North Atlantic (Bernoulli and Jenkyns 1974).

1.1 Palaeogeographic domains

The Apulia palaeo-margin (Fig. 10.2B) is bounded to the east by the shallow marine carbonate platform of Friuli deposited from Triassic to Cretaceous time, i.e. during the pre-, syn- and post-rift phases. The area is somewhat symmetric with the Causses–Jura platform of the inner European margin. To the west of the Friuli platform, Early and Middle Jurassic rifting differentiated four major palaeogeographic domains. These are:

− The Belluno Trough − infilled with calcareous turbidites, which were derived from both Friuli to the east and Trentino to the west.

− The Trentino, or Venice, submarine platform − covered with shallow marine limestones.

A system of faults or syn-sedimentary flexures oriented N10°E to N20°E marks the boundary between the subsiding basin and adjacent platform. The system was active from the Permian to the Jurassic with a 2 km

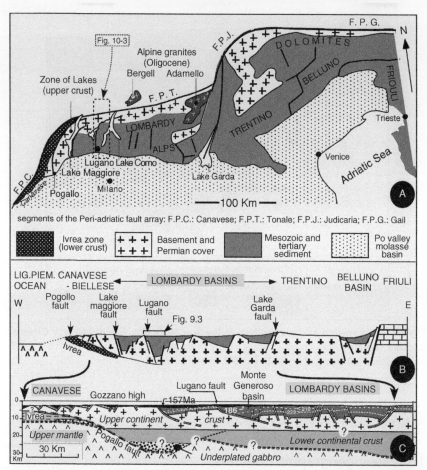

Figure 10.2 *Map and cross section of the Jurassic margin in the Southern Alps.*
(A) The map shows in bold lines the Peri-adriatic fault system. FPG: Gail segment;
FPJ: Judicaria segment; FPT: Tonale segment; FPC: Canavese segment (the latter
should not be confused with the palaeogeographic domain with the same name)
and the main syn-rift faults. South of the Peri-adriatic fault system, crosses indicate
basement and Palaeozoic cover; dark grey: Mesozoic and Tertiary sediments (except
molasse); light grey: Molasse Basins of the Po valley. (B) The corresponding cross
section is a time slice at the end of rifting and onset of post-rift sedimentation.
Post-rift sediments are not shown. In the Lombardy Basin and further west,
radiolarites mark the onset of post-rift deposition, to the east, in Trento and
around Belluno, nodular, *Ammonitico Rosso* limestones are the first post-rift
sediments. LM: Lake Maggiore fault; Lu: Lugano-Monte Grona fault (Fig. 10.3); Ga:
Lake Garda faults. See also Figs 2.9 and 4.7. (C) Tentative reconstruction of the
Lombardy rift. *In part after Winterer and Bosellini 1981; Bertotti et al. 1993;
Manatschal 2004; Manatschall et al. 2007, Fig. 6, p. 300. Modified and completed
from Lemoine et al. 2000, Fig. 11.6, p. 125.*

thickness of Permo-Triassic on one side of the fault array and 4–5 km on the other. In the Dolomites, this family of extensional faults is associated with another array of transfer faults oriented N70° to N90°E (type: Valsugana line; Fig. 5.3B), which are both transtensional and transpressional, and associated with the regional extension.

– The Lombardy basin, a wide and complex domain bounded by horsts and grabens with well-preserved syn-rift faults.

– The Biellese and Canavese zones. These areas are poorly known because of poor, sparse outcrop in low-lying areas. In addition, Alpine deformation is intense due to their original proximity to the former continent–ocean boundary of the major Canavese (Peri-adriatic) fault trend.

The Liguro-piemontais Ocean was present to the west of the area as documented by some ophiolite outcrops within western Canavese units.

1.2 Extension in the proximal parts of the Apulia margin in the Southern Alps: the Lombardy rift and the Tethyan break

The Lombardy rift (Fig. 10.2B and C), which developed between Late Triassic and Late Liassic times, was located between the Lugano and Lake Maggiore faults, and the Lake Garda fault. The Lugano and Maggiore faults downthrow eastward towards the Apulia–Africa continent while the Lake Garda fault downthrows westward towards the nascent Tethys. The Lombardy rift does not predate the future Tethyan break-up line which was located west of the Canavese as shown below.

1.3 The Lugano fault in the Lombardy rift system

The north–south Lugano fault is a major palaeofault largely unaffected by Alpine deformation that can be followed for 30 km (Fig. 10.3). It has great scientific importance because it is a rare example of an exposed fault in the Alps that allows study at outcrop of the continuation in depth of a major listric fault located at the boundary or transition between the ductile lower crust and brittle upper crust.

In the Monte Generoso basin, immediately adjacent to the Lugano fault, thousands of metres of hemipelagic sediments of Rhaetian to early-Middle Liassic age are interbedded with sedimentary breccias derived from the adjacent fault scarp. In contrast on the footwall, west of the fault, Rhaetic and Liassic beds are only a few tens of metres in thickness. These differences in thickness are also present in the Norian. The Lugano fault is sealed by

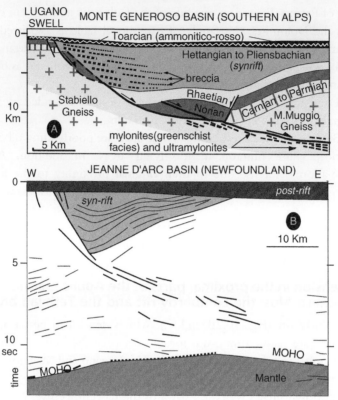

Figure 10.3 *(A) Reconstruction of the Lugano - Monte Grona syn-rift fault*. The thickness of Rhaetian and Liassic sediments exceeds several kilometres. The Lugano fault became active during deposition of Norian dolomites which vary in thickness across the half graben. Above the syn-rift sediments, the nodular, "Ammonitico Rosso" limestones of Late Liassic (Toarcian) age are not displaced by the fault. *After Bertotti in Bernouilli and others, 1990; modified from Lemoine et al., 2000, fig. 11.7, p. 127.* **(B) Comparison with the Jeanne d' Arc Basin (Newfoundland margin)** Line drawing of Lithoprobe deep seismic line 85-4. The Murre fault soles at about 22km (7sec.). Conspicuous features include a slight rise in an otherwise flat Moho, intracrustal detachment, rollover of crust and Mesozoic stratigraphy into listric extensional faults and a relatively thin post rift megasequence. *After Tankard and Welsink, 1989; Manatschal and Bernoulli, 1998, fig. 4b, p. 376.*

post–rift pelagic and hemipelagic sediments that include the *Ammonitico Rosso* nodular limestone of Toarcian age (190–180 Ma).

The characteristics of this fault were first documented as early as 1964 by Bernoulli prior to the recognition of tilted blocks on present–day passive margins. However, its geodynamic significance as an eastward–dipping fault separating the footwall from the half-graben to the east was not understood until some years later. The Lugano fault is a basement-involved listric fault as shown by more recent work which has defined the deeper part of the fault trace in Triassic sediments as well as in the underlying crystalline basement (Fig. 10.3). Within the basement, the fault trace is characterized by narrow mylonitic bands affected by retrograde green schist metamorphism. At about 10 km below the base of the Upper Liassic, the fault curves to become almost parallel to the sedimentary strata, notably to the Jurassic–Triassic boundary and the contact between the crystalline basement and overlying Triassic sediments. The lower trace of the fault was therefore approximately horizontal at the time of formation.

An elegant comparison can be made between the observed and restored structure of the Monte Generoso half-graben with the deep structure of the half-graben underlying the Jeanne d'Arc basin off Newfoundland interpreted from seismic profiles. In the two examples, the bounding major listric fault approaches the horizontal towards 10–15 km depth, i.e. towards the middle of the crust. The seismic profile also shows elevation of the Moho above the stretched zone (Fig. 10.3A and B). Comparable structures are known on the European margin in the Sub-alpine domain. However, the level reached by erosion is insufficient to expose the deeper traces of the normal faults. Their qualification as listric normal faults is based only on geometrical reconstructions constrained by the structure observed at outcrop (see Fig. 6.2, Ornon and Fig. 13.1).

1.4 Extension in the distal part of the Apulia margin: Zones of ductile shear in the basement west of the Southern Alps

Within the basement, two rock assemblages are present representing the upper and lower continental crust (Fig. 10.2C). The upper continental crust consists of pre-Permian gneisses and micaschists known as the 'Zone of Lakes' unit on the Central Alps transect (Chapter 2). The lower crust is composed of granulite-facies-grade gneisses and, towards the base, i.e. close to the palaeo-Moho, of gabbros and peridotites. This lower crust corresponds to the Ivrea zone of geologists, also called the First Dioritic-kinzigite zone that outcrops south of the Peri-adriatic fault line (Fig. 2.7). Zones of ductile shear intersect the continental crust at several places of the Lombardy, Biellese and Canavese Alps (Fig. 10.2C). The most important shear zone on the distal margin is the Pogallo (Fig. 10.2A and C), which transects the upper crust to curve at greater depth and then merge with the upper to lower crust boundary. It plunges eastward, i.e. beneath the Apulia margin itself.

Ductile shear is accompanied by a local, weak metamorphism which typically post-dates the (Hercynian) main metamorphism and deformation. Formation of the post-Hercynian shears at depth is shown by their association with green schist facies metamorphism and locally by amphibolite facies, as in part of the Pogallo shear. Thermo-barometric results show the fault was active during Liassic time. Likewise, Rb/Sr and K/Ar radiometric dating of other early Mesozoic shear zones within the Ivrea have yielded some Permian ages but mainly Triassic and Liassic ages, i.e. from 220 to 170 Ma. The structural, metamorphic and geochronological signatures of rifting are therefore recorded within the lower continental crust.

The effect of the Pogallo fault has been to reduce the thickness of the continental crust by about 10 km solely by stretching of the lower continental crust without apparently affecting the upper crust. Even if the ages are the same, the mechanism of stretching is clearly different for the proximal and distal parts of the margin.

2. THE GRISONS (=GRAUBÜNDEN) TRANSECT IN EASTERN SWITZERLAND

The Austro-alpine nappes outcrop mainly to the east in the Graubünden (=Grisons) within the Eastern Alps. This area, situated close to the Swiss–Austria border, lies at the western erosional limit of the Austro-alpine nappes.

2.1 Palaeogeographic domain

Within the Engadine window in the Inn valley region, the following Austro-alpine nappes are distinguished from base to top (Fig. 10.4B): the Err (mainly basement with cover remnants), the Samedan (cover), the Julier and Bernina nappes (basement), the Ela and Ortler and some Engadine dolomites (mainly cover with minor basement) and the Silvretta and Otztal (mainly basement). The entire nappe pile overthrusts the Platta ophiolitic nappe that was derived from the Liguro-piemontais Ocean.

The Err and Platta nappes provide a unique and important record of the early phases of rifting on the distal parts of the margin and involve well-preserved structures resulting from the exhumation of sub-continental mantle.

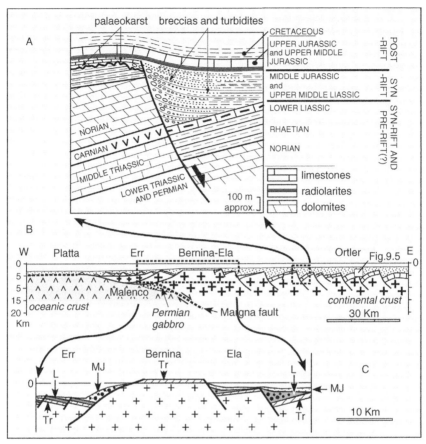

Figure 10.4 (A) Synthetic and theoretical Triassic to Cretaceous stratigraphy of the Austro-alpine nappes of the Grisons in relation to a syn-rift palaeofault. Along the arrays of palaeofaults, syn-rift deposits vary in thickness and facies; shallow marine carbonates deposited on fault block highs commonly rest on an eroded surface cut by tectonic fissures and perhaps karstified. Hemipelagic sediments interbedded with breccias derived from fault escarpments infill the half-grabens. Fault throws are towards the west and the ocean in the Err nappe, which originated close to the Liguro-piemontais Ocean. *Modified from Lemoine et al. 2000, Fig. 11.1, p. 121*. **(B) and (C) Cross section of the Jurassic margin in the Grisons (Central Alps)** *simplified from Manatschal and Nievergelt (1997); Manatschal and Bernoulli 1999, Fig. 4b, p. 1105.*

2.2 Stratigraphy

Pre-rift: basement and Triassic. Above the crystalline basement and Palaeozoic sediments, the pre–rift succession consists of:

- Early Triassic sandstones.
- Middle Triassic platform carbonates deposited in very shallow marine environments.

- Late Triassic beds composed of Carnian evaporites, overlain by Norian dolomites (*Hauptdolomit* or *Dolomia Principale* facies) which were deposited over a broad, tidally dominated area. The overlying Rhaetian beds (latermost Norian) were deposited in an open but shallow sea. However, there is no evidence of active Late Triassic faulting in the Grisons.

Syn-rift: Middle to Late Liassic and Middle Jurassic in part. The succession consists of sediments deposited either on the raised edge of tilted blocks or in the adjacent half-graben. Sediments infilling the half-grabens include olistoliths, breccias and turbidites derived from the fault scarps of the tilted blocks. Successive phases of rifting are dated Hettangian, Sinemurian, and probably Early Pliensbachian (?) and Late Liassic (?) (Fig.10.7). The age of these episodes compare well with those recorded on the European margin (Fig. 6.19).

Post-rift: Late Middle Jurassic and Cretaceous. The lithostratigraphy compares well to that of the European margin:

- radiolarites and light-coloured pure *Aptychus*-bearing limestone dated as Late Middle Jurassic and Late Jurassic,
- alternating shales and limestones dated as Early Cretaceous that are closely comparable to the Replatte formation of the Western Alps and the *Palombini* of the Apennines, and
- pelagic calcareous shales dated as Late Cretaceous.

2.3 Extension in the proximal parts of the Apulia margin as seen in the Graubünden region: normal faults, listric faults and tilted blocks

Extension in the proximal parts of the margin is mainly due to normal fault movements in the brittle upper part of the crust and ductile stretching in the lower part following the classical model mentioned above for the Southern Alps (Fig. 10.3) and Western Alps (Fig. 6.2A), and developed from seismic profiles across present-day passive margins (Fig. 1.8).

In these areas, the majority of the syn-rift faults have either been reactivated or sheared by thrusts (Fig. 14.1: Ortler and Campo Nappes). However, several have been preserved within nappes such as the Ortler (Figs. 10.5 and 10.6: Livigno area). Intercalated clastic beds, breccias and then turbidites further basinward in half-graben that also include mega-breccias at the foot of palaeofaults provide additional evidence for syn-rift faulting (Figs. 10.4A and 10.6). The distribution of these syn-rift sediments allows reconstruction of the orientation and displacement of the syn-rift

Livigno Lake

Figure 10.5 *The palaeo half-graben of Il-Motto near Livigno (Grisons, Central Alps)*. The very well-preserved palaeofault (F), virtually unaffected by post-Liassic deformation, dips S or SE, i.e. towards the right. Displacement took place during the Hettangian (1, 2) and Early Sinemurian (3) and marks the first phase of rifting known on the European margin. Emphasized by a narrow couloir of fallen rocks, the throw of the normal fault is more than 300 m. At the southern or southeastern foot of the fault escarpment, breccias and mega-breccias are intercalated within ammonite and belemnite-bearing limestones and marls (L) of the half-graben. These breccias and mega-breccias are due to collapse of the fault scarp. T: Triassic (Norian) dolomites; L: syn-rift Liassic limestones and marls with intercalated breccias, dated by ammonites. 1, 2: Hettangian; 3: Early Sinemurian; 4: Late Sinemurian. *Stratigraphic results after Eberli 1988. Adapted from Lemoine et al. 2000, Fig. 11.2, p. 123.*

faults: the throws of syn–rift faults, from direct observation or reconstruction, are eastward and towards the Apulia–Africa continent in the Bernina, Ela and Ortler nappes (Fig. 10.5). In addition, major listric normal faults have been documented from field evidence.

2.4 The presence and ages of distinct phases of rifting

The timing of syn-rift faulting is documented mainly from the age of turbidites derived from submarine fault escarpments (Figs. 10.4, 10.5 and 10.7). Each phase of fault movement triggered deposition of a sequence of coarse, thinning upward turbidites that is also widely observed on the European

Figure 10.6 *Syn-rift palaeofault in the Western Ortler nappe: Piz Chaschauna sector near Livigno.* This well-preserved palaeofault dips NE (towards the right). To the left, pre-rift Triassic carbonate (T); to the right, dark coloured marls and limestones, intercalated with breccias, are syn-rift Liassic (L) deposits. *After Froitzheim and Eberli 1990. Photo N. Froitzheim. Adapted from Lemoine et al. 2000, Fig. 11.3, p. 123.*

margin (Figs. 6.5, 6.6 and 6.19). The first phase of Early Liassic age (Hettangian–Sinemurian, about 200 Ma: 1a and 1b in Fig. 10.7) is well dated by ammonites. The later phases are of Middle Liassic (Early Pliens-bachian: 1c) and Late Liassic–Middle Jurassic (2) age but are poorly dated, especially the latter. Again, these ages are similar to those documented on the European margin (Chapter 6).

2.5 Deformation during the start of extension on the distal parts of the Apulia margin in the Graubünden region

Those structures resulting from extension on the distal parts of the margin (Fig. 10.4) are very different from those on the proximal parts (Figs. 10.3 and 10.4B). They are due to displacement along low-angle detachment

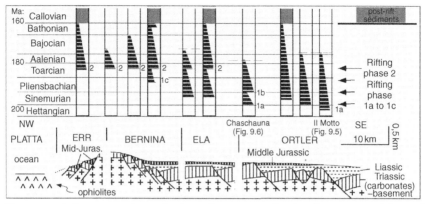

Figure 10.7 *Reconstruction of the African margin in the Austro-alpine of Graubünden at the syn-rift–post-rift boundary and ages of the principal phases of rifting*. The successive phases are: 1a: Hettangian; 1b: Sinemurian; 1c: Early Pliensbachian (?); and 2: Late Liassic (?). Those correspond in the half-graben to megasequences composed of turbidites whose bed thickness and grain size decrease upward (compare also Figs. 6.3 and 6.5). Initiation of each megasequence is considered to mark the start of a rifting phase. See the migration with time of the onset of rifting from the proximal to the distal part of the margin, as shown by Manatschal et al. (2007). *After Eberli 1988; Lemoine and Trumpy 1987. Adapted from Lemoine et al. 2000, Fig. 11.4, p. 124.*

faults leading to stretching of the lower crust but without any notable stretching of the brittle upper part of the crust during the start of extension.

The Margna fault (Figs. 10.4B and 10.8B) is one of the main detachment faults situated in a distal position. It plunges eastward beneath the continental Apulia block. It separates the brittle upper crust from the ductile lower crust. Extension along the fault has led to significant thinning of the continental crust. Thermo-barometric results from the footwall and hanging wall show a difference of 0.3–0.4 GPa indicating about 10 km of crustal thinning during the Liassic. Comparable results have been obtained from the Pogallo fault in the Southern Alps (see Fig.10.2C).

The constant thickness of syn-rift sediments, though small due to starvation, suggests that there was no extensional fault with a notable throw in the upper continental crust. A half-graben appears to be absent in a distal position, in contrast to what is clearly observed on the proximal part of the margin. The transition from platformal sediments to pelagic sediments with cherts from the start of rifting shows, however, that the subsidence of the whole margin began early at the start of rift-driven subsidence.

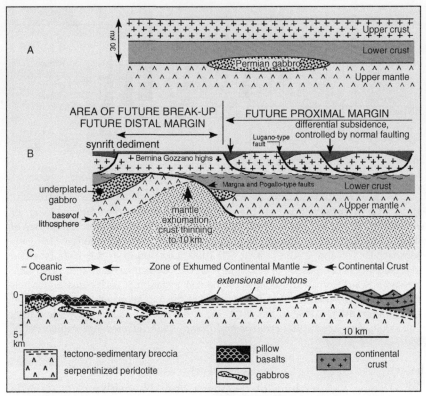

Figure 10.8 *Tentative reconstruction of the development of the Apulia-African passive margin on both the Southern Alps and Grisons transects (Northern Italy and SE Switzerland)*. *Simplified from Manatschal 2004, Fig. 4, p. 439 and Fig. 8, p. 447.*

2.6 Deformation during the final phase of rifting and continental break-up: low-angle detachment faults and 'extensional allocthons'; the Platta, Err and Tasna nappes

The succession of syn–rift deposits in the Err nappe (Fig. 10.4) comprises from base to top: (1) hemipelagic limestones with cherts of Middle Liassic age, (2) a submarine hardground of Pliensbachian age, (3) poorly dated deep marine hemipelagic sediments interbedded with clastic horizons and (4) radiolarian cherts of Middle Jurassic age.

These sediments rest horizontally in places on the surfaces of the detachment faults, which are characterized by a fringe of crushed granite topped by a well–defined 'black gouge horizon' which outcropped on the sea floor. These relationships show that the youngest syn–rift sediments were deposited during or just after the detachment and exhumation of the sub-continental mantle. Following Manatschal (2009), the absence of ductile deformation in crystals shows that these faults were active at temperatures lower than 300°C, i.e. in the upper 10 km of the crust, with a major role being played by fluids in localizing extensional faults in the ductile part of the crust.

It is important to strongly emphasize that the detachment surfaces are very clearly established as being quite distinct and different to Alpine thrust surfaces.

The detachment fault of the Err nappe can be followed in the field semi-continuously for 18 km parallel to the direction of transport. The hanging wall comprises block faults, fragments of upper continental crust constituting *extensional allocthons* with a basal truncation that corresponds to the detachment surface (Figs. 10.8 and 12.9). It appears that the blocks belonging to the hanging wall were separated or '*boudinaged*' at the time of detachment and now rest on top of the footwall. Their existence is the signature of low-angle detachment faults.

Similarly, in the mainly ophiolitic Platta nappe (Figs. 10.8 and 12.9), a system of detachment faults separates serpentinized peridotite beneath from basement blocks derived from the adjacent continental crust above. The blocks themselves are covered by syn-rift sediments succeeded by those of the post-rift. They are also interpreted as extensional allocthons lying tectonically on exhumed sub-continental mantle.

The zone of faults composed of foliated serpentinite, cataclastic serpentinite and serpentinite gouges records a system of deformation and hydration in conditions evolving between amphibolite-grade facies to those at outcrop on the seafloor. It marks the denudation of the upper lithospheric mantle which is accompanied by modest magmatic activity.

3. SUMMARY AND CONCLUSION

Each of the Apulia-African and European margins was about 250–350 km wide. Prior to spreading the whole rift between the western (Causses–Jura) and eastern (Friuli) carbonate platforms was at least 500–700 km wide. However, the rift was probably asymmetric because of the presence of the elevated, several hundred kilometres wide, SBR (Briançonnais) block on only part of the European margin. Neither the Trentino platform, which remained submerged, nor probably the Canavese zone, poorly known due to sparse outcrop, seem to have played a comparable role on the conjugate margin.

Stretching began at the same time on the proximal and distal parts of the margin. Indeed, independent results show that in the Southern Alps, the Lugano listric fault and the low-angle Pogallo fault were active approximately at the same time during the Liassic. In addition, the ages of the rift phases in the Grisons area are broadly contemporaneous with those observed on the European margin (Figs. 6.19 and 6.20).

On the proximal part of the two conjugate, European and Apulia-African margins, the fabric of reconstructed or preserved palaeofaults and the associated syn-rift sediments provides a classic image of tilted blocks and associated half-grabens. Ductile shear zones observed in the basement of the proximal part of the margin are the lower parts of curved listric syn-rift faults that bounded the half-grabens (Fig. 10.3).

On the distal part of the margin, structures resulting from extension are quite definitely different from those on the proximal part.

Movement along low-angle faults achieved the reduction in the thickness of the continental crust on the distal part of the margin by about 10 km. Such faults ensure decoupling between the upper crust and mantle and affect only the lower crust (Fig. 10.8B).

At the end of rifting and during the start of continental break-up, movement along detachment faults exhumed the sub-continental mantle and lower crust with the emplacement of extensional allocthons composed of fragments of continental crust covered by thin syn-rift sediments succeeded by those of the post-rift.

The consequence of the difference in styles of extension between the distal and proximal parts of the conjugate margins is that the domain of upper crustal thinning (for example, the Lombardy rift or Rhone valley) and the line of future break-up are shifted relative to each other,

To explain the location of the line of break-up, the role of structural inheritance may also be taken into account. Thus, under the Magna fault (Fig. 10.4), gabbros were emplaced at the crust–mantle boundary during the course of one of the episodes of extension or transtension accompanying Permian magmatism. These gabbros cooled slowly, in isobaric conditions over about 50 Ma, but without the thickness of the crust being reduced from more than 30 km. A possible conjecture is therefore (Fig. 10.8) that the presence of this intrusive body might have induced the line of break-up because of the heterogeneity affecting the base of the crust and the top of the mantle.

CHAPTER ELEVEN

Liguro-piemontais Ophiolites and the Alpine Palaeo-Ocean

Contents

The Western Alps, from Rift to Passive Margin to Orogenic Belt, Volume 14
ISSN 0928-2025, DOI 10.1016/S0928-2025(11)14011-0

© 2011 Elsevier B.V.
All rights reserved.

Summary

The ophiolites of the Alps, Apennines and Corsica were mainly derived from the Liguro-piemontais Ocean, a branch of the Mesozoic Tethys formed during Middle to Late Jurassic times, simultaneously with the Central North Atlantic.

Closure of the ocean began during the Late Cretaceous resulting in disappearance of oceanic crust and mantle mainly by subduction during end Cretaceous and/or earliest Tertiary times.

The oceanic crust of the Liguro-piemontais Ocean is characterized by the absence of dyke complexes and only minor gabbros and basalts with respect to abundant serpentinized peridotites. Ophicalcites are commonly underlain by serpentinites or gabbros and overlain by pillow basalts or sediments. The normal stratigraphic contact (i.e. not deformed by Alpine tectonism) between serpentinites and pelagic sediments is interpreted as sub-continental mantle tectonic denudation. In effect, the characteristics of a significant part of Alpine ophiolites are those of the continent–ocean transition zone comparable to the 'oceanic' basement of the continental margins of the North Atlantic between Iberia and Newfoundland. Moreover, the various ophiolites of the Western Alps including the Chenaillet and Monviso share common characteristics with slow-spreading mid-ocean ridges such as those of the Central North Atlantic and Indian Ocean.

The sediments resting on the oceanic crust of the Liguro-piemontais Ocean comprise the 'Schistes Lustrés' or 'Calcschisti' or 'Bundnerschiefer' of classical French, Italian and German authors respecitvely. Their stratigraphy is now much better known and is dated as mainly Cretaceous with fewer Jurassic dates.

1. OPHIOLITES

1.1 Geological setting of structural units derived from the Liguro-piemontais Ocean

Alpine units originating from the Liguro-piemontais Ocean are composed of ophiolites derived from the oceanic basement of the Liguro-piemontais Ocean (Chapter 3) and also from their overlying sedimentary cover of pelagic sediments deposited in deep marine environments. They were thrust onto the European continental margin of the Alps and Corsica and onto the Adria block in the Apennines.

The small ophiolite bodies of the Valais Zone are an exception since they were very probably derived from Palaeozoic oceanic basement (Chapter 7).

The Liguro-piemontais ophiolites now outcrop as small, disconnected bodies as a consequence of Alpine deformation. In comparison, the ophio-lites of the Eastern Mediterranean fold belts, such as those of Turkey, form major nappes some hundreds to thousands of metres in thickness. The strongly deformed and metamorphosed sediments overlying the ophiolites of the Alps and Corsica are the classical *Schistes Lustrés* or *Calcschisti* or *Bündnerschiefer* of classical French, Italian and German authors respectively.

Several types of structural setting can be distinguished that also reflect the characteristics of Alpine deformation and metamorphism (Fig. 11.1A):

– Ophiolite units including serpentinized peridotites, gabbros and basalts without high-pressure metamorphism that have therefore escaped subduction. Examples include the upper unit of the Chenaillet massif (Fig. 11.2), the Platta nappe (Grisons, Central Alps) and internal Ligurian units of the Northern Apennines.

– Ophiolite units with serpentinized peridotites, gabbros, basalts and a thin film of sediment, metamorphosed to eclogite grade; they occur in the nappes thrust on the Dora Maira massif of the Western Alps such as Monviso, the Voltri Group in the Ligurian Alps, the Zone of Zermatt-Saas Fe in the Valais and Val d'Aoste.

– Ophiolite units with serpentinized peridotites, gabbros, basalts and their sedimentary cover, metamorphosed to blue schist-grade facies and therefore subjected to subduction, but at depths less than the eclogite-grade ophiolites above (Fig. 14.5). They occur in the Sestri-Voltaggio Zone of the Ligurian Alps, the Queyras area and also form the lower unit of the Chenaillet ophiolite complex (Figs. 11.1 and 11.2).

– Blocks and olistoliths of ophiolitic material such as basalts, gabbros and ophicalcites together with clastics of continental origin involved in sedimentary complexes (olistostromes) at the base of Late Cretaceous flysch units. Examples include the nappes of the Ubaye valley (near the village of Serenne), the Wildflysch of the Gets nappe (Prealps). They are best developed in the external units of the Ligurian Alps (Fig. 11.1).

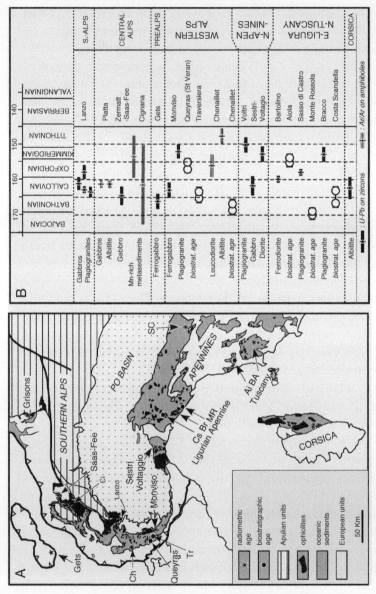

Figure 11.1 (A) Location of radiometric and biostratigraphic age determinations in the Central and Western Alps, Corsica and the Apennines. Ai: Aiola; BA: Bartolina; Br: Bracco; Ch: Chenaillet; Montgenevre; Ci: Lago di Cignana; Cs: Costa Scandella; Mau: Maurienne; MR: Monte Rosscla; SC: Sasso di Castro; Tr: Traversiera (Acceglio area). (B) U/Pb and 40Ar/39Ar dates of magmatic rocks and Mn-rich meta-sediments in Alpine and Apennine ophiolites. Comparison with biostratigraphic ages in radiolarian cherts associated with ophiolites. Biostratigraphic dates are from radiolaria. Location of samples on Fig. 11.1A. Sources: Lombardo et al. 2002; Kaczmarek et al. 2008; Manatschal and Müntener 2009. For the Montgenèvre ophiolite, see Cordey and Bally 2007.

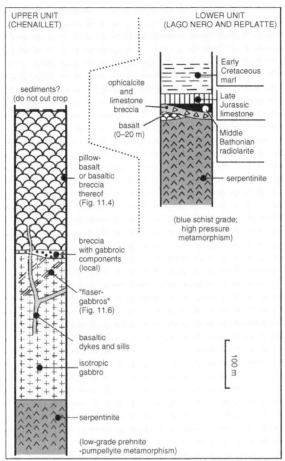

Figure 11.2 *The ophiolite successions of the two superposed units of the Chenaillet massif (Montgenevre).* The Chenaillet massif consists of two stacked ophiolite nappes derived from different parts of the ocean and not subjected to the same history of subduction and collision. The lower unit was subjected to high-pressure, blue schist-grade metamorphism. The first sediments are represented by ophicalcites, which everywhere rest directly on the serpentinites. Locally, at Mt. Corbioun, a conglomeratic horizon with granite pebbles of continental origin is interbedded with the ophicalcites. Above, granitic and Triassic dolomite clasts are found locally in Early Cretaceous sediments. The lava pillows of the upper ophiolite unit of the Chenaillet were subjected only to low-grade prehnite–pumpellyite metamorphism, which took place on the floor of the actively spreading Liguro-piemontais Ocean. The unit is unaffected by Tertiary Alpine syn-metamorphic deformation. In particular, the pillow basalts are nearly undeformed unlike those of the lower unit; see Fig. 11.4. *Adapted from Lemoine et al. 2000, Fig. 12.1, p. 130.*

1.2 Dating of Alpine ophiolites

1.2.1 Brevity of Jurassic magmatic activity

Available radiometric and biostratigraphic dates (Fig. 11.1B) directly demonstrate magmatic activity in the Liguro-piemontais Ocean extending over an interval of 15–20 Ma between the Bathonian (170 Ma, Gradstein time scale) and the Kimmeridgian (150 Ma, *Ibid*). The two most common available dates are Callovian (165–160 Ma) and Kimmeridgian (155–150 Ma).

The oldest date, around the Bajocian–Bathonian boundary, is from radiolaria contained in sediments resting on pillow lavas in Tuscany (Monte Rossola). The youngest date in the Western and Central Alps is Tithonian (149 +/−2 Ma), a radiometric age determined on an anorthosite (now transformed into albitite) from the Chenaillet massif. Albitization is tentatively described as the result of early hydrothermal alteration.

There are few available radiometric dates younger than the Tithonian from Alpine ophiolites (Fig. 11.1B). An exception is the 93 Ma date obtained from the Chiavenna ophiolites (Liati et al. 2003, 2005).

Nonetheless, the Early Cretaceous *Argile a Palombini* resting in normal stratigraphic contact on pillow lavas shows that oceanic basement outcropped on the seafloor at this time (Fig. 11.10). Examples occur in the Monviso massif and in the Ligurian Apennines.

The range in ages opens the question as to whether greater part of the oceanic basement of Cretaceous age disappeared during subduction together with the corresponding geological record during this interval. This may imply that only the older parts of the Tethyan oceanic basement are preserved today.

1.2.2 Episodic magmatic activity during Jurassic times

Magmatic activity in the Ligurian-piemontais Ocean seems to have occurred episodically over relatively short intervals of geological time.

Dates from the Platta nappe in the Grisons record this brevity (Figs. 1A and B). In this unit, zircons from a poorly differentiated Mg-gabbro, a more differentiated Fe–Ti gabbro, a dioritic dyke and an albitite fragment (extremely differentiated) in a breccia with elements of pillow lavas have yielded mean ages of 160.9+/−0.4 Ma. These ages suggest that the magmatic event can be dated to 161+/−1 Ma (Callovian?) and that its duration was not longer than 2 Ma.

The succession of relatively brief magmatic episodes results in the occurrence of apparently coeval basalts, poorly differentiated gabbros and other highly differentiated igneous rocks (for example albitites derived from anorthosites) at different points in the Liguro-piemontais Ocean.

In a given ophiolite massif, the most differentiated members of the tholeiitic suite, such as anorthosites, often yield older ages than those less differentiated. However, the presence of differentiated end members does not imply a given age for that part of the host palaeo-oceanic basement.

1.3 Development of ideas on Alpine ophiolites

In the 1960s, ophiolites were interpreted as fragments of fossil oceanic crust exhumed during orogenesis.[1]

The seminal 1972 GSA Penrose conference resulted in the following classical model of the oceanic crust: it typically comprises from top to bottom, basalts (1 km in thickness), dyke-complexes (1 km) and gabbros (3–5 km in thickness). Pillow basalts comprise the seafloor except near transform faults where gabbros and serpentinites may also outcrop. Two models of ophiolite complexes drawn from the Troodos massif of Cyprus and the Semail nappe of Oman also became classical in the geological literature as illustrating oceanic crustal accretion processes at fast-spreading mid-ocean ridges explored during this period by the JOIDES programme.

Alpine geologists adhered to the notion that the ophiolites of the Alps could be derived from Tethyan oceanic crust. But they knew immediately that the Penrose model was inappropriate to the point that various authors questioned the presence of true oceanic crust in Alpine ophiolites. In effect, Alpine ophiolites are not composed of sheeted dyke complexes. In contrast, they consist in volume of 90–95% serpentinized peridotites. While gabbros and basalts are present, they occur only in the form of small, rare and dispersed bodies. Gabbros are intrusive into serpentinized peridotites.

Research therefore began into the present-day nature of mantle rocks outcropping on the seafloor that might provide possible analogue characteristics to those of Alpine ophiolites.

Work on mid-ocean ridges, in particular the Central North Atlantic (Fig. 11.1A) as well as the Southwest Indian Ocean ridge and the Gakkel ridge of the Arctic Ocean, has shown the presence of very large areas of

---

[1] Ophiolites were named by Alexandre Brongniart (1813) from the Greek ophis: serpent because of the resemblance between the alteration patterns in serpentinite and the skin of a snake.

seafloor composed of denuded oceanic mantle poor in magmatic rocks gabbros and basalts.

Cited here by way of example (Fig. 11.14B), the oceanic crust, younger than 5 Ma shows along the Mid-Atlantic Ridge near 15°20′, the juxtaposition of:

(1) 'amagmatic' segments composed essentially of more or less serpentinized ultrabasites with rare gabbros and basalts, and

(2) 'magmatic' segments composed of gabbros and basalts (typical oceanic crust).

Such surface segmentation may result from vertical heterogeneities in the thermal regime prevailing in the crust and irregularities in depth reached by the asthenosphere (Fujiwara et al. 2003).

The 'amagmatic' strips (Fig. 11.14B) are respectively located to the south and north of the 15°20′N transform zone and are oriented near parallel to it. The strips have an extremely rugose surface, linked to an irregular fault pattern. Relatively low amplitude, irregular to discontinuous magnetic lineations probably reflect the thinness of the basaltic layer and its tectonic dislocation during the past 5 Ma. Average full-spreading rates have been 2.5 cm/Ma for the past 5 Ma, the rate of a slow-spreading ridge.

Located in one part or another of the strip of amagmatic segments, those segments with basalts and gabbros show long and linear abyssal hills. Magnetic anomalies are strongly linear, of high amplitude and symmetrical across the ridge axis. This pattern is consistent with normal crustal accretion and gravity results confirm the interpretation proposed here.

The geophysical interpretations have been tested using samples obtained by dredging during submersible dives (Fig. 11.4B) and also by deep-sea drilling (Fig. 12.6) particularly (1) the length of transform faults (for example Bonatti and Honnorez 1976), (2) the length of normal faults associated with slow-spreading ridges (for example Lagabrielle and Cannat 1990) and (3) the deep marine ridges at the foot of the West Iberian margin discovered during ODP Leg 103 (Boillot et al. 1980 and 1987).

It is therefore appropriate to identify possible relics of these three types of structural situation in Alpine ophiolites. Their different origin could be as follows:

− typical oceanic crust, comprising gabbro and basalt,
− 'oceanic' crust, could be comprising gabbro together with basalt and/or exhumed oceanic mantle and
− transitional crust, mainly constituted of exhumed sub-continental mantle. This is spatially located between thinned continental crust and 'oceanic' crust.

1.4 Alpine ophiolites and the continent-ocean transition zone of the North Atlantic

In the North Atlantic between the margins of Iberia and Newfoundland, the seafloor, located between anomalies M 17 (?) and M0 of Berriasian to Barremian age, is neither strongly attenuated continental crust nor typical oceanic crust; it is transitional crust resulting from exhumation and denudation of the pre-rift sub-continental mantle. Work on Alpine ophiolites shows that mantle ultramafic rocks may also, and in large part, be derived from the denudation of sub-continental lithospheric mantle (Manatschal and Müntener 2009). However, the difficulty lies in identifying among the ultrabasites of Alpine ophiolites those of denuded continental mantle origin and those derived from denuded oceanic mantle.

1.4.1 Mantle-derived serpentinized peridotites and gabbros

The peridotites exhibit generally a characteristic strong foliation that is indicative of high-temperature flow in the asthenospheric mantle. Serpentinized lherzolites are predominant. The small proportion of harzburgites is characteristic of low partial melting in the mantle peridotites, which is compatible with the small quantity of magmatic basic rocks. They are intruded locally by gabbroic and dolerite dykes (Figs. 11.3–11.5).

In places, the gabbros are cut by ductile shear zones ranging in thickness from tenths to tens of metres (Figs. 11.5 and 11.6). The shear-related foliation superimposed on the gabbros was accompanied by metamorphism resulting in replacement of pyroxenes by brown amphiboles at high temperatures (800°C) and low pressure (at a few kilometres depth).

Figure 11.3 **(A) and (B) Examples of gabbroic bodies cut by dolerite dykes (Chenaillet Massif).** (See colour plate 12) *Photographs: Roberts 2004.*

Figure 11.4 *Tubular and pillow basalts (Chenaillet Massif).* In the upper unit of the Chenaillet (Figs. 11.1 and 11.2), the pillow lavas have completely escaped Alpine deformation and metamorphism. (A) Cliff of Collet Vert: upper surface of an intact flow reoriented from horizontal to vertical by Alpine folding. Diameter of basalt tubes is 0.5–1 m. Note the conjugate normal faults related to late Neogene extensional collapse of the Alps (Chapter 15). (B) Same locality. Detail of lava tubes. Note striae parallel to the elongation of the tubes demonstrating plastic stretching of the glassy carapace during eruption of the lava flow. (See colour plate 13) *Reproduced from Lemoine et al. 2000, Fig. 12.3, Table IX.*

Figure 11.5 *Gabbro subject to high-temperature–low-pressure ductile deformation cut by a dolerite dyke (lower unit in Chenaillet massif).* (See colour plate 14) *Photo: Roberts 2004.*

Unmetamorphosed dolerite dykes that cut the foliated gabbros show that the metamorphism predated dyke intrusion (Fig. 11.5). In ophiolite bodies affected by high-pressure–low-temperature metamorphism, shear zones resulting from tectonic denudation of the mantle can be difficult to distinguish from those caused by later Alpine deformation.

Figure 11.6 *Gabbro subjected to high-temperature–low-pressure metamorphism and ductile deformation (upper unit in Chenaillet Massif).* The white and dark bands are felspathic and amphiboles, respectively. Large pyroxene crystals are still visible surrounded by dark amphiboles. Scale of photograph: 15 cm. (See colour plate 15) *Reproduced from Lemoine et al. 2000, Fig. 12.2, Table IX.*

Manatschal and Müntener (2009) divide the rocks resulting from denudation of the mantle into three main types which are, according to their model:

1. Spinel lherzolites associated with pyroxenites. They are present in the Tasna nappe, the upper Platta nappe (Fig. 11.8), the Malenco valley and the external Ligurides (Fig. 11.1). A singular case is the gabbroic body of Val Malenco (Central Alps) intruded into mantle rocks. It is composed of many generations of amphiboles of which the oldest is of Permian age. The assemblage and notably the gabbro and the rocks to which it is welded are interpreted as representing evidence of sub-continental pre-rift mantle denuded during rifting (Figs. 10.2 and 12.4).

2. Serpentinized lherzolites associated with rare harzburgite and dunites. They are present in the lower Platta nappe (Fig. 11.7) and in the upper Chenaillet unit (Fig. 11.2). They were subject to pervasive impregnation by tholeiitic melt after partial melting of Permian age, but later or just contemporaneous with the start of denudation.

3. Depleted peridotites related to MOR magmatic rocks present only in the ophiolite of Monte Maggiore in Corsica (Fig. 11.1A).

1.4.2 The rugose submarine topography of exhumed sub-continental and oceanic mantle

In the ophiolites of the Western and Central Alps, post-rift sediments of Late Jurassic or Early Cretaceous age are observed to rest in direct

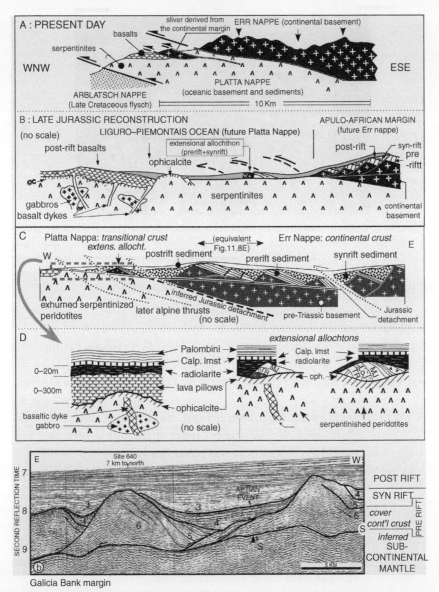

Figure 11.7 *Section across the ophiolites of the Platta nappe in the Grisons of the Central Alps.* The Platta nappe is composed of ophiolites and sediments derived from the Liguro-piemontais Ocean. It rests on units belonging to the SW European craton and is covered by the Austro-alpine nappes. The overlying oceanic sediments rest either on serpentinites or generally thin pillow basalts. (A) Present-day section. (B) Reconstruction of the structure of the 'oceanic' crust.

—(Continued)

Figure 11.7 —*(Continued)* This reconstruction is inspired by drilling results obtained from the foot of the present day Atlantic margin of Galicia during the 1990s. The oldest seafloor of the Liguro-piemontais Ocean adjacent to the Apulia–Africa continental margin (now involved in the Austro-alpine Err nappe) is composed of serpentinites. In the younger part of the ocean beyond generated at the axis of a newly formed ridge, thin basalts rest on mantle serpentinites injected by gabbroic intrusions but a dyke complex is absent. According to the notion of transitional crust proposed by Tucholke et al. (2007) and Sibuet et al. (2007), the basalt may be qualified as syn-rift and not post-rift as indicated on the figure of which the original is dated 1995. *Simplified after G. Manatschal, Thesis, Zurich 1995. Adapted from Lemoine et al. 2000, Fig. 12.11, p. 139.* *(C) Detail of the reconstruction of the boundary between thinned continental crust and transitional crust.* This figure illustrates in particular the flat detachment surfaces one of which has allowed withdrawal of slices of continental crust above serpentinized peridotite during its denudation. *(D) Detail of the stratigraphic succession just after denudation.* The succession of radiolarian cherts (=radiolarites)–Calpionnellid limestones (=Calp. Lmst)–Palombini shales is characteristic of the lower part of the sedimentary cover of Tethyan oceanic basement. It rests as well on pillow lava flows and serpentinized peridotites themselves (Fig. 11.21), and on the distal part of the adjacent thinned continental crust (Fig. 11.22). Its position in stratigraphic contact on the extensional allochtons shows that these were emplaced between the phase of denudation of the sub-continental mantle and the deposition of the first oceanic sediments. The blocks defined as extensional allochtons are not the same as the klippes dismembered by Alpine thrusting. TR: Triassic carbonates; MJ: Middle Jurassic layers of the extensional allochtons; Radiolarite Formation: Middle to Late Jurassic; Calp. Lmst: Calpionnellid and Aptychus-bearing mudstone, dated Berriasian here (Earliest Cretaceous); Palombini Formation: alternating mudstone and shale dated Valanginian to Barremian approximately (Early Cretaceous). *Simplified from Bernoulli and Jenkins 2009, Fig. 27, p. 174.* *(E) Section of ISE 2 multi-channel reflection profile across the western margin of Galicia Bank.* The section has been taken from the proximity of the boundary between continental and transitional crust. Its equivalent is located in the middle of Fig. 11.7C. The tilted blocks are composed of pre-rift continental crust. The first peridotite ridge is located only 10 km west of the left edge of the figure. The S reflector marks one of the major detachment surfaces that allow the denudation of sub-continental mantle present beneath. Imagine that the denudation of the continental mantle has been pushed much further: the block marked 'site 640' could be the site of an extensional allochton. Seismic sequence 6: Pre-rift Tithonian to Berriasian platform carbonates. Seismic sequence 5: Syn-rift Valanginian–Hauterivian layers, with oblique and divergent reflectors conveying extension and block rotation during denudation. Seismic sequence 4: Late syn-rift Barremian–Aptian, not faulted. The Aptian sediments are mainly turbidites. Seismic sequence 3: Post-rift Aptian black clay-stones. The strong reflector at the Aptian–Albian boundary (seismic sequence 3: post-rift) is interpreted as recording the break-up unconformity in the Newfoundland–Iberia rift. Note that the dips and divergence between reflections are exaggerated due to the increase in velocity with depth. The undulations of the S reflector may result from distortions due to variations in acoustic impedance along the traverse. Location on Fig. 12.2. *Reproduced from Tucholke et al. 2007, Fig. 6b, p. 20.*

stratigraphic contact on ophicalcitic breccias that veneer serpentinites or gabbros (Figs. 11.21–11.23). Most common are radiolarian cherts and more rarely *Calcaires à Palombini* assigned to the Early Cretaceous using facies analogues. With regard to the latter, examples occur in the Monviso massif (Fig. 11.9) and in the Ligurian Apennines. The existence of syn-sedimentary topography explains the presence of ophiolitic blocks included in sediments of Early Cretaceous age visible in the landscape of the Queyras valley (Fig. 11.9).

The age of the radiolarian cherts ranges between the Bajocian–Bathonian boundary and the Oxfordian–Kimmeridgian boundary. Their thickness is widely variable (Figs. 11.2, 11.8, 11.21 and 11.22). The age of the first sediments resting in stratigraphic contact on the crust is therefore spread over 30 Ma at least if not more than about

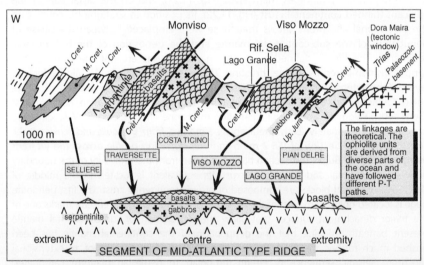

Figure 11.8 *Schematic section across the Monviso: stacking of MP (magma-poor) and MR (magma-rich) units.* The Monviso ophiolites reach a height of 3940 m (Fig. 16.7) and form a N–S oriented zone, 35 km long and 9 km wide. They were subjected to eclogite-grade metamorphism at sub-surface depths ranging from 50 to 90 km. In the remnant sedimentary cover, depending on locality, the oldest sediments are of Late Jurassic, Early or 'middle' Cretaceous age in so far as the ages are identifiable from the facies. The presence of both MP and MR ophiolitic types offers a good comparison with those known on the Mid-Atlantic Ridge (Fig. 10.19). However, these units do not prove mantle denudation of the same segment because each has been subjected to different metamorphism paths and the cover sediments have different ages. *Adapted from Lemoine et al. 2000, Fig. 12.10, p. 138.*

Figure 11.9 *Ophiolitic olistoliths included in Early Cretaceous sediments (Aigue Agnelle valley of Queyras).* The ophiolitic olistoliths occur in an alternation of the Palombini facies shales and limestones characteristic of the Early Cretaceous oceanic succession. The same sediments also contain limestone blocks of Late Jurassic age (such as the large lens on the upper left of the figure), serpentinite blocks and intercalated breccias with ophiolite clasts. The chaotic breccias are confined to specific localities in the Queyras valley and Corsica. The ophiolite fragments intercalated with pelagic sediments were derived by collapse of faults that controlled ocean bottom topography. It is not known if the faults correspond to transform fault escarpments or to ridge axis faults. *Reproduced from Lemoine et al. 2000, Fig. 12.8, p. 136.*

70 Ma. This is without taking into account the relationship between the distality or proximality of the dated sediments to the boundary between continental and transitional crust. The identified differences in ages may also result from progressive onlap of the initial sediments onto the rugose topography of the surface of the sub-continental mantle undergoing denudation. Therefore, the age of the oldest sediments probably do not uniquely allow dating of exhumation at each place. However, the age of the oldest sediment, Bajocian–Bathonian in the Apennines, provides a reasonable estimate for the start of denudation. The proposed interpretation herein results from comparison with the foot of the Galicia margin of Iberia where results from ODP leg 102 (Boillot et al. 1980) and later drilling show that the recently exhumed continental mantle has a rugose topography of ridges and troughs. In a similar manner, ridge flank crust near the 15°20′N Atlantic fracture zone (Fig. 11.14A) is characterized by irregular and blocky topography that is interpreted as resulting from irregular fault patterns (Figs. 11. 14B; Fujiwara et al. 2003).

1.4.3 Top–basement detachment faults and extensional allochtons

The flat top–basement faults observed on the Apulo-African border of the Tethys are considered by Manatschal and Müntener (2009 and previous cited publications) as the most characteristic structures of the continent–ocean transition zone. They are best preserved in the Platta nappe (Grisons; Fig. 11.7). The surfaces of these flat faults are surmounted by extensional allochtons, by tectonosedimentary breccias, by sediments, in principle post-rift, and, in a distal position, by basalts.

The extensional allochtons (Figs. 11.8C and D) are blocks detached from ultrastretched continental crust during exhumation. Their reality is reinforced by comparison with seismically imaged structures at the foot of the Galicia margin of Iberia. Their dimensions range from the order of a kilometre to tens of metres. They are often associated with mixed syn-sedimentary breccias composed of mantle and continental debris. The presence of either is a sign of proximity to the contiguous continental crust. Both are known from the foot of the Galicia margin (Site ODP 1068; Manatschal et al. 2009). However, in contrast to the Platta nappe (Figs. 10.8 and 11.7), it has not been possible to recognize the presence of extensional allochtons at the foot of the European margin in the Alps though rare mixed breccias have been described from the Queyras area (Fig. 11.1) and the lower unit of the Chenaillet (Lago Nero). There, they overlie pillow lavas and the first sedimentary cover. Such sequences show on one hand that mixed breccias do not rest directly on top–basement detach-ment faults and on the other that magmatic activity existed not far from the distal extremity of the continental margin.

1.4.4 Evolution of the characteristics of the ophiolites as a function of reconstructed distance to the adjacent palaeo-margin

The Platta nappe in the Grisons (Fig. 11.8) is appropriate for reconstructions of the initial structure because it has escaped strong Alpine deformation. The Platta nappe consists of two superposed thrust units themselves surmounted by another unit, the Err nappe, derived from the contiguous Apulo-African continental margin (Fig. 10.4). Removal of the effects of Alpine thrusting allows demonstration of the evolution of proximal to distal structures.

In the Platta nappes, the ophiolite units show the same sedimentary succession and metamorphic evolution as the immediately overlying Err nappe of Apulian continental origin. In addition, the first radiolarian cherts deposited on the most distal units of the Apulia–Africa continental margin and on the underlying ophiolites have the same age. The ophiolites may therefore derive from a proximal position with respect to the adjacent Apulia-African margin.

In consequence, the lower continental unit of the Err nappe and the upper Serpentinite unit of the immediately underlying Platta nappe may represent the remnants of the transition zone between continental and 'oceanic' crust.

The Grisons ophiolites (Fig. 10.8) are uniformly composed, like other Alpine ophiolites, of serpentinized spinel lherzolites and harzburgites both intruded locally by gabbroic bodies, then by doleritic dykes.

Serpentinites interpreted as having a proximal origin with respect to the margin mainly originate from non-depleted peridotites (lherzolites). In contrast, serpentinites interpreted as having a distal origin with respect to the margin derive either from non-depleted or from largely depleted peridotites (harzburgites) depending on location. However, the relative proportion of harzburgites remains modest.

In a correlative sense, the relative abundance of basaltic intrusions and pillow basalts increases with distance from the adjacent margin. In a distal position, they are aligned and apparently controlled by syn-magmatic faults. Elsewhere, however, the presence of basaltic lava flows close to the edge of the continental crust is testified by the existence, at least in one place, of a conglomerate of granitoid blocks underlying a pillow lava flow (Queyras valley; Fig. 11.11).

Thermo-barometric results obtained from two pyroxenes and an ortho-pyroxene sampled only from the peridotites of the Platta nappe show an increase in temperature from an equilibrium temperature of 950°C plus or minus 50°C for 0.9–1.2 GPa to more than 1000°C from proximal to distal. This may illustrate the rise of a thermal dome as the mantle was progressively denuded (Fig. 11.7).

1.4.5 Ophicalcitic breccias

Ophicalcites are almost ubiquitously intercalated between flow basalts or sediments and the underlying serpentinites or gabbros (Figs. 11.12 and 11.13). The ophicalcites are breccias composed of serpentinites cemented by calcite or dolomite.

They have given rise to a variety of interpretations, in particular that they are the products of contact metamorphism after the intrusion of peridotites (Steinmann 1905–1927, cited by Bernoulli and Jenkyns 2009). Today, there is general agreement that they mark the traces of low-angle normal faults along which the sub-continental mantle was denuded to outcrop on the seafloor.

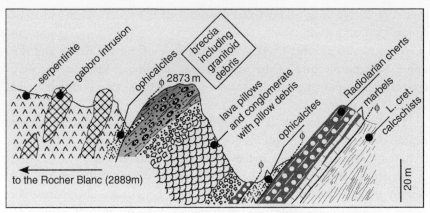

Figure 11.10 *Example of a breccia with elements derived from granitoids at the base of a pillow lava flow (Queyras valley near Saint Veran).* The breccia at the 2873 m summit is composed of basaltic blocks, fragments of pillow lavas, gabbros, serpentinites and granitoid blocks. This outcrop demonstrates European continental clastic material and thus the proximal position of the underlying 'oceanic' crust. It is therefore consistent with the conjugate observations from the Platta nappe whose upper units were also in a proximal position with respect to the Apulo-African margin (Fig. 11.8). Location of the Queyras valley on Fig. 11.1A. *Simplified after Caby et al. 1971, Fig. 2, p. 999.*

Two types of ophicalcites, OC1 and OC2, can be distinguished (Figs. 11.12 and 11.13).

The OC1 ophicalcites correspond to a fringe of serpentinites some metres in thickness cut by a network of veins (Fig. 11.12C).

These veins are typically composed of sparry calcite or dolomite commonly occurring as large crystals associated locally with minor talc. Between the veins, the sedimentary infill consists of calcite and serpentinite grains. These ophicalcites show 'puzzle-type' structures. The latter are interpreted as hydraulic breccias produced by fracturing induced by overpressured fluids perhaps active along normal faults and the extensional detachment surfaces that contributed to the uplift and exhumation of mantle rocks on the seafloor.

The OC2 ophicalcites are clastic sediments, debris flows or grain flows (Fig. 11.12B) deposited on the seafloor as mass-flow deposits. These typically comprise cataclasites derived from sedimentary breccias associated with faults which cut the serpentinites and accompanied tectonic denudation of the mantle. Some of these ophicalcites grade laterally to breccias that include reworked gabbroic elements together with granitoid blocks derived from the adjacent continental crust (Fig. 11.11).

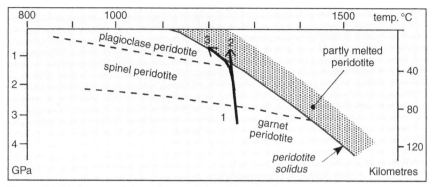

Figure 11.11 *P–T conditions for partial melting in the upper mantle.* The diagram illustrates the different equilibrium conditions for different types of peridotites in the pressure–temperature field. Trajectory 1–2: Rapid expansion and elevation of the asthenosphere. The conditions are nearly adiabatic, as temperatures did not re-equilibrate during decompression due to the relatively rapid rise of the asthenosphere. Partial melting takes place towards 40 km depth. The volume produced is elevated to 20–30% of the initial volume. Residue of melting: depleted peridotite (harzburgite). Trajectory 1–3: Slow expansion. Cooling which accompanies the decompression as a result of the slower relative rise of the asthenosphere limits the rate of partial melting to 10% at maximum. Residue of melting: fertile peridotite (lherzolite). This evolution has been experienced by peridotites as shown by the study of mantle pyroxenes across the Platta nappe of the Grisons (Müntener et al. 2000, 2002 and 2004; Desmurs et al. 2002). *Simplified after Nicolas 1999. Adapted from Lemoine et al. 2000, Fig. 3.2, p. 36.*

1.5 Alpine ophiolites and slow ridges of the Central North Atlantic

Outcrops or groups of outcrops composed of serpentinites, gabbros, dolerite veins, pillow basalts and associated sediments (Figs. 11.1 and 11.4) can be interpreted as having been formed at the axis of a slow-spreading ridge comparable to those known from the Central North Atlantic from dredging, drilling and observations from submersibles (Fig. 11.14). The model of Lagabrielle and Lemoine (1997), adopted by Lemoine et al. (2000; Fig. 11.15) for the Alps, has suggested possible reconstructions of many ophiolite bodies in the Queyras (Monviso, Cascavelier, Roche Noire; Figs. 11.9 and 11.16).

1.5.1 Magmatic rocks of the Alpine ophiolites

Alpine gabbros, dolerite dykes and basalts have the characteristic MORB geochemical signature of mid-ocean ridge basalts.

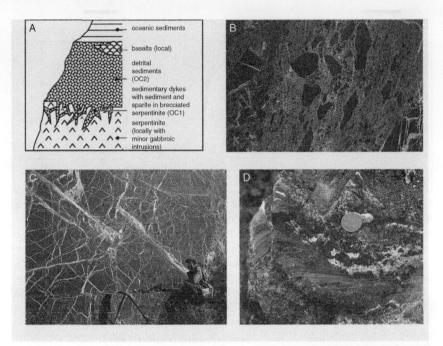

Figure 11.12 *The two types of ophicalcites.* (A) Schematic section showing the relationship between the serpentinites, OC1 and OC2-type ophicalcites, and oceanic sediments, i.e. radiolarites, limestones and marls or clays. OC1-type comprises hydraulic serpentinite breccias with puzzle-type structures intersected by a network of carbonate veins. OC2-type comprises mass-flow sedimentary deposits that include serpentinite or gabbroic clasts. (B) and (C) OC2- and OC1-type ophicalcites (quarry in the Ligurian Apennines north of La Spezia). (D) Geopetal-type infilling of a cavity in a type OC1 ophicalcite by finely stratified and laminated limestone. This image shows that ophicalcite outcropped on the seafloor from the moment of infilling of the cavity. Coin diameter: c. 1 cm. *In: Weissert and Bernouilli 1985.* (See colour plate 16) *Photograph: D. Bernouilli. Adapted from Lemoine et al. 2000, Fig. 12.5, Table X.*

The gabbros were intruded into serpentinized mantle peridotites as irregular bodies tens of metres to kilometres in width, and tens of metres in thickness. The primary contacts with the serpentinites have often been tectonized. Gabbroic bodies are generally boudinaged within their serpentinite matrix.

The gabbros are mainly Mg-rich rocks, showing isotropic structures commonly including magmatic breccias, such as in the Chenaillet Massif and in the Queyras. In the thicker gabbros of the Platta nappe, compositions range from olivine gabbros to differentiated Fe–Ti gabbros. The lower half of the upper Chenaillet unit (Fig. 11.2) exhibits Mg-rich gabbros plus some olivine gabbros at the very base.

Figure 11.13 *Example of mantle serpentinized lherzolites directly covered by ophicalcites and pelagic sediments (south face of the Roche Noire, between the Queyras and Ubaye valleys).* From east to west, i.e. from right to left on the photograph: (1) dark green serpentinites whose upper few metres are transformed into OC1-type ophicalcites; (2) light-coloured OC2-type sedimentary ophicalcites, affected by Alpine cleavage; and (3) Late Jurassic pelagic limestones. Adjacent to this outcrop, 5–10 m thick radiolarites are intercalated between horizons 2 and 3. (See colour plate 17) *Reproduced from Lemoine et al. 2000, Fig. 12.4, p. 133.*

Throughout the Alpine ophiolites, gabbros intruded into the serpentinized peridotites do not occupy more than 5% of the total volume. The small quantity of rocks of magmatic origin is indicative of a very slow rate of partial melting in mantle peridotites.

The gabbros are cut locally by dolerite dykes (Fig. 11.3), and more rarely by diorites and by albitites belonging to the plagiogranite suite. Albitite intrusions also occur in the peridotites.

Where present, the basalts are not more than a few hundred metres in thickness. Where basalts are absent, often over large areas, sediments rest directly on serpentinites or locally on the gabbros.

The majority of the basalts in the Alps, Apennines and Corsica exhibit pillow or tubular structures (Fig. 11.4) indicative of relatively viscous and thus low-temperature magmas. In consequence, magma conduits bringing basalts to the surface would have cooled rapidly during the interval between eruptions of successive flows.

1.5.2 Temporal succession of the magmatic emplacement of gabbros and basalts

The radiometric ages of the less–differentiated gabbros and differentiated albitites group narrowly around 160 Ma between the top of the Middle Jurassic and the base of the Late Jurassic (Fig. 11.1B). Movement on the detachment faults exhuming the mantle was therefore contemporaneous with the emplacement of gabbros there and at outcrop on the seafloor, all of which took place within a relatively short time.

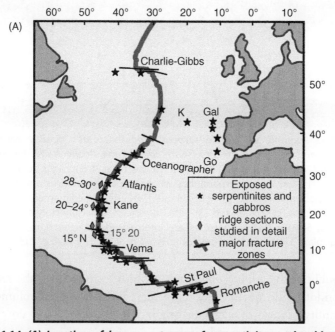

Figure 11.14 *(A) Location of known outcrops of serpentinites and gabbros in the Equatorial and Central Atlantic.* The serpentinite and gabbro outcrops have been located either by drilling, dredging or *in situ* sampling from submersibles. K: King's Trough; G: Galice; Go: Gorringe Bank. *After Cannat 1993; Tucholke and Lin 1994. Adapted from Lemoine et al. 2000, Fig. 3.4, p. 40.* *(B) Shaded relief bathymetry from multi-beam survey of the Mid-Atlantic Ridge in the vicinity of the 15–20 Fracture Zone.* White solid lines mark the 15°20′N Fracture Zone. Red discontinuous lines mark bathymetric lows in the axial valley. N1, N2, N3 and S1, S2, S3 identify spreading segments. Rock samples are from dredging and from submersible dives. S1, S2, N1 and N2 segments that are located on both sides of the fracture zone are magma-poor. Note the recovered ultramafites in these segments at distance from the ridge. They separate magmatic segments N3 from S3 that show regularly elongated hills and correspond to classical oceanic crust. Basalts have been sampled at distance from the ridge in segment S3. (See colour plate 18) *Reproduced and adapted from Fujiwara et al. 2003, Fig. 1, p.3.*

(B)

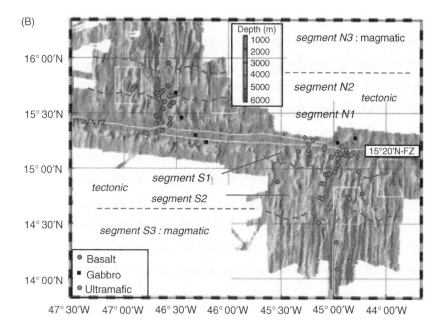

Figure 11.14 (Continued)

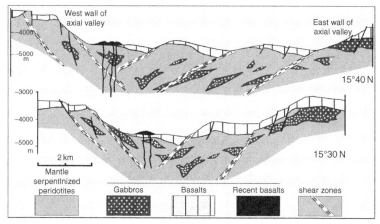

Figure 11.15 *Two schematic sections in the region of the 15°20′N transform fault showing the relationship between mantle serpentinites, intrusive gabbros and basalts.* Note the absence of a dyke complex which has not been found in any part of the area by drilling or submersible. This model implies that the basalts were extruded onto a seafloor composed of serpentinites and gabbros. The same relationship is also present in Alpine ophiolites. Location: Fig. 11.14A. This model of Lagabrielle et Lemoine (1997) was used by Lemoine et al. (2000) for the Alps and is an adaptation of that published for the Atlantic by Lagabrielle and Lemoine 1997 (1998). *Simplified after Cannat et al. 1997. Adapted from Lemoine et al. 2000, Fig. 3.6, p. 42.*

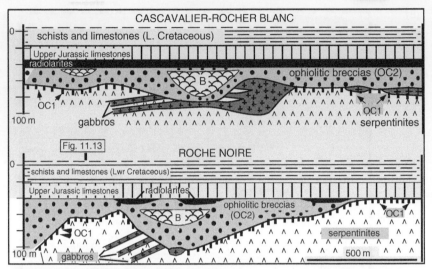

Figure 11.16 *Reconstruction of the relationships between ophiolites and sediment in the Roche Noire and Cascavelier massifs of Queyras.* In the Roche Noire massif, sediments rest directly on serpentinized mantle lherzolites. Reconstructions of the Roche Noire and neighbouring Cascavelier massif prior to Alpine deformation are possible because of the continuity and quality of the outcrops at high altitude. The outcrop geology pictures the extremities of a segment of the accretionary ridge developed in the Liguro-piemontais Ocean, by analogy with such features on the Mid-Atlantic Ridge (Fig. 11.14). OC1 and OC2: type OC1 and OC2 ophicalcites; B: pillow basalts. *Adapted after Tricart and Lemoine 1983 and Lemoine et al. 2000, Fig. 12.6, p. 134.*

Chilled margins along some of the dolerite dykes feeding the basalts show that these dykes cut across solidified gabbros although both share a co-magmatic, MORB, origin. In places, the pillow basalts were erupted directly onto the gabbros, which then comprised the seafloor. While these observations show that eruption of the basalts postdates intrusion of the gabbros, the two events are often too closely spaced to be distinguished in geological time. In some localities, such as in the upper unit of the Chenail-let, a relatively complex succession of events is shown by magmatic breccias composed of fragmented fine-grain dolerites and basalts included in a coarse-grain gabbroic matrix. At other localities, sediments are absent between pillow basalt flows and the underlying substrate of gabbros or serpentinites. Given that magmatic emplacement of a part of the gabbros and basalts may have been quasi-contemporaneous at each point, variation in ages from place to place can thus be attributed directly to the nature of mantle tectonic denudation and/or spreading processes.

At other localities, discrete sedimentary breccias comprised reworked gabbros and serpentine debris underlie pillow basalt flows. However, true pelagic sediments are exceptional. Accepting that the duration of deposition of the breccias is unknown, the phase of partial melting resulting in gabbroic intrusion might also have been caused by underplating during the syn–rift phase. Such gabbros could therefore be appreciably older than the pillow basalt flows. But this possibility is again the subject of debate.

Faults affect the serpentinite bodies and are sealed by basaltic flows (Fig. 11.17) or by sediments and escarpment breccias along faults which exhume mantle rocks. In the Queyras and Ubaye valley areas, and in Corsica, sands, conglomerates, as well as olistoliths, up to 200 m in size, composed of basalts, gabbros or serpentinites form interbeds in the pelagic formations overlying the ophiolitic bodies (Figs. 11.9 and 11.23). While their precise origin is unknown, the submarine relief sourcing these clastic deposits may be represented by faults bounding

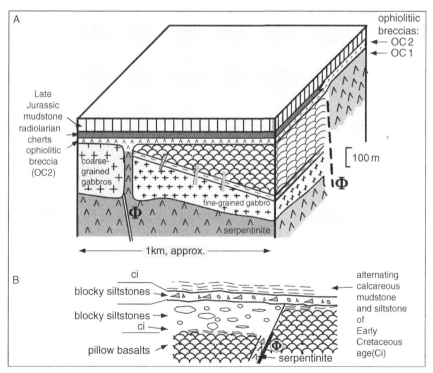

Figure 11.17 *Geometric relationship between ophiolites and sediments in the Chabriere massif (upper Ubaye valley).* Submarine faults have created escarpments, whose surfaces are onlapped by sediment. *Adapted from Tricart et al. 1985a and b* and *from Lemoine et al. 2000, Fig. 12.9, p. 136.*

the spreading axis. However, they may also be associated with transform faults offsetting the ridge axis or later faults which record the widespread Late Jurassic to Early Cretaceous extension which affected a large part of the West European craton and its margins (Chapter 7).

1.5.3 Comparison of spreading processes on the Mid-Atlantic Ridge and in the Liguro-piemontais Ocean: Magmatic and tectonic accretion

Along strike segmentation characterizes the Mid-Atlantic Ridge in common with all mid-ocean ridges (Fig. 11.14A). The discontinuities separating the segments differ in order or rank. The first-order discontinuities correspond to major transform faults, spaced some hundreds of kilometres apart. Second-order discontinuities separate ridge segments 20–100 km in length that are horizontally offset by a few kilometres at most.

Each second-order segment (Fig. 11.18) consists of a central area with a relief of 1000–500 m, which decreases towards the end of the segment. Basalts outcrop in the centre of these segments. At the extremities or tips of those segments, serpentinites and locally gabbros outcrop while basalts are rare.

Geophysical studies show that basalts and gabbros are relatively thick at the centre of the segments where the lithosphere is thin and the asthenosphere is shallow. In contrast, the tips of the segments are characterized by the minor presence or absence of basalts and by the relative thickness of the oceanic lithosphere. Submersible observations made on the Vema transform fault (Fig. 11.14A) have provided a vertical section through the oceanic crust along a fault scarp. From base to top, the section comprises of: (a) serpentinites, (b) about 1 km of gabbros, (c) a dyke complex of about 1 km in thickness and (d) basalts of unknown thickness. The Vema transform is the only locality on Atlantic Ocean crust where a dyke complex has been recognized to date.

Other submersible observations at segment extremities made near the Kane and 15°20′N transforms (Fig. 11.14A) have found serpentinites cut by dykes and gabbroic intrusions overlain by apparently thin basalts (Fig. 11.15).

Observations made on different sectors of the Mid-Atlantic Ridge and on a 400 km strip of the ridge system have distinguished a pattern of uplifted lozenge-shaped compartments, culminating in depths of 2500 m, and depressions lying in greater than 3000–3500 m water depths that trend obliquely to the ridge axis (Fig. 11.19).

The uplifted compartments, in which basalts outcrop, consist of relatively thick oceanic crust composed of basalts and gabbros above thin lithospheric mantle. These high areas are *magma-rich* (MR) segments.

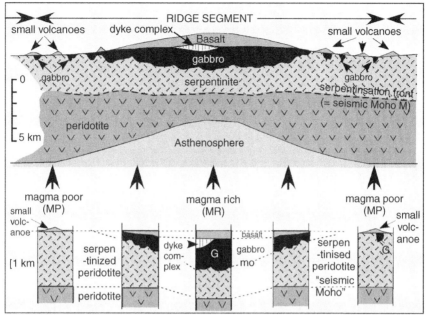

Figure 11.18 *Schematic longitudinal section of a segment of the Mid-Atlantic Ridge.* Columns situated at the base of the figure show the types of succession encountered on the ridge. The 'MR' or *magma-rich* sections comprise a thick mafic crust with gabbros, basalts and rarely a dyke complex. The 'MP' or *magma-poor* sections show mantle serpentinites outcropping directly at seabed without a mafic crust. An intermediate-type section is shown between the two. Horizontal scale: between 20 and 100 km. G: gabbro. The 'seismic Moho' is marked by a jump in P wave velocity between serpentinized and unserpentinized peridotites. The petrologic Moho (mo) is not defined at segment extremities. In the case of rapid-spreading ridges where mafic crust is well developed, the seismic and petrologic Moho coincide everywhere because of the difference in P wave velocity between mafic (basalts and gabbros) and ultramafic (peridotite) rocks. In such a case the serpentinite fringe is quite thin. *Modified from Lemoine et al. 2000, Fig. 3.5, p. 41.*

In contrast, the depressions are floored mainly by outcrops of serpentinites, and more rarely gabbros with sparse basaltic volcanoes. The lithospheric mantle is relatively thick. These low areas are *magma-poor* (MP) segments.

1.5.4 Slow-spreading ridge processes: Alternation of phases of magmatic and tectonic accretion

Along the mid–ocean ridge axis, two types of seafloor can be recognized and are composed respectively and mainly of basaltic material and mantle-derived serpentinites. The two types appear to alternate at once in time

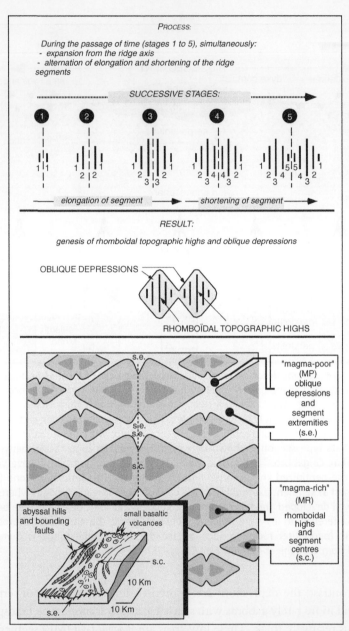

Figure 11.19 *Theoretical sketch of the 'Atlantic patchwork'.* The ridge is segmented along its axis into elevated basalt-rich segment centres and depressed extremities with rare basalts. The evolution of these structures during spreading results in elevated lozenge-shaped areas with abundant basalts separated by oblique depressions composed of outcropping gabbros and mantle serpentinites. The combination of spreading along the ridge and along axis segmentation results in rhomboidal-shaped high areas and oblique depressions which occur at the extremities of the segments. In the box, a sketch of part of a segment is shown in perspective. No scale. *After Durand et al. 1995; Gente et al. 1995; Lagabrielle and Lemoine 1997. Modified from Lemoine et al. 2000, Fig. 3.7, p. 43.*

and space (Fig. 11.19) so that each point of the ridge passes alternatively through one of the two stages.

During periods of magmatic accretion, magmatic activity results in basaltic volcanism with probable gabbroic emplacement at depth. A dyke complex may form locally, as at the Vema transform.

During periods of tectonic accretion, magmatic activity is reduced to zero. Basaltic volcanism is rare to absent. Horizontal traction on the lithosphere leads to the exhumation on the seafloor of serpentinites and gabbroic intrusions.

1.5.5 'Magma-rich' and 'magma-poor' successions in Liguro-piemontais ophiolites as in the Atlantic

In Alpine ophiolite bodies, two types of structural units are recognized from the main lithological associations:

- units with gabbros and relatively thick basalts of the MR type. Examples include the upper unit of the Chenaillet (Fig. 11.2) and most units constituting the Monviso (Fig. 11.9) and
- units constituted mainly of serpentinites with or without rare basalts and gabbros of the MP type. Examples include the lower unit of the Chenaillet (Fig. 11.2), most units in Queyras valley and other units in the Monviso massif.

It is thus possible that spreading in the Liguro-piemontais Ocean resulted in segmentation comparable to that described from the present Mid-Atlantic Ridge. The MR-type successions would have been derived from the centres of the rhomboidal bathymetric highs and the majority of the MP-type successions from the segment tips or from oblique depressions separating the highs.

1.6 Serpentinite, serpentinization and hydrothermal activity

In the serpentinites of those units unaffected by intense Alpine metamorphism, low-temperature chrysotile and lizardite are the main constituents. These minerals are the characteristic products of the hydrothermal serpentinization of mantle peridotites found on both exhumed sub-continental mantle and present-day mid-ocean ridges.

Serpentinization results from two different processes in the submarine environment; one is due to the pervasive circulation of juvenile waters and the other to contact with seawater (Fig. 11.20). This is demonstrated by oxygen isotope profiles from serpentinites drilled in two wells during the ODP programme (Skelton and Valley 2000).

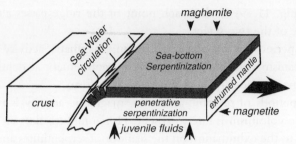

Figure 11.20 *Serpentinization.* This model results from a comparison of the ophiolites of the Err and Platta nappes (Fig. 11.7) with those of the Iberian abyssal plain. It proposes a suite of reactions with seawater causing serpentinization in two stages: (1) Penetration of seawater to react with peridotites as a consequence of deformation and thermal expansion, both due to the exhumation of the sub-continental mantle. The depth reached under the seafloor is of the order of 2–3 km. (2) As soon as the mantle rocks were exhumed on the seafloor, another phase of serpentinization resulted from direct contact with seawater. The thickness of the altered rock would be some tens of metres. The temperatures of the reactions would be above 150°C for the first and below 50°C for the second. The ferromagnetic minerals released during this process would be magnetite and titanomagnetite for (1) and maghemite for the altered part (2). These different minerals are oriented according to the prevailing direction of the terrestrial magnetic field during these stages of serpentinization. The recording of the corresponding magnetic anomalies in the rocks permits kinematic reconstructions of the nascent ocean during the period of creation of the transitional crust. *Redrawn after Sibuet et al. 2007, Fig. 7.*

The first reaction takes place at temperatures above 170°C and releases magnetite. The second takes place at temperatures below 50°C releasing maghemite, a low-temperature mineral formed by the oxidation of spinels containing ferrous iron. The presence of these minerals is recorded by magnetic anomalies in the transitional crust off Iberia (Sibuet et al. 2007).

Associated sulphide minerals

In rare localities, such as the type example at the Saint Veran mine in the Queyras valley (Hautes Alpes), copper sulphide and also native copper ore deposits occur precisely at the boundary between the serpentinites and the overlying cover of radiolarites and pelagic limestones of Jurassic age. This mineralization is interpreted as the expression of fossil hydro-thermal activity at the seafloor that may be similar to the 'black smokers' known at present-day active spreading centres (see Tricart 2003 for references).

2. OCEANIC SEDIMENTS

2.1 The sedimentary succession of the ophiolitic nappes of the Western Alps, Corsica and the Apennines

In these areas, units derived from the Liguro-piemontais Ocean consist of pelagic sediments of Middle Jurassic to Late Cretaceous age that directly overly the ophiolites or the ophicalcites where present. The most complete succession (Figs. 11.8, 11.21 and 11.22) comprises from base to top:

- radiolarian cherts ranging in age from Middle Jurassic to Late Jurassic (Late Bajocian to Kimmeridgian) depending on locality,
- pelagic limestones of Late Jurassic to earliest Cretaceous age (? 145–165 Ma). Where radiolarian cherts are absent, the limestones rest directly on ophiolites or ophicalcites (Fig. 11.13),
- argillaceous sands with intercalated micritic limestones of Early Cretaceous age (140–120 Ma) known locally as the Replatte formation in the Western Alps, the Erbajolo Formation in Corsica and the '*Argile a Palombini* Formation' in the Apennines,
- Black Shales of 'middle' Cretaceous (115–95 Ma) age,
- the Late Cretaceous succession includes calcschists (originally foram – nanno oozes), deposited in the remnant, non-subducted part of the Liguro-piemontais Ocean, or calcareous flysch (type: Helminthoid flysch) deposited in trenches above the recently initiated subduction zones as shown earlier (see Chapter 14 and Fig. 14.5).

2.2 Reconstruction of the sedimentary succession in the Tethyan oceanic domain: The Schistes Lustrés

Authors from the end of the 19th century and for most of the 20th century have grouped, under the names of *Schistes Lustrés, Bundnerschiefer* or *Calcschisti*, comparable sedimentary formations that outcrop in the internal Alpine zones frequently, but not always, in association with ophiolites. These sedimentary successions were then considered to be lithologically monotonous and particularly thick. Their interpretation remained enigmatic because their original bedding and fossil content had been largely destroyed by multiple phases of deformation and metamorphic recrystallization due to Alpine subduction and collision. Additional confusion in interpretation also arose from the fact that, while the lower parts of the succession are quite different, the post-Middle Jurassic successions are comparable on a stage-by-stage basis (Figs. 11.21–11.23).

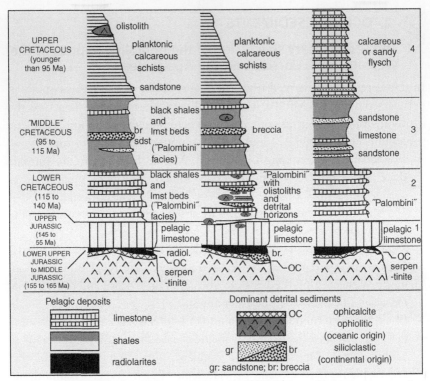

Figure 11.21 *Examples of the Jurassic and Cretaceous sedimentary succession resting on Liguro-piemontais ophiolites.* The sections represent the different types of sedimentary succession studied in the Queyras area and Chenaillet massif. The sediments are mainly pelagic with some clastic material. The pelagic sediments are sometimes siliceous or argillaceous–siliceous and were deposited below the CCD (episodes 2 and 3); where calcareous, these sediments were deposited above the CCD (episodes 1 and 4). The clastic material is most often of local, oceanic origin. Very large olistoliths occur locally and were derived from the collapse of submarine fault scarps. The original thicknesses of the beds have not been preserved because of Alpine deformation. The estimated original thickness of the Late Jurassic limestones is about a few tens of metres while that of the Cretaceous shales and calcareous shales is a few hundreds of metres only. *M. Lemoine 2002, unpublished.*

It is now known also that the apparently large thickness of the *Schistes Lustrés*, as defined by previous authors, results from the superposition of multiple phases of thrusting and isoclinal folding.

Between 1900 and 1910, the first ammonites, of Liassic age, were discovered in external Alpine units assigned to the *Schistes Lustrés* (Figs. 11.22 and 11.23). These discoveries led to the assignment of an Early Jurassic age to

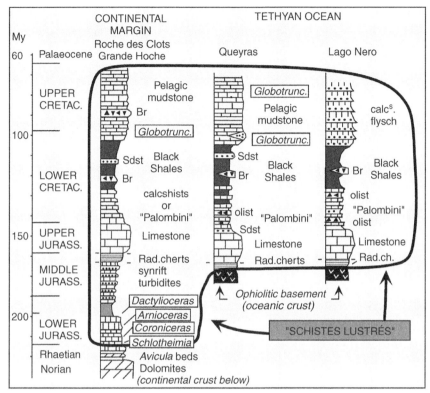

Figure 11.22 *The different successions of the Schistes Lustrés* on the Tethyan continental margin and ocean in the Central Western Alps. *M. Lemoine 2002, unpublished.*

the whole of the *Schistes Lustrés*. However, the effects of Alpine schistosity on the original lithology were not deciphered until after 1960. It now appears that most of the *Schistes Lustrés* is composed of calcschists of Late Cretaceous age.

Latterly, a succession of different lithological units (Figs. 11.22 and 11.23) has been progressively recognized despite the destructive effects of metamorphism. Identification of radiolaria of Middle and Late Jurassic age has permitted dating of the supra–ophiolitic radiolarites. Rare palynomorphs show the Black Shales to be middle Cretaceous in age. Other rare forms, interpreted as planktonic foraminifers (Globotruncanids), identify the calcschists as Late Cretaceous in age.

Finally, the same type of lithological succession has been recognized in units derived from the Tethyan Ocean in all parts of the Alpine fold belt

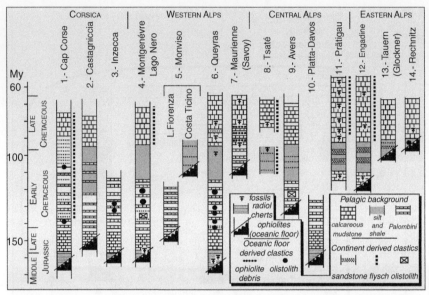

Figure 11.23 *Selected reconstructed stratigraphic columns of the Schistes Lustrés between Corsica and Rechnitz (Eastern Alps).* The long-distance stratigraphic correlations presented here result from (1) very few and loose biostratigraphic markers dispersed within the Alpine metamorphic terranes and (2) the existence of typical lithostratigraphic markers. For example the development of pelagic limestones at the Jurassic–Cretaceous boundary in the Tethyan realm records both the bloom of calcareous nano-plankton at the end of the Late Jurassic and a major maximum marine regression dated as latest Jurassic. These lithostratigraphic markers are characteristic of specific periods recording the principal phases of the biological evolution and major geodynamic global events such as long-term transgressive/regressive cycles. The correlation is dramatic and of wide importance to the correlation of similar sections in folded belts elsewhere. See Fig. 9.1 for location of Corsican sections, Fig. 11.1 for Montgenevre, Monviso, Queyras and Maurienne, Figs. 10.4 and 11.7 for Platta nappe, Fig. 7.2 for Prätigau and Fig. 7.1 for the Eastern Alps sections. *M. Lemoine 2002, unpublished.*

from Cap Corse as far as the Eastern Alps in Hungary (Fig. 11.23). This regional correlation has validated the stratigraphic reconstruction proposed for the *Schistes Lustrés (= Bundnerschiefer* or *Calcschisti).*

2.3 Clastic material interbedded in the Schistes Lustrés

Two types of clastic sediments are present.

2.3.1 Clastic interbeds of continental origin

Remnants of granitoids and associated arkosic material outcrop in the Chenaillet massif and Rocher Blanc massif, south of the Queyras valley

(Fig. 11.1A). In the latter, they are associated with gabbroic and basaltic debris in a conglomerate which is interbedded between a flow of pillow basalts and underlying ophicalcites, which rest in turn on serpentinized lherzolites, as shown above (Fig. 11.11). In a comparable manner, a bed of arkoses resting on Tithonian limestones can be followed for many kilometres in the region of the Col Agnel in eastern Queyras (Tricart 2003). These siliciclastics of cratonic origin demonstrate the proximity of the adjacent continental margin.

2.3.2 Clastic interbeds of ophiolitic origin: Intra-oceanic deformation of Late Jurassic–Early Cretaceous age

In the Queyras and Ubaye valley areas, and Corsica, sands, conglomerates, as well as olistoliths, up to 200 m in size, composed of basalts, gabbros or serpentinites form interbeds in the pelagic formations mentioned above (Figs. 11.21–11.23). While their precise origin is unknown, the submarine relief sourcing these clastic deposits may be represented by rarely recognized faults bounding the spreading axis. However, they may also be associated with transform faults offsetting the ridge axis or later faults which record the widespread Late Jurassic to Early Cretaceous extension which affected a large part of the West European craton and its margins (Chapter 7).

3. SUMMARY AND CONCLUSIONS

3.1 The main common characteristics of the 'oceanic' crust of the Liguro-piemontais and Central Atlantic Oceans

The variation in degrees of metamorphism observed between individual ophiolites, and often in different units in the same ophiolite body (Fig. 11.2), shows that the original 'oceanic' crust was dismembered from the onset of subduction through to final thrust emplacement in the nappe stack and later exposure by Alpine exhumation. Reconstruction of the original structure and development of Tethyan 'oceanic' crust is therefore particularly problematic.

During the period 1960–1970, it was difficult to conceive that part of the oceanic basement could have been formed outside the axis of accretion on mid-ocean ridges. Nonetheless, the possibility of oceanic basement or denuded sub-continental mantle (already expressed by Decandia and Elter 1972) was implied, if not present, in the publications of this period. Since then, the unification of results obtained on the transitional crust between Newfoundland and Iberia on the one hand with the mechanisms of slow spreading on mid-ocean ridges in the Central North Atlantic and Indian Ocean have provided important points of comparison.

We propose after others that the development of the Tethyan Ocean began by tectonic denudation of mantle ultramafites and was followed by the rise of the asthenosphere and formation of a slow-spreading ridge. The Alpine ophiolites are composed of both types now dispersed by thrusting during orogenesis.

The Atlantic transition zone between continental and 'oceanic' crust results from tectonic denudation of the sub-continental mantle. Its characteristics are as follows:

- The denudation surface, or at least its remains, is marked by tectono-sedimentary breccias. The surface locally supports fragments of continental crust or 'extensional allochtons' (Fig. 11.8) abandoned during denudation.
- The ultramafic rocks resulting from tectonic denudation of the mantle are mainly serpentinized peridotites derived from spinel lherzolites (less depleted or non-depleted), and in modest proportion, harzburgites (depleted), both intruded by gabbroic bodies and basaltic dykes.
- The relative proportions of depleted peridotites, gabbros and basalts increases with distance from the edge of the margin. However they remain modest in volume.
- The emplacement of gabbros and basalts was localized, episodic and each episode was of short duration.
- The equilibrium melting temperature of the ultramafic rocks increases oceanward in relation to the progressive creation of a spreading axis.
- The rate of creation of the seafloor by mantle denudation in the transitional zone of the Newfoundland–Iberia rift was 6–12 mm per year during the Early Cretaceous (Sibuet et al. 2007). In comparison, rates of fast-spreading ridges are 60–200 mm per year.

Similar characteristics are found in the 'oceanic' crust of the Liguro-piemontais Ocean and are as follows:

- Dominance of serpentinites intruded by gabbros, with or without overlying basalt flows. Gabbros and basalts are minor in volume compared with serpentinites. Dyke complexes are absent. Development of layered gabbros is absent or weak, more or less differentiated at the base of isotropic gabbroic bodies.
- Peridotites and serpentinites have been impregnated by magma along localized 'magma-rich' segments (Fig. 11.18 and 11.19).

- The frequent occurrence of OC2-type ophicalcites (debris flows containing carbonates and serpentinite; Fig. 11.13A) underlain by serpentinites and gabbros and overlain by pillow basalts or sediments.

The first sediments deposited on the ocean floor, interpreted as resulting from tectonic denudation, are dated to the Middle Jurassic around 170 Ma (radiometric age, Bathonian–Bajocian boundary; Fig. 11.1).

One of the difficulties is to make a clear distinction at outcrop between an ophiolite sequence emplaced in the context of transitional crust or at a slow-spreading ridge axis. For example it can be reasonably agreed that the ophiolites of the Chenaillet and Monviso (Figs. 11.2 and 11.9) were formed at the initiation of a slow-spreading ridge. In effect, it is possible to reconstruct from rocks at outcrop segments of a slow 'magma-poor' ridge. Another supporting reason is the presence in ophiolite bodies of extremely differentiated rocks such as anorthosites that compare well to the plagiogranites of the present-day oceans. The anorthosites may have been transformed into albitites as a result of (early?) serpentinization processes.

3.2 Two key conclusions: The limited life and narrowness of the Tethyan Ocean

The dated range in magmatic activity between the Bathonian and the Tithonian (165–145 Ma) implies that it did not last much longer than about 20 Ma (Fig. 11.1B). On the other hand, the characteristics of Alpine ophiolites support comparison with North Atlantic oceanic basement between Iberia and Newfoundland. The spreading rate there did not exceed 5–10 mm per year, with a velocity increasing to 5–12 mm per year according to the time scale used by Sibuet et al. (2007). These clues lead to an estimation of the width of the Ligurian Tethys of the order of 200 km at least, if the magmatic activity designating spreading did not exceed more than 20 Ma. However, if the spreading lasted until the middle of the Cretaceous at the same rates over 60 Ma (although this is not actually known), the width of the Alpine Tethys could have reached 350–650 km. The value of these simplistic estimates is perforce limited by the disappearance of key parts of the record due to subduction or to the erosion of the nappes at the summit of the pile.

Comparison of the Tethyan 'oceanic' crust involved in the Alpine nappes and 'oceanic' crust of the present-day Atlantic crust off Galicia has been valuable in understanding the transition from rifting to earliest spreading. It remains possible that a part of the Tethyan Ocean did not pass much beyond the stage of transitional crust between hyper-stretched continental crust and typical oceanic crust.

FURTHER READINGS

Concerning the Grisons ophiolites: Desmur et al. 2002; Manatschal et al. (2007); Manatschal and Müntener (2009). Concerning the oceanic crust: Cannat et al. (2006) Tucholke et al. (2007); Sibuet et al. (2007); Peron-Pinvidic and Manatschal (2009); Lagabrielle (2009). Concerning the ophiolite concepts: Bernoulli et al. (2003).

CHAPTER TWELVE

Recapitulation and Comparisons: Oceans and Continental Margins in the Alps, an Overview

Contents

Summary

All palaeogeographic syntheses of the Liguro–piemontais Ocean and its passive margins are beset with serious difficulties because of the fragmentary geological record. Much of the sedimentary succession has been lost through subduction of vast areas of oceanic and continental lithosphere. In addition, most of the remnants incorporated in the former superstructure of the fold belt have been lost by erosion during the Tertiary and Quaternary. Reconstructions of the conjugate margins of Europe and Apulia–Africa thus remain conjectural while estimates of the original width of the ocean basins prior to convergence rely on indirect reasoning. The latter rests, on the one hand, on the accumulation of direct observations and, on the other, from comparisons with regions that have experienced comparable events during their history such as the North and Central Atlantic, and the Indian Ocean.

In the preliminary phase of the birth of the Tethys Ocean and the North Atlantic, the continental crust was stretched over a large area due to listric

The Western Alps, from Rift to Passive Margin to Orogenic Belt, Volume 14
ISSN 0928-2025, DOI 10.1016/S0928-2025(11)14012-2
© 2011 Elsevier B.V.
All rights reserved.

normal faulting and then thinned due to the effects of detachment faulting. The result was to pull apart the continental crust leading to tectonic denudation of the sub-continental lithospheric mantle along the length of a well-defined line of rupture. Finally, the ascent of asthenospheric mantle and the separation of the continental lithosphere into two plates resulted in the creation of a very slow-spreading axis. As a result, the total width of the Tethyan oceanic crust probably did not exceed more than a few hundred kilometres.

Two branches composed the Jurassic and Cretaceous part of the Tethys. One opened in the Middle Jurassic and consisted of two segments elbow-shaped in plan view. Of these, the Liguro-piemontais segment, from which the Western Alps originated, was oriented approximately SW–NE. The second, oriented E–W and corresponding to the Grisons (=Graubünden)–Tauern line of the future Central and Eastern Alps, linked the Liguro-piemontais Ocean to the Neotethys. The Late Jurassic–Early Cretaceous Valais basin, which is grafted onto the Grisons–Tauern segment, is partly superimposed on the earlier European margin of the Liguro-piemontais Ocean.

All the margins, together with the adjacent oceanic crust, were trans-ected by transform faults. These divided the rifted margin into segments each characterized by somewhat different histories.

1. A SHORT PALAEOGEOGRAPHIC REMINDER

Successive episodes of rifting and spreading, from Late Triassic through to Early Cretaceous times, created a set of oceanic basins extending from the Caribbean via the Atlantic to the Eastern Mediterranean and beyond (Fig. 3.1). These basins include the Central Atlantic, the Liguro-piemontais Ocean and, to the east, the western part of the Neotethys already open in Triassic time (Fig. 3.2).

The history of the different parts of the Alpine Tethys resulted from the relative movements of Africa, North America, Eurasia and the small Apulia and Iberian plates. The development of the Liguro-piemontais Ocean, from which the Alpine ophiolites are derived, was linked to, and contempora-neous with the Central North Atlantic (Fig. 12.2) in the Late Triassic and Jurassic (Fig. 3.4). It led to the separation of the Eurasia–Iberia plate in the north from the Apulia-African plate in the south.

During the Early Cretaceous, the drifting of the southern part of the North Atlantic and specifically the area between the Iberia and Newfound-land margins together with the Bay of Biscay resulted from the relative movements between Iberia, North America and Eurasia (Fig. 3.5). This rift was contemporaneous with that known in Provence and the Valais basin.

The major difference that distinguishes the evolution of the Atlantic sector of the Tethys from the Alpine Tethys is that the Atlantic has continued spreading to the present day while the Alpine sector closed during the course of the Cretaceous and Tertiary (Fig. 3.7). The processes generating the oceanic crust and continent–ocean transition zone of the Atlantic and Alpine sectors of the Tethys were almost similar as has been shown from comparative studies of ophiolites and mid-ocean ridges.

2. RIFTING: COMPARISONS AND REFLECTIONS

Extension, which resulted in the progressive disintegration of Pangaea and particularly of the post-Hercynian platform, affected the European craton from the Late Carboniferous–Early Permian (c. 300 Ma) to the middle of the Cretaceous (c. 100 Ma). This long phase led to the formation of subsiding intra-cratonic Mesozoic sedimentary basins not only in the future Alpine and Pyrenean domains but also in other areas such as the Aquitaine and the London–Paris basins. The main Jurassic and Early Cretaceous rift episodes were not accompanied by magmatism in contrast to those of the Late Palaeozoic and Triassic. This is why the pre-Tethyan rift and likewise the Newfoundland–Iberia rift are termed *non-volcanic or amagmatic.*

The schematic section (Fig. 12.1) of the main features of the pre-Tethyan rift shows three branches preserved after the Alpine orogeny: the Dauphine–Helvetic–Valais branch on the European side, the Piemontais–Canavese–Err branch in the centre and the Lombardy–Ortler–Ella branch on the Apulia–Africa side. The width of the NW and SE branches respectively is estimated at between 300 and 400 km. The width between the platforms represented by the Jura-Cevennes and Friuli shallow water carbonates thus could have exceeded 600 km.

The evolution of each of the branches of the rift was relatively different: the Dauphine–Helvetic–Valais branch is separated from the central branch by the 100 plus km wide Saint Bernard–Monte Rosa or Brianconnais block ('SBR' block) in the Central and part of the Western Alps (Fig. 6.13). On the Apulia–Africa margin, there are no equivalents and both the Trentino (Fig. 10.2) and Bernina highs (Fig. 10.4) are considerably less wide; the pre-Triassic basements do not have the same characteristics as the SBR block vis-à-vis the neighbouring continental block.

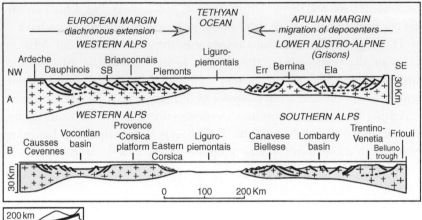

Figure 12.1 *Schematic reconstruction of the Liguro-piemontais Ocean and its margins.* Many reconstructions are possible according to the initial arrangement of the conjugate margins. This model supposes that part of the European margin in Corsica was conjugate to the Apulia–Africa margin of the Southern Alps. Crosses show continental crust. The scale is approximate and the width of the ocean is arbitrary. SB: Subbrianconnais. The migration with time of the main areas of deformation in the direction of the line of continental breakup is well observed on both the European and Apulo-African margins (Figs. 6.17 and 10.7). It is also widely known on the Iberian margin (Wilson et al. 2001). *Modified from Lemoine et al. 2000, Fig. 13.4, p. 148. Inspired from Manatschal (2004), Manatschal et al. (2007).*

2.1 The Tethys and Newfoundland–Iberia: a general comparison

Both the Tethyan and Iberian margins experienced two rift events separated by long, 40–50 Ma, periods of quiescence (Figs. 6.19, 6.20, 10.7 and 12.3). Both were subject to early diffuse rifting from the Late Triassic to the Early Jurassic, the same time as in the contiguous central North Atlantic, and also most of the major European basins.

In the case of the better-documented Tethys, rifting was not a continuous process and it comprises a succession of discrete episodes, all spaced between 5 and 10 Ma at most, between the end of the Triassic (210 Ma approx.) and the Middle Jurassic (160–165 Ma approx.), i.e. an interval of about 50 Ma. These episodes can be correlated quite well in time from the European to the Apulian margin of Tethys (Figs. 6.19

and 10.7). In the case of Western Iberia, this early rift was aborted unlike that of the Tethys and contiguous Central North Atlantic which evolved to ocean spreading.

In contrast, Late Jurassic–Early Cretaceous rifting (Tithonian to Aptian) off Iberia resulted in the formation of oceanic crust west of Iberia and in the Bay of Biscay (Fig. 12.2). The associated rift formed part of a system extending through the Bay of Biscay, Pyrenees/Aquitaine, and Provence to the European margin of Tethys including the Valais basin (see Chapter 7).

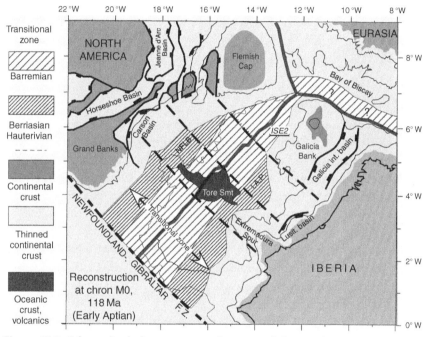

Figure 12.2 *Schematized plate reconstruction around the Newfoundland–Iberia rift at chron M0 (118 Ma; Early Aptian).* This reconstruction shows the transitional oceanic zone generated by the two-steps exhumation of the sub-continental lithosphere. During the first step the southwestern part of the Newfoundland–Iberia rift separated Grand Banks from Iberia margins. The northeastern part between Flemish Cap and Galicia Bank ruptured during the second step. The Tore seamount is an exception in this panorama. It resulted from melt intrusion above a plume in the south part of the rift during the Barremian. There is no (again) equivalent in Tethyan ophiolites. Eurasia is supposed fixed. ISE 2: seismic line shown on Fig. 11.7E; IAP: Iberian Abyssal Plain; NFLB: Newfoundland Basin. Thick black hyphen lines are M25–M0 flow lines deduced from bathymetric, gravimetric and magnetic data. The thick solid line is the trace of the future rupture of the continental lithosphere dated as at the Aptian–Albian boundary. *Simplified from Sibuet et al. 2007, Fig. 1 and Tucholke et al. (2007a, Fig. 1).*

In the latter two areas, this rift phase did not lead to oceanic spreading although the kinematic linkages are poorly understood.

West of Iberia rifting took place in three main episodes (Fig. 12.3). The first two date to the end of the Berriasian and Hauterivian respectively. These led to two main stages in the separation of continental crust in two sections separated by a transfer zone (Fig. 12.2). The third episode is dated to the Aptian–Albian boundary (112 Ma approx.). It is recorded as the breakup unconformity and is marked by a prominent seismic reflector associated with a major unconformity (Groupe Galice 1979; Montadert et al. 1979). This unconformity marks the end of development of transitional crust, the complete rupture of the continental lithosphere and the onset of oceanic crustal accretion *sensu stricto* (Tucholke et al. 2007; Sibuet et al. 2007).

2.2 The syn-rift phase

The syn-rift phase is generally considered to be defined as the time interval between the onset of rifting and the end of rifting. The former is widely considered to be marked by a rift onset unconformity and the latter by the breakup unconformity or its correlative conformity. Observations from the Tethyan margins and Iberia discussed below suggest that this definition may be too restrictive in terms of what is now known of the development in space and time of the rift phase in these areas.

Manatschal et al. (2007) have shown that during each episode of rift development on the future Apulo-African margin (Fig. 12.1), the depocentres migrated progressively towards the distal parts and the line of future continental breakup (Fig. 10.7).

However, the same simple evolution cannot be applied easily to the European margin. The proximal Dauphinois, in the west, is mainly a major half-graben beneath the Rhone Valley infilled by thick sediments throughout, all post-Middle Triassic in age (Figs. 5.6, 6.17 and 12.1). With regard to the Piemontais units, which are located in a distal position, active subsidence only began from the Middle Jurassic. Moreover, it was never accompanied by thick sedimentation because of starved conditions caused by distance from the source of material and also the role played by the Brianconnais platform as an effective barrier (Fig. 6.14).

Successive episodes of extension became progressively younger and younger from the proximal to the distal parts of the future European margin. For example (Fig. 12.1A), the Jurassic rift appears to have been

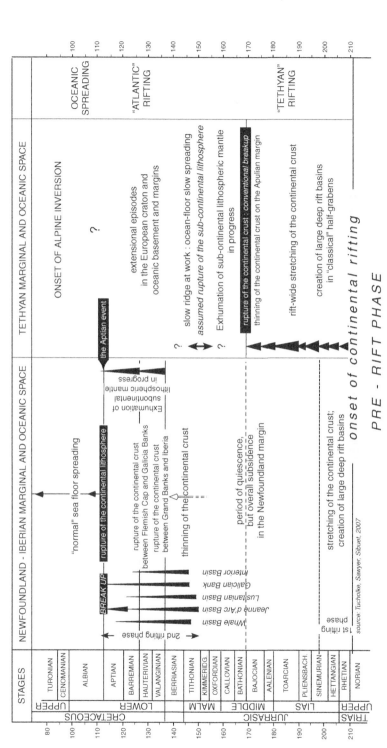

Figure 12.3 Rifting events and development of oceanic basement in the Tethys and the North Atlantic between Newfoundland and Iberia. *Inspired from Sibuet et al. (2007) and Tucholke et al. (2007) for the Atlantic part. Ages from Gradstein et al. (1994).*

active earlier in the western Dauphine segment than the central branch. Supporting evidence is provided from drilling, which shows an approximately 600 m thick earliest Jurassic section beneath the Rhone Valley (proximal situation), compared with the much smaller, less than 100 m (?) thickness of Hettangian–Sinemurian outer-shelf (now multi-folded) sediments at outcrop in the distal Piemontais domain.

In the Dauphinois, Bajocian and above all Late Bathonian strata are post-rift and post-date extensional faults of the Tethyan rift phase. The succession is regionally transgressive as far as the Hercynian basement including the Massif Central. The transgressive surface corresponds to the 'Mid-Cimmerian' unconformity well known at the scale of the West European craton (Jacquin and Graciansky 1988).

In contrast, in the distal Piemontais domain, sediments transported by gravity, mainly turbidites of the same Middle Jurassic age, are syn-rift. They were deposited in a relatively deep environment at the foot of the master fault bordering the east side of the Brianconnais platform (Fig. 6.14; Fig. 6.19, event J3, Piemontais column).

In the case of the Apulian margin, Manatschal et al. (2009) describe syn-rift sediments as younging from the proximal part of the rift to the distal, paralleling the migration of depocentres noted above (Figs. 10.4 and 10.7). The thinning of the crust along major mylonitic faults such as the Pogallo is post-Late Sinemurian–Pliensbachian (184 Ma). However, strata of this age seal the extensional faults on the proximal parts of the margin (Il Motto for example; Fig. 10.5). These observations further demonstrate migration of extensional activity towards the distal part of the margin to be finally concentrated on the line of oceanic opening.

The Iberian margin is similar in that the main phases of rifting in the Interior Basin pre-date rifting under the future Iberian abyssal plain to the west (Figs. 12.2 and 12.6; Wilson et al. 2001).

While the onset of rifting seems to be contemporaneous over wide areas, the end of rifting becomes younger from the proximal to the distal parts of the margin. The hiatus associated with the breakup unconformity thus conceals the diachroneity of the correlative conformity marking the end of rifting in various parts of the margin. Care must be taken to avoid oversimplification of the significance of the breakup unconformity.

Extension of the lithosphere during the phases of rifting is expressed by successive types of deformation that affected the continental crust leading to continental rupture and then exhumation of the upper mantle (Figs. 12.4 and 12.5).

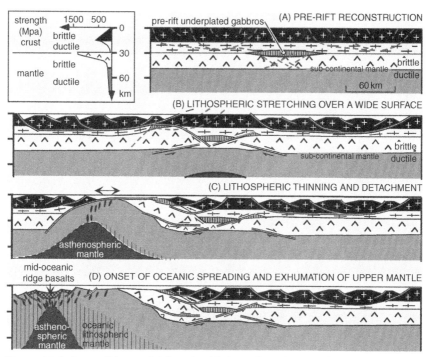

Figure 12.4 *A possible model for the birth and evolution of a slow-spreading ridge.* This conceptual model is supposed to explain the development of the 'MP' or *magma-poor* oceanic crust situated at the foot of the Galicia margin related to the fixed Iberian continental edge. It can be applied to the development of Alpine ophiolites. (A) Pre-rift situation. The model involves a supposed four-layered lithosphere. The continental crust is shown here to be locally thickened by a pre-rift underplated gabbroic body. (B) Necking of the ductile upper mantle beneath the gabbros allows the asthenosphere to rise. According to one classic model for rifts and passive margins, ductile stretching of the lower crust is accompanied in the upper crust by displacement along listric faults bounding tilted blocks. At the end of the rift phase the crust is reduced to about 10 km thickness. As a result, the classic pattern of tilted blocks and half-graben are missing on the line of future breakup. (C) Lithospheric thinning by localized detachment faults as a consequence of asthenospheric rise. Sub-continental mantle denudation at a very low spreading rate gave rise to the observed 10–20 km wide ribbon of outcropping serpentine associated with sparse gabbroic bodies and small basalt flows (double arrow on figure). Concave downward faults are shown on the model even they have not been imaged. Such geometries differ from typical listric faults. These could be due to redistribution of the stress field resulting from the rising magma and high thermal gradients above the asthenospheric dome. (D) Rising of the asthenosphere close to the surface. Intrusion and local extrusion of mid-oceanic-ridge basalts into and onto the sub-continental mantle. Onset of oceanic spreading. *Simplified from Whitmarsh et al. 2001, Fig. 3, p. 152 and Manatschal 2004, Fig. 6, p. 444.*

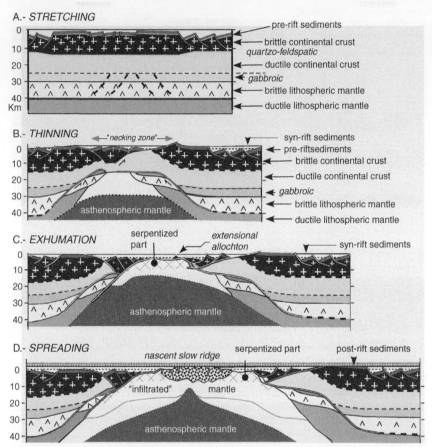

Figure 12.5 *A conceptual model to illustrate the evolution of the Mesozoic rift between Newfoundland and Iberia.* (A) Stretching. Weak rate of large-scale regional stretching of the continental crust by listric normal faulting associated with classical subsiding half-graben. The listric faults are concave upward. (B) Thinning. Thinning of continental crust in a localized corridor succeeding in separation due the effect of conjugate detachment faults. This deformation accommodates the exhumation of the sub-continental mantle. The faults are concave downward. (C) Exhumation. Serpentinized peridotites of the sub-continental mantle reach outcrop on the seafloor due to the effect of detachment faults. 'Extensional allochtons' are abandoned by detachment of the continental crust. (D) Spreading. Start of mid-ocean ridge accretion. Breakup of the continental lithosphere takes place at this moment. *Simplified from Peron-Pinvidic and Manatschal 2009, Fig. 5.*

2.3 Distributed lithospheric stretching

This first phase is characterized by the formation of large subsiding sedimentary basins localized in classical half-grabens such as those of the Rhone

Valley (Figs. 6.16 and 6.17), Bourg d'Oisans in the Western Alps (Figs. 6.2 and 6.5), Monte Generoso in the Central Alps, the Jeanne d'Arc basin (Fig. 10.3) and Harseshoe Basins on the Grand Banks of Newfoundland, the Galicia Interior and Lusitanian Basins of Western Iberia margin (Wilson et al. 1989; Figs. 12.2, 12.4 and 12.5).

The initial extensional deformation was not limited to the locus of future continental breakup but, on the contrary, affected a much wider area. It resulted from (i) ductile stretching of the lower crust and (ii) movement along listric faults separating tilted blocks in the upper crust (Figs. 10.8 and 12.4B).

The listric faults moved discontinuously due to successive phases of pulsed extension affecting the upper crust. The duration of each episode was almost instantaneous in terms of geological time and of the order of the time represented by a single ammonite sub-zone (Fig. 6.20). In addition, on the same transect, the periods of movement and peaks in the rates of movement on different faults were not simultaneous (Fig. 6.20). The faults in general become younger from the proximal to the distal parts of the margin.

2.4 Focussed lithospheric thinning

According to observations on the Apulian margin (Fig. 10.8B) and those from the present-day margins of Iberia and Newfoundland (Figs. 12.2–12.4), the second phase affects mainly the distal part of the future margin. The zone of lithospheric thinning is localized but is not dispersed over a large area as was the zone of stretching. This phase is the least well documented on the European margin because the distal parts of the Tethyan margins are rarely preserved due to Alpine deformation compared with their proximal parts.

The lithosphere was strongly stretched and thinned in a 'necking zone' where the thickness of the crust decreased rapidly from proximal to distal from about 15 km to zero (Figs. 12.4 and 12.5). The line of future continental rupture would appear within this zone of extreme thinning.

On the European passive margin near Briancon, the distal parts of the margin are involved in Piemontais units originally located beyond the master fault controlling the east flank of the Brianconnais platform (Figs. 6.17 and 12.1). Always however, the pre-Triassic basement is missing due to detachment of the overlying cover by Alpine thrusts rooting in Triassic evaporites. That said, it seems probable that the crust was thinned as strong subsidence is shown by the deep marine environments in the late syn-rift and early post-rift. In Switzerland, where the basement is preserved, the distal part of the margin is involved in the Mont Fort and Mont Rose nappes (Table 1 of Chapter 4). However, the imprint of Alpine subduction is too strong to allow credible reconstructions of original sediment thicknesses. In consequence, Fig. 12.7 presents only thinning on the Apulian margin with a pronounced structural asymmetry between the two margins. The models of Figs. 12.4 and 12.5 also offer this direction.

Concerning the Apulian margin, deformation of the crust was concentrated along major ductile shear zones, such as the Margna and Pogallo faults, which pre-conditioned the line of oceanic rupture (Figs. 10.2, 10.4 and 10.8). The zones of major ductile shear allowed accommodation of the relative movement of the upper crust with respect to the lower crust. The model (Fig. 12.4C) shows that the lower crust alone was thinned but not the upper crust whose thickness remained approximately the same.

The difference in styles between the distal and proximal parts of the margin can be interpreted in terms of relative differences in the magnitude of the extension factor (bêta), which is least on the proximal parts of the margin and very much higher in the distal extremities (Fig. 12.7).

3. CONTINENTAL RUPTURE AND EXHUMATION OF THE UPPER MANTLE

Alpine subduction of the European margin has removed all evidence of exhumation. Perforce, ideas on exhumation must be sought from the conjugate Apulian margin and by comparison with the undeformed margins of Iberia and Newfoundland (Manatschal et al. 2009; Figs. 12.2, 12.4 and 12.5).

The structural organization of the passive margin of the Liguro-piemontais Ocean can be compared to that of the Atlantic passive margins west of Iberia and Ireland where Galicia Bank and Porcupine Bank may have played roles analogous to the 'SBR' block emphasized by the absence of oceanic crust in the basins situated between these banks and the adjacent platforms of Iberia and Ireland (Fig. 12.2).

However, in the rifting basin east of the SBR block, only a narrow and elongate area would evolve into the spreading Liguro-piemontais Ocean (Fig. 12.4D). In a palaeogeographical and structural context, continental rupture was probably located between the Err nappe to the east (Fig. 12.1) and the most distal Piemontais units to the west. In effect, the remains of the continent–ocean transition are preserved in the Err and Platta nappes on the Apulia–Africa side. In contrast, on the European side, there are no visible remains of the most external parts of the margin and the continent–ocean transition. Pre-rift Hercynian structure may have controlled the locus of continental rupture thereby demonstrating the importance of structural inheritance in the development of rift structures (Chapter 4).

A now classical view, developed since 1990, considers that spreading was initiated by tectonic denudation of the sub-continental mantle. The exhumation of mantle rocks (Fig. 12.4C and D; Fig. 12.6) resulted from much higher extension than is observed in intra-continental rifts (Fig. 12.7). Slices of continental crust torn apart by extension are nonetheless preserved on mantle rocks as 'extensional allocthons' (Figs. 10.8C, 12.4 and 12.5), which mark the oceanic crust adjacent to the neighbouring margin as a true continent–ocean transition zone. There, mantle denudation was immediately accompanied by the production of basaltic magma in modest amounts.

In the southern part of the Newfoundland–Iberia rift, the continental crust ruptured around the Jurassic–Cretaceous boundary and in consequence the sub-continental mantle began exhumation (Welsink et al. 1989; Fig. 12.2). The continental lithosphere separated near the Aptian–Albian boundary, at the 'Aptian event' (Groupe Galice 1979). A time-span of 14–15 Ma lies between these two events.

In the case of the Tethys, rupture of the continental crust and denudation of the sub-continental mantle began in the Middle Jurassic, probably close to the Bajocian–Bathonian boundary (about 170 Ma) at the foot of the European margin and perhaps a little earlier in the Apennines (Fig. 11.1).

The correlative issue is the age of the rupture of the continental crust and the beginning of slow mid-ocean ridge accretion. Acid-differentiated plagiogranites of Kimmeridgian–Tithonian age provide an answer to this question since the ophiolite bodies correspond to the remains of the mid-ocean ridge dismembered by Alpine thrusting. The duration of the interval between the Bajocian–Bathonian boundary and the Early Kimmeridgian is about 15 Ma and of the same order as the homologous interval in the Newfoundland–Iberia rift. But there is no known trace of a major discontinuity in the Tethys comparable to the Aptian event that records at distance the rupture of the lithosphere in the North Atlantic.

 ## 4. THE NOTION OF THE BREAKUP UNCONFORMITY: DISCUSSION

The breakup unconformity according to classical views dates the onset of plate separation and oceanic crust accretion (Chapter 1) and is a

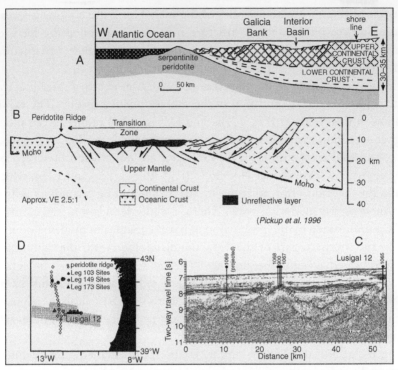

Figure 12.6 *A model for the initiation of a slow-spreading ocean: the serpentinite seafloor at the extremity of the West Iberia (Galicia–Portugal) passive margin. (A) Interpreted section of the continental margin off Iberia* at the latitude of the Galicia Bank. *Redrawn from Lemoine et al. 2000, Fig. 2.3, p. 29. (B) Transition zone between the Galicia Bank block and the peridotite ridge. (C) Seismic diagram across the transition zone and location of wells. (D) Location diagram. Reproduced from Pickup et al. 1996, Fig. 5, p. 1082.* At the foot of the margin, cut into tilted blocks, the ocean basement of the transition zone consists throughout of serpentinites with accessory gabbros either at outcrop or covered by sediments over a width of 10–20 km or less. Across this zone, the available data is provided only from the interpretation of seismic profiles and some deep-sea drilling sites. Note especially that there were two rifts. The first rift, between Galicia Bank and the Iberia continent, did not lead to spreading. The second rift, west of the Galicia Bank horst, led to complete rupture of the North Atlantic thereby separating Galicia Bank from North America. A similar disposition is found in the Tethys with many separate rifts of which only one evolved to complete rupture and spreading.

major discordance. It separates syn-rift sediments below from post-rift sediments above (Fig. 1.8). In principle, it corresponds to the end of extensional deformation on newly formed continental margins and the onset of oceanic crust accretion. The newly formed oceanic crust continues to widen in area due to accretion at the young mid-ocean ridge axis.

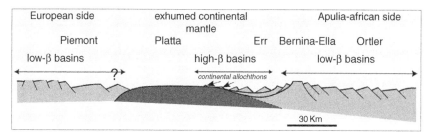

Figure 12.7 *Variation in extension (b) factors across two conjugate passive margins.* *Modified from Manatschal 2004, Fig. 14a, p. 459.*

With regard to the domains of Newfoundland–Iberia and the Bay of Biscay, the ambiguity in the definition of the breakup unconformity comes from the existence of a large *transitional zone* corresponding to the denudation of sub-continental mantle. This zone is neither continental crust nor 'normal' oceanic crust composed of basalts and gabbros as well as peridotites of asthenospheric origin. In terms of the geography of the seafloor, transitional crust separates thinned continental crust from 'normal' oceanic crust. A major discordance reported as the breakup unconformity, dated as Late Aptian–Early Albian has been characterized as recording the end of formation of transitional crust and the rupture of the continental lithosphere as already outlined above (Figs. 12.2–12.4). The Aptian event therefore marks the end of rifting between Newfoundland and Iberia.

The definition and nature of the breakup unconformity can be posed as a question: if it is associated with the start of denudation of the sub-continental mantle, it dates to the end of the Berriasian or at least to the oldest identified magnetic anomaly. Moreover, the structures of the Iberian margin were essentially acquired in the Berriasian, exhumation of the mantle absorbing the main extension with respect to the thinned continental crust (Tucholke, Sibuet and Sawyer 2007).

If the rifting continued until the end of the Aptian, with important extension in the transition zone and modest extension in the adjacent continental crust, then the breakup unconformity dates to the end of the Aptian. It thus records the separation of the continental lithosphere that occurred 15–20 Ma later than the rupture of the continental crust as indicated above (Tucholke et al. 2007; Sibuet et al. 2007).

For this reason there are no criteria in this book that allows escape from the definition of the breakup unconformity as recording the appearance of true ophiolitic material on the seafloor, in place of denuded sub-continental mantle. For the Tethys, it is dated as Middle Jurassic and it truly records the onset of subsidence of both margins associated with a large regional transgression on the adjacent continental areas. In this way, there is no contradiction with our previous publications on this subject (Lemoine et al. 2000) that might otherwise introduce confusion. However, we are conscious of an apparent ambiguity that exists in comparison with the Newfoundland–Iberia rift.

5. OCEANIC ACCRETION IN THE LIGURO-PIEMONTAIS OCEAN

With regard to the Alpine Tethys, part of the ophiolites may result from denudation of sub-continental mantle, a point of view supported by Manatschal and his co-workers. Another part of the ophiolites, such as the Chenaillet (Fig. 11.2) and Monviso (Fig. 11.9) formed by accretion at a slow-spreading ridge, later dismembered by Alpine thrusting (Lemoine et al. 2000). One subject in debate is identifying among the ultrabasites of Alpine ophiolites those of denuded continental mantle origin and those derived from denuded oceanic mantle (Lagabrielle 2009). No easy distinction can be made between underplated gabbros of sub-continental lithosphere (Fig. 10.8) and those formed at a slow-spreading ridge axis (Fig. 11.14B). In addition, there is no major unconformity analogous to the Aptian event of Newfoundland–Iberia which might record the rupture of the lithosphere between Europe and Apulia.

The interpretation of Alpine ophiolites has been greatly aided by an understanding of the accretion mechanisms at slow- and ultra-slow-spreading ridges in the Central Atlantic and the southwestern Indian Ocean. These studies have provided comparative models.

The composition and structure of Alpine ophiolites demonstrate formation at a mid-ocean ridge (Fig. 12.4D) with a very slow spreading rate of 5–20 mm per year. Taking into account the 50–60 Ma duration of spreading of the Liguro-piemontais Ocean, its width could not therefore have exceeded a few hundred kilometres. These figures, which provide an upper limit for the width of oceanic crust in the area, clearly cannot be compared with the present width of the Atlantic. The difference arises because the Atlantic has not undergone Alpine shortening and the ocean has continued to widen by spreading since its inception in Jurassic time.

6. BRANCHES OF THE TETHYS IN THE FUTURE ALPINE DOMAIN

6.1 The Liguro-piemontais branch of the Tethys Ocean

In the reconstruction proposed here, the Jurassic segment of the Tethys was elbow-shaped with two segments representing the upper and lower arms (Fig. 12.8A). The present-day Central and Eastern Alps are approximately parallel to the Grisons (=Graubünden) segment between Grisons and the Tauern window. Reconstruction of the orientation of the Graubünden–Tauern segment is uncertain because of Alpine deformation. Nonetheless, the present-day orientation of this segment suggests that it may correspond to a major zone of transtension linking the Neotethys in the east to the Ligurian Tethys in the West (Fig. 12.8).

The Liguro-piemontais segment *sensu stricto* was probably parallel to the elongation of the tilted blocks and the strike of the normal faults bounding the blocks, i.e. along the Cevennes or Belledonne trend that is approximately NE–SW oriented (Fig. 12.8). It coincided in position with the Western Alps and Corsica in pre-Miocene time. Modest Alpine deformation has effectively allowed preservation of the original orientation of the faults so that those in the External Crystalline Massifs (Fig 6.16) are as well preserved as those on the eastern border of the Massif Central.

Structural trends inherited from rifting occur widely on the rest of the adjacent continental margins. However, the original trends are not preserved in the nappes because of translation and rotation during thrusting.

6.2 The Valais domain and the Valais branch of the Tethys: a subject of debate

Valais units (=Lower Penninic or North Penninic, Figs. 2.7 and 7.1) have been classically attributed in Alpine geology to a phase of Late Jurassic and Early Cretaceous rifting and spreading that led to the formation of the Valais trough or small oceanic basin. These units only occur close to the NE boundary of the Western Alps, in the Tarentaise valley at the Petit Saint Bernard pass, in the Central Alps as far as the Grisons and also further east in the Eastern Alps. In the Central Alps, Valais units correspond in large part to the continental basement of the Simplon–Ticino nappes (Fig. 7.2) and their more or less detached cover of metamorphosed Mesozoic sediments, traditionally known as Valais *Schistes Lustrés* or *Bundnerschiefer* (Fig. 7.3). Ophiolites related to the Valais domain and their overlying sediments are rare and of modest volume where present in the Central Alps. However, they are more important in volume in the Glockner nappe of the Eastern Alps.

Late Jurassic and Early Cretaceous extension was particularly concentrated along two corridors (Figs. 7.10 and 12.8). The first extended from the Bay of Biscay through the North Pyrenean zone and via the Languedoc into southern Provence. In this corridor, only the Bay of Biscay evolved to spreading. The second corresponds to the Valais domain which also could

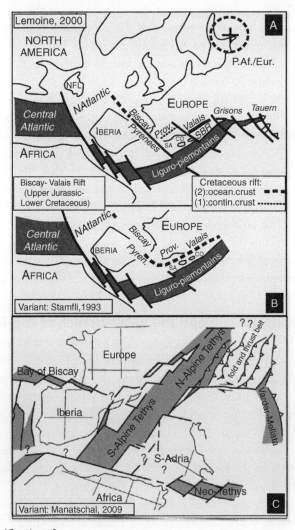

Figure 12.8 *(Continued)*

have evolved to spreading especially in its eastern part although this is no longer agreed by the Alpine geological community (Chapter 7). The two corridors were linked geographically via the Vocontian Basin, part of the Sub-alpine zone of the Western Alps, where the extension was relatively modest and did not evolve to spreading.

The traditional, classical and agreed Alpine view is that spreading in the Valais branch of the Tethys began in the Early Cretaceous and therefore later than in the Liguro-piemontais branch of the Tethys (Antoine 1968; Antoine 1971; Antoine et al. 1973; Antoine et al. 1992). The prior

---

Figure 12.8 *Three palaeogeographic models of the Liguro-piemontais Ocean and Valais trough in the Early Cretaceous. (A) Transfer faults divided the Tethyan continental margin into separate compartments.* These compartments differ in structure. Following the model depicted here, the Bay of Biscay–Valais rift system comprises the North Atlantic margins, the North Pyrenean zone (Pyrenees), Provence and the Sub-alpine chains of the Western Alps (Provence), the Valais, Graubünden and the Hohe Tauern in the Eastern Alps. The Bay of Biscay parts of the rift system have evolved to spreading; in contrast the other parts have not evolved to spreading. The existence of spreading in the Valais has become much less probable following recent new results. The Liguro-piemontais and Valais basins merge in the East where the SBR block pinches out. *Adapted from Lemoine et al. 2000, Fig. 13.1, p. 142. (B) Prolongation of the hypothetical Valais Ocean towards the Bay of Biscay* through a hypothetical arm of the Tethyan Ocean located between Provence and the Corsica–Sardinia block (Stampfli 1993). Iberia, Sardinia, Corsica and the Brianconnais collectively form a unique block, mobile with respect to the European craton, but displaced laterally along the line joining the Valais to the Bay of Biscay. While this hypothesis reconciles the major differences in the structure and history of the basement of the Helvetic and Brianconnais respectively (Chapter 7), there are, however, no results available to show how these two domains came to face each other. In addition, the existence of a hypothetical ophiolitic suture beneath the present-day Mediterranean cannot be proven. This model also does not take into the account the existence of NW–SE transfer faults, which provide a severe constraint on reconstructions of the origin of continental blocks. Oceanic crust is shown in black: Central Atlantic, Gibraltar transform, Liguro-piemontais segment and Grisons transform. Co: Corsica; Sa: Sardinia block; SBR: Saint Bernard–Monte Rosa block; P.Af/Eur.: approximate position of the pole of rotation for Africa–Europe during the Middle and Late Jurassic (see Fig. 6.16). *Simplified from Stampfli 2001 in Lemoine et al. 2000, Fig. 13.1, p. 142. (C) Model of the Valais oceanic arm but without oceanic continuity with the Bay of Biscay.* This model does not integrate the work of (Masson 2002; Masson et al. 2008) which ties the Valais ophiolites to the Hercynian cycle and not to the Tethys Ocean. As in the model (A) above it does not figure oceanic basement in the south of the Western Alps and the Pyrenees and it shows the start of closure of the Tethys Ocean at the end of the Early Cretaceous in the Eastern Alps. *Reproduced from Manatschal and Müntener 2009, Fig. 4, p. 8.*

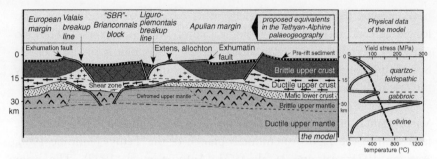

Figure 12.9 *Conceptual model for the symmetrical thinning of crust in the case of a magma-poor oceanic margin.* The figure shows the result of numerical modelling of the thinning of a crust composed of many layers of different rheological properties. Thinning is accommodated in this model by the exhumation of both middle crust and the mantle. The development of the model shows the existence of two families of faults concave towards the base, which will be used to exhume the mantle. The two groups of faults are symmetrical with respect to a block of brittle upper crust which they isolate and then separate the two previously adjacent margins. This model is designed to be applied to the opening of the Liguro-piemontais Ocean between the Apulia-African and European margins, and to the opening of the Atlantic between the Newfoundland and Iberian margins. The block of isolated continental crust is comparable and similar here to the 'SBR' block of the Alps. The two groups of faults are respectively compared, on the right to the trace of opening of the Liguro-piemontais Ocean and that on the left to the opening trace of the hypothetical Valais branch. One result of the model is to show that the two groups of faults are active simultaneously and not successively. Consequently, it is an open question as to whether the spreading of the Liguro-piemontais and Valais domains was contemporaneous so that both are of Middle Jurassic age contrary to traditional Alpine geological wisdom. *Simplified from Lavier and Manatschal 2006.*

argument was based on the observation that no sediments older than Early Cretaceous (*Bundnerschiefer*; Fig. 7.3) are associated with the Valais ophiolites while sediments dated to Middle or Late Jurassic rest in stratigraphic contact on the Liguro–piemontais ophiolites (Chapter 11). The Atlantic and Liguro–piemontais segments of the Tethys were thus said to have shared a contemporaneous evolution beginning in the Middle Jurassic. In contrast, the Valais branch and Bay of Biscay and Newfoundland–Iberia rift (Fig. 12.2) would have commenced spreading simultaneously but, much later, at the start of the Cretaceous only.

A different view results from modelling exercises founded on a realistic rheological model of the lithosphere (Manatschal et al. 2004–2006). These results suggest that continental breakup could have been simultaneous, and therefore of Middle Jurassic age, in the Valais and Liguro–piemontais branches of the Tethys and symmetrical with respect to the

'SBR–Brianconnais' block. According to the modelling results, Biscay–Pyrenean extension would have been limited to the corresponding part of the Atlantic and its margins. Valais 'spreading' would then be quite separate from that of the Bay of Biscay. The pattern of opening of the Bay of Biscay, which is dated to the Early Cretaceous, has therefore been erroneously linked by Alpine authors to that of the Valais through the extensional fault system of Cretaceous age in the Pyrenees and Sub-alpine zones (Figs. 7.10 and 12.8).

New field observations by (Masson 2002; Masson et al. 2008) have lead to a re-interpretation of the ideas progressively formulated since 1950 (Fig. 7.5A). The classical views have been opened to question by new radiometric dating of gabbros from the Versoyen ophiolites: zircons from the gabbros are of Early Carboniferous (Visean) age and not Early Cretaceous. Revision of the geological map around the ophiolites has shown that they are overlain by a sedimentary succession of Triassic to Jurassic age (Fig. 7.5B). These two data points show that the Versoyen ophiolites belong to the Hercynian basement and they were not derived from the Tethys Ocean of either Jurassic or Cretaceous (Valais) age. During later Alpine orogenesis, the assemblage comprising the ophiolites and their Triassic to Jurassic sedimentary cover was partly eroded to provide the clastic sediments at the top of the Valais flysch to then be thrust over the Valais basin so concluding the history of this basin. The proposed new interpretations concern the Valais domain situated at the boundary of the Western and Central Alps; it is not known if they can be extended to the Eastern Alps. The new results raise very important questions that will require a full debate on a potential controversy that lies at the heart of Alpine geology. As a result, these questions are left open in this book.

7. SEGMENTATION OF THE OCEANIC LITHOSPHERE AND ADJOINING CONTINENTAL MARGINS

The Alpine Tethys, like the Central Atlantic (Fig. 12.2), furnishes beautiful examples of segmentation of oceanic structures inherited from the start of their history.

The NE–SW array of palaeofaults, belonging to the Cevennes-Belledonne trend in the Western Alps, represents extensional faults dating to the Early Jurassic (Liassic) rift phase. The NW–SE faults of the Argentera trend lie parallel to the main extension direction of the rifting and cut the principal tilted blocks approximately perpendicular to their strike. Active as strike-slip faults, they divide the rift along the strike into segments whose

syn-rift evolution differs from segment to segment. As shown above, the
faults are parallel to the small circles defining the rotation of Africa relative
to Europe in the Jurassic (Figs. 6.16 and 12.8A). These faults are thus
interpreted as classical transfer or transform faults, which offset continental
margins.

The option chosen here (Fig. 12.8A and C) gives priority to the
hypothesis that transfer faults divide the Tethyan continental margin into
separate compartments differing in structure.

The continental 'SBR' block was cut into several compartments each
with a slightly different history from its neighbour (Chapter 7, Section 4 and
Fig. 12.10). The block was approximately bounded to the SW by the
Pelvoux–Argentera transfer zone (Figs. 6.16, 6.17, 12.8 and 12.10).

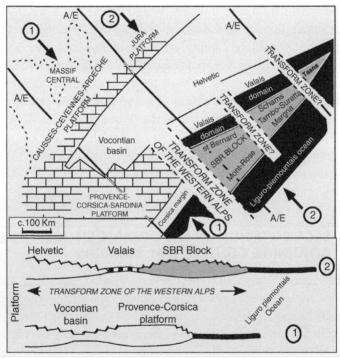

Figure 12.10 *Sketch of Alpine palaeogeographic domains in the Early Cretaceous.*
Oceanic crust, dark grey. Saint Bernard–Monte Rosa block, light grey. The fact that
reconstructed sections 1 and 2 cannot be superposed leads to the hypothesis of a
transform fault zone in the future Western Alps. *Modified from M. Lemoine et al. 2000,
Fig. 13.3, p. 146*

The southern limit of the Pelvoux and Argentera line of External Crystalline Massifs separates two segments of the rift in which the syn-rift structures and also Alpine structures are different (Figs. 6.16 and 6.17). It is interpreted as a typical transfer fault. In particular, this line corresponds to the southwestward termination of the 'SBR' block. Although clearly expressed, this transfer has less importance than the transfer which separates the Central and Western Alps and marks the disappearance of the Valais domain to the SW. However, it provides a comparison between structures of Jurassic age in the Central Alps with those along a NW–SE (Argentera) transect from the Cevennes to Provence and Corsica (Fig. 12.10, sections 1 and 2). NW–SE displacement during the Mesozoic was constrained by the orientation and position of the transfer faults. During Jurassic and Early Cretaceous times, the SBR block, which comprises collectively the Sub-brianconnais, Brianconnais and Piemontais palaeogeographic domains, moved progressively southeastward away from the European craton towards the Apulia margin. This movement reflects composite rifting of Early and Middle Jurassic age (Helvetic) followed by the Late Jurassic–Early Cretaceous (Valais) rift phase.

These observations show that the boundaries of the palaeogeographic domains from which the Alpine nappes originated were defined by a network of faults forming a grid-like pattern (Fig. 12.8A). One of the directions of the grid parallels that of the syn-rift extensional faults while the second parallels the transfer faults and their prolongation onto newly formed oceanic crust. Application of this geometric model leads to the possibility that the Grisons segment of the Jurassic Tethys may correspond to a major transtensional or strike-slip zone linking the Neotethys to the east to the Ligurian Tethys to the west as noted above (Fig. 12.8A).

Similarly, the transfer fault between the future Central and Eastern Alps limits the opening of the Valais basin towards the SW. The reality of oceanic spreading in the Valais is, however, questionable as noted above and earlier in this book. Further to the SW, the zone of rifting of Late Jurassic–Early Cretaceous age which affected the Sub-alpine domain, Provence and the North Pyrenean zone certainly did not evolve to spreading.

There is good evidence for faults cutting the palaeo-oceanic crust exemplified by preservation of small fault surfaces in the Chabrieres massif (Fig. 11.17, Haute Ubaye valley). Additional evidence from Queyras and Corsica includes breccias composed of abundant large ophiolite fragments that are interpreted as fault scarp breccias (Fig. 11.10). However, given the small size and distribution of ophiolite bodies in Alpine structures, it remains unknown whether these faults are associated with the ridge axis or oceanic transform faults, or indeed any another type of fault. To reconstruct the fabric of the spreading ocean, reliance has to be placed on the trends of structures preserved on the rifted margins using the concept that the oceanic transform faults lie in principle on the continuation of rift transfer faults.

FURTHER READINGS

On reconstitutions of past oceans and continental margins: Schlee and Klitgord (1988); Rosenbaum et al. (2002); Bernoulli and Jenkins (2009).

PART THREE

From the Tethys to the Alpine Fold Belt

The history of the Alps can rightly be said to have begun during Cretaceous times when the motion of Africa relative to Europe changed from divergence to convergence. This marked the start of the closure of the Tethys Ocean.

Due to the progressive convergence between the two margins of Tethys, the European plate began to subduct beneath the Apulia-african plate accompanied by sub-horizontal shear at the scale of the lithosphere. In this subduction system, Europe comprised the lower plate and Apulia-african the upper plate (Chapter 14).

From the start of shortening of the Ligurian Tethys, an 'oceanic' accretionary wedge formed during subduction of the Tethyan oceanic lithosphere beneath the Apulia-african margin. Later, following complete subduction of the oceanic lithosphere, a collisional wedge progressively incorporated thrust sheets derived from the European continental plate (Fig. 14.2). A 'lid' composed of Austro-alpine nappes, originating from the Apulia-african continental margin, then the upper plate, was emplaced above the latter wedge.

During these events, which took place over more than 60 Ma, two major phases of metamorphism took place successively. Firstly, high-

267

pressure–low-temperature metamorphism led to the formation of blue schists or eclogites or even coesite-bearing white schists (Fig. 2.10, P–T path 3). Later, medium pressure and increased temperatures due to thickening of the continental plate by accretion of the wedge led to green schist metamorphism (Fig. 2.10, P–T path 2).

As the European and African continents continued convergence, the African continent pushed the Apulia microcontinent before it, resulting in thickening of the collisional prism and growth of major topographic relief or, in other words, the formation of the Alpine mountain belt as clearly seen today. However, both relief and altitude evolved constantly. The dynamic topography and its evolution reflects the equilibrium between the rate of uplift, estimated at a few millimetres per year, and the rate of opposing effects which have tended to lower the relief as uplift continues. These opposing processes, active today, combine superficial erosion of different parts of the fold belt and crustal thinning resulting from the activity of late tectonic extensional faults (Figs. 15.6 and 15.7) some of which are seismically active (Chapter 15).

CHAPTER THIRTEEN

Birth of the Western and Central Alps: Structural Inversion and the Onset of Orogenesis

Contents

Summary

The onset of Alpine orogenesis resulted from the reversal or 'inversion' of the deformation field, which changed completely to one of shortening during the Late Cretaceous. Extensional structures inherited from the Tethyan rift phases as well as those formed at the start of the compression were, for the most part, destroyed in the internal zones because of shortening and metamorphism. In contrast, rift phase structures are well preserved in the external part of the Alpine fold belt so that the expression of the onset of Alpine shortening can be readily understood.

The Western Alps, from Rift to Passive Margin to Orogenic Belt, Volume 14
ISSN 0928-2025, DOI 10.1016/S0928-2025(11)14013-4
© 2011 Elsevier B.V.
All rights reserved.

In the Sub-alpine domain, the fold patterns were complicated due to the change with time of the principal shortening direction from the initial N–S phase dominant in Provence (called Pyrenean–Provence) in the south to the E–W direction dominant in the north (Alpine phases *sensu stricto*).

From the start of the orogeny, large sub-horizontal displacements along regional decollement thrust surfaces accommodated Alpine shortening. Certain of these surfaces were created during the Tethys rift phase and bounded the main palaeogeographic domains of the passive margin. These same surfaces were reactivated during the Alpine cycle as major reverse faults. This phenomenon accounts for the long-established and key observation that the Alpine palaeogeographic domains coincide with main structural units and zones.

1. THE CONCEPT OF STRUCTURAL INVERSION

In common with all passive margins, the structural fabric of the Tethyan passive continental margin was largely acquired during the rift phase responsible for the stretching and rupture of the continental lithosphere. In contrast, the construction of the Alpine fold belt resulted from crustal thickening due to shortening of the original passive margin (Chapter 2). In addition, the rift structures controlling the sedimentary domains later involved in Alpine folding from its initiation correspond exactly to those of Alpine structural inversion. There is thus a clear staging point that corresponds to a reversal of the field of deformation affecting the lithosphere; initially, extension and associated syn- and post-rift subsidence created accommodation space infilled in part by sediments, subsequently shortening and crustal thickening characterize the start of Alpine orogenesis.

As a generality, the formation of the Alpine fold belt can be considered to result from the closure of the Liguro-piemontais Ocean and the inversion of its margins in response to the displacement and convergence of the Apulia-african continent with respect to Europe. In conventional usage, the term 'inversion' describes modestly deformed short-wavelength structures at most tens of kilometres in size, formed by the reversal of throw along normal faults bounding half-grabens. Such structures are normally of interest to petroleum exploration as for example in the southern North Sea or SE Asia. However, the application of the concept of inversion to folded belts facilitates an understanding of how structures formed during extension might have influenced early structures formed by compression.

Alpine deformation occurred sequentially from the internal zones (E and SE of the fold belt), and then outward to the external zones (W, NW and N of the fold belt). Structures resulting from inversion are therefore best seen in the external zones where relatively modest Alpine shortening has allowed preservation of Tethyan rift structures. In the internal zones by contrast, nappe units were stacked to form the orogenic prism. First to be deformed, they record successive phases of tectonism and were subject to the most intense deformation in a mainly ductile regime as shown by their metamorphic assemblages. Although some structures inherited from rifting are preserved in the internal zones, there are few field observations that allow reconstruction of the onset of inversion.

2. INCREASING INTENSITY OF ALPINE DEFORMATION FROM W TO E IN THE WESTERN ALPS

In the most external part of the Western Alps, normal faults and the adjacent tilted blocks have been preserved nearly intact due to minimal deformation during orogenic compression. This is notably the case along the eastern border of the Massif Central, especially in the Ardèche (Les Vans: Fig. 5.4, and Uzer: Fig. 6.20). Towards the east, a transect parallel to the latitude of the city of Valence provides a good example of inversion in the classical sense (Figs. 13.1 and 13.2). Here, part of the

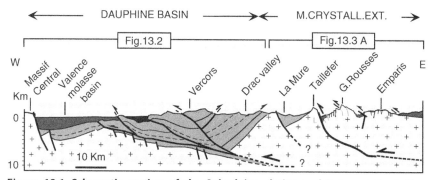

Figure 13.1 *Schematic section of the Sub-alpine chains and External Crystalline Massifs at the latitude of the Vercors.* Dauphine section modified after Mascle et al. 1996, Fig. 13, p. 15. External Crystalline Massifs section simplified after Chevalier 2002.

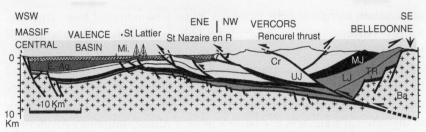

Figure 13.2 *Schematic section of the Vercors (Northern Sub-alpine zone).* The concept of reactivation, due to Alpine deformation, of major listric normal faults inherited from the Tethyan rift phase allows understanding of how the Vercors, previously located in a subsiding basin, is now in an elevated position. The present-day high elevation of the basement of the External Crystalline Massifs can also be similarly understood. Note that the small extensional faults that affect the basement beneath the Dauphine basin have preserved their original geometry beneath the regional detachment surface located in Triassic evaporites. Ba: Hercynian basement; Cr: Cretaceous; E.-Aq: Eocene to Aquitanian; LJ: Lower Jurassic (Liassic); Mi: Marine Miocene molasse; Mj: Middle Jurassic (Dogger); TR: Trias layers including evaporites; UJ: Upper Jurassic (Malm). *Modified after Mascle et al. 1996, Fig. 13, p. 15.*

Dauphine basin located below the Vercors (including the Vercors massif) was actively subsiding during the Mesozoic. Due to the effect of the combined throws on a network of Alpine inversion faults, the Vercors is now found in an elevated position, notably with respect to the Valence basin which remained in subsidence. In this area, extensional faults of Mesozoic age of minor throw affecting the basement are effectively preserved as shown from seismic profiles. Alpine shortening in the basement is apparently modest, at least at depths accessible to direct or indirect observation. In the cover, Alpine shortening has been facilitated by detachments in incompetent strata.

The Valence transect also shows the increase in Alpine deformation from west to east from the undeformed Massif Central via the non-inverted molasse basin beneath the Rhone Valley to the inverted Vercors (above). Further east on the Valence transect (Fig. 13.1), the Belledonne massif has been elevated by thrusting along a deep crustal detachment whose role and throw increases from west to east. The increasing intensity of Alpine deformation shown successively by the Bourg d'Oisans inversion and the internal zones is a simple way of summarizing the more detailed treatment below.

The Early Jurassic sediments of the 'syncline' of Bourg d'Oisans, located between the Taillefer and Grandes Rousses crystalline massifs (Figs. 13.1 and 13.3A and B) represent the wedge-shaped infill of a half-graben. Inversion of this half-graben has been achieved by partial expulsion of its sedimentary infill ('wedging'), with almost vertical ductile stretching of the sediments, towards the crest of the fault block.

Further to the east again, only normal faults of minor throw are preserved and the most obvious faults are the major overthrusts and minor thrusts such as the Meije or Combeynot (Figs. 6.2D and 13.3).

The later stage of the effects of 'wedging' was complete or almost complete expulsion of the graben infill inherited from the rift phase along detachment surfaces leading to the formation of overthrust nappes. These are characteristic of the internal zones (see below) or the Helvetic nappes to the north in the Central Alps (Fig. 13.4). However, this phase is no longer inversion but Alpine tectonism in the generally understood sense.

 ## 3. MULTIPHASE INVERSION IN THE SUB-ALPINE DOMAIN

In the Sub-alpine domain, the initial development of Alpine structures was partly a response to the inversion of several older extensional basins of different ages. These are respectively late Hercynian (age: Permo-Carboniferous), Tethyan (age: mainly Early Jurassic – Chapter 6 – then Early Cretaceous – Chapter 7), and Oligocene age, the latter representing the system of rifts of this age that affected Western Europe. Furthermore, the Sub-alpine domain was located in both the Alpine and Pyrenean forelands (Chapter 7). The main direction of Late Cretaceous shortening in the Pelvoux and Diois areas and then Pyrenean shortening, of end Cretaceous to Eocene age in Provence (*ibid*), was oriented principally N–S. In contrast, Cenozoic Alpine shortening was oriented E–W at the north of the Pelvoux Massif and ESE–WNW in the south, resulting in a relatively complex system of superimposed folds (Fig. 13.5). Where the folds intersect, particularly in the Diois area south of the Vercors, sections oriented E–W (Figs. 13.1 and 13.6) and N–S (Fig. 13.7) exhibit the characteristic profiles of inversion structures.

Multiphase inversion can also be illustrated by the evolution with time of those faults that separate the Manosque basin from the Valensole plateau in the southern Sub-alpine zone (Fig. 13.8).

The simple picture of an east–west gradient of compressive deformation from the internal to the external zones of the Western Alps is valid for the northern Sub-alpine chains where Alpine shortening is best

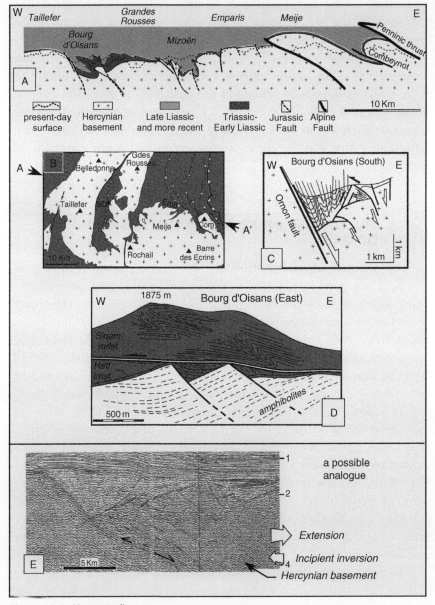

Figure 13.3 *(Continued)*

expressed by E–W verging structures (Fig. 13.1). However, this concept of an E–W gradient is inappropriate for Provence which formed part of the Pyrenean foreland where the folds verge from south to north (e.g. Fig. 13.9).

4. REACTIVATION OF EXTENSIONAL FAULTS DURING INVERSION

The susceptibility to inversion of extensional faults inherited from Tethyan rifting depends on the rate of Alpine shortening as well as their size and orientation with respect to the main shortening direction.

For example, many extensional faults situated in the most external part of the Sub-alpine chain, such as the Ardèche border of the Massif Central, have almost escaped inversion because of their small size and the modest Alpine shortening in this area (Fig. 6.10).

In other situations, the reactivation was modest so that the initial geometry of the extensional faults and micro-faults were preserved; examples can be found in the Sub-alpine domain (Fig. 13.2), in the hanging wall of the Bourg d'Oisans half-graben (Figs. 6.2A and B and

Figure 13.3 **The tilted block and half-graben of Bourg d'Oisans. (A) Actual cross section**. *Courtesy of Thierry Dumont, unpublished 2006.* **(B) Location map.** A–A' is the trace of section A. BO: Bourg d'Oisans Village; Com: Combeynot. **(C) Detailed schematic section** across the Bourg d'Oisans 'syncline'. Alpine cleavage with vertical stretching lineation is well developed here. Alpine inversion faults cut across earlier structures, notably Tethyan extensional faults. *Simplified after Chevalier, unpublished 2002.* **(D) The Rochers d'Armentiers section above the village of Bourg d'Oisans**. The decollement surface close to the upper surface of the basement separates two types of structures: above, Alpine deformation is accommodated by tight folds in the alternating marls and limestones of Sinemurian age; below, the effects of Alpine deformation are not visually obvious, though small Tethyan extensional faults have retained their geometry as at Alpe d' Huez (Fig. 6.7). *Simplified from Gillchrist et al. 1987, Fig. 12, p. 16.* **(E) Possible present-day analogue of the structures across the Bourg d'Oisans 'syncline'.** Movement along the bounding fault of the half-graben to the left has been partly reversed and is accompanied by reverse faulting which has thrust the graben infill to the right. The reverse movements on these faults have elevated the breakup unconformity (1.6 s TWT on the right of the section) above regional producing the characteristic inversion antiform. *Seismic data courtesy of TOTAL 2010.*

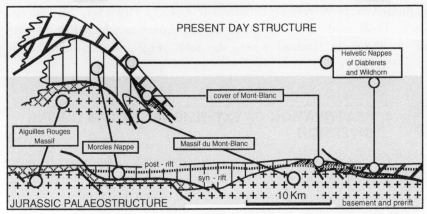

Figure 13.4 *The External Crystalline Massifs (Mont Blanc and Aiguilles-Rouges) and the Morcles (Helvetic) nappe: extreme inversion.* During Early and Middle Jurassic times, the Aiguilles-Rouges massif and the internal part of Mont Blanc formed the raised edge of tilted basement blocks, which were not covered by sediments until the beginning of the post-rift phase. The syn-rift sediments were deposited in the half-graben separating the two structural culminations. These sediments were expelled from the half-graben during Alpine deformation along decollement surfaces that cut the sedimentary succession. These sediments are now found in the Morcles nappe situated at the base of the stack of Helvetic nappes. *Modified from Lemoine et al. 2000, Fig. 14.2, p. 153.*

13.3C) or within the overthrust nappes of the internal zones. In this case, the main shortening has been accommodated along minor decollement surfaces or the major ones discussed below (Section 5.2). The sedimentary section or volume between decollement surfaces has therefore often been protected from pronounced Alpine shortening. In this way, the common preservation of small normal faults displacing the basement and overlying Triassic sandstones can be easily understood because the main regional detachment lies stratigraphically higher in Triassic evaporites (Figs. 13.1 and 13.6).

4.1 Easy reactivation and simple geometry: main shortening near perpendicular to the strike extensional faults

When the shortening direction is near parallel to the plane of a pre-existing fault, reversal of throw takes place easily whether the dip of the fault is steep or flat. The Maritime Alps provide many examples of steeply dipping Tethyan normal faults that later became Alpine strike-slip faults with a

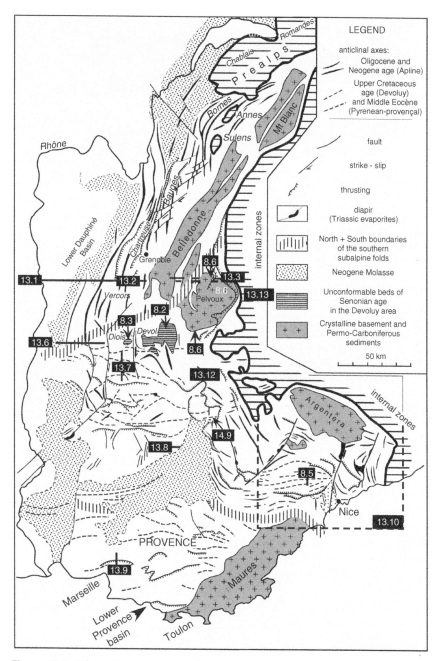

Figure 13.5 *Schematic structure of the Sub-alpine chains (external Western Alps).* The other figures are located by the numbers of 13-1 to 13-10. *Redrawn and complemented after Lemoine 1972 and Lemoine et al. 2000, Fig. 14.8, p. 165.*

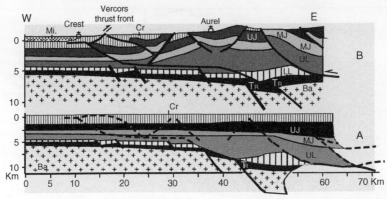

Figure 13.6 *Example of inversion in the 'Terres Noires' basin, south of the Vercors.* (A) Pre-inversion. (B) Post-inversion. Location: Fig. 13.5. Present-day section parallel to the Drome valley at the Southern foot of the Vercors plateau and end Cretaceous reconstructed section. This E–W section illustrates the control on inversion structures of lateral changes in lithofacies and bed thicknesses. The thickness of the Mesozoic section is about 8 km in the axis of the Dauphine basin in the east, but less than 4 km beneath the Rhone Valley. The main decollement surface rises from east to west level by level from Triassic evaporites up to Early Cretaceous marls to die out, probably in a triangle structure. The shortening is accommodated by several backward-verging reverse faults. The anticline of the Aurel ramp is located vertically above the point where the decollement surface climbs from Triassic evaporites to Late Liassic shales and is coincident with a rapid variation in the thickness of the Liassic shales. The structure allows understanding of the existence of broad outcrops of Late Jurassic (UJ) Terres Noires (black shales) eroded as deep as the Late Bathonian next to the village of Aurel as well as the existence of repeats in the Liassic recognized from deep exploration wells. Ba: Hercynian basement; Cr: Cretaceous strata including Early Cretaceous marlstones and middle Cretaceous 'Urgonian' platform carbonates; LL: Early Liassic limestone; UL: Late Liassic Black shales; Mi: Miocene molasse; MJ: middle Jurassic beds: Tr: Triassic strata including evaporites; UL: Late Jurassic strata, including Terres Noires (Black Shales) and Tithonian pelagic limestone. *Simplified after Roure and Coletta 1996, Fig. 12, p. 199.*

small oblique component (Fig. 13.10). Note that the reactivation of previous extensional faults is not inversion in the strict sense.

In a comparable way, the minor dip of the listric faults bounding tilted blocks on the distal part of the margin facilitated inversion during the start of orogenesis. This notion is applicable to the interpretation of the evolution of the Ornon listric fault during inversion (Figs. 6.2 and 13.3). The Austro-alpine nappes of Graubünden also provide another good example (Fig. 13.11).

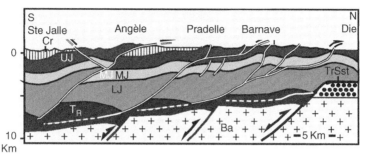

Figure 13.7 *Normal faults, tilted blocks and Late Cretaceous inversion in the Jurassic of the Diois (Western Alps).* Late Cretaceous inversion of structures inherited from Tethyan rifting has led to the development of two decollement surfaces located respectively in Triassic evaporites and Oxfordian black shales. Beneath the surface of the lower decollement, faults, which displace the basement, have been only partly inverted and again exhibit normal fault geometry. Within the ground between the two decollement surfaces, extensional faults and parts of tilted blocks are preserved – see for example, the offset in the middle Jurassic surface beneath Pradelle and Barnave. E–W fold axes assigned to Late Cretaceous shortening interfere in this area with N–S folds due to Alpine Cenozoic folding: compare Figs. 13.1 and 13.5. Die is the main small town of the Diois area. Ba: Hercynian basement; Cr: Cretaceous strata; LJ: Liassic limestone and black shale; MJ: middle Jurassic beds; Tr: Triassic evaporites; TrSst: Triassic sandstone on the Vercors basement high. Location: Fig. 13.5. *Simplified from Roure et al. 1994.*

4.2 Complex reactivation: main shortening oblique to the strike of extensional faults

Standard observations of listric faults imply that their fault surfaces become horizontal with increasing depth (Fig. 10.3): they are therefore likely to be reactivated probably without particular complication during Alpine deformation. However, the situation is more complex along the higher, up-dip parts of listric faults where the fault plane becomes vertical towards the surface. A good insight is again provided by the Bourg d'Oisans half-graben. The major Ornon listric fault (Figs. 6.2A and 13.3) has been straightened with respect to its initial position by Alpine deformation. The fault has had a buttressing effect forcing extrusion of the half-graben infill towards the high. The buttressing effect can also lead to back-thrusts (Fig. 13.9, Sainte Victoire). While effective in producing structural elevation, it, however, only results in minor, local shortening.

In the case of the crests of the tilted blocks, the accommodation required for shortening is a well-known product of a shallow dipping footwall cut off

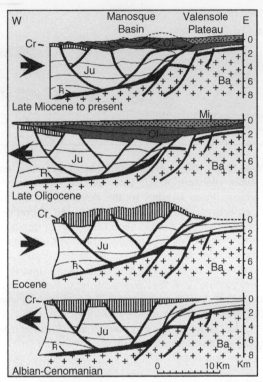

Figure 13.8 *Reconstruction of multiphase structural inversion along the Durance fault corridor.* The Durance fault corridor was formed during two phases of extension respectively associated with the Tethyan rift phase and that of the Early Cretaceous. These faults separated the subsiding Manosque block to the west from the much less subsident Valensole block to the east. During the Pyrenean deformation phase (end Cretaceous–Eocene), the hanging wall was subject to partial inversion, due to transpression along the fault corridor, followed by erosion of the uplifted area. Rifting of Oligocene age differentiated the Manosque basin and sediments of this age partly seal the Durance faults. Alpine deformation from Late Miocene time resulted in a new phase of inversion of the Manosque basin. The Oligocene infill outlines an antiform partly displaced towards the east. The crestal elevation of the antiform, which is higher than the Miocene of the Valensole, demonstrates the reality of Alpine inversion. The geometry of the antiform is partly controlled by conjugate inversion faults that root in Oligocene salt beds. Ba: Hercynian basement; Cr: Cretaceous strata; Ju: Jurassic section; Mi: Miocene molasse; Ol: Oligocene molasse; TR: Triassic strata including evaporites. Location: Fig 13.5. *Simplified after Roure et al. 1992* and *Roure and Coletta 1996, Fig. 11b, p. 197.*

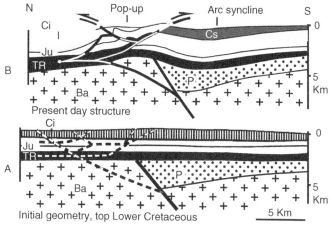

Figure 13.9 *The Sainte Victoire pop-up structure in Provence.* (A) Pre-inversion. (B) Post-inversion. The Sainte Victoire massif, located just east of Aix-en-Provence, belongs to the domain of structures mainly inverted during the Pyrenean deformation. The Triassic and Jurassic succession is more or less constant in thickness from N to S thereby excluding the presence of extensional faults of Jurassic age. Permian sediments exceeding 1000 m in thickness outcrop beneath the eastern end of the Arc syncline. In contrast, Triassic beds rest directly on the basement in the north. It is therefore probable that the Permian basin is defined by an extensional fault that passes beneath Sainte Victoire. The Triassic evaporites and Jurassic black shales give rise to two decollement surfaces throughout the Sub-alpine domain. The interpretation proposed here demonstrates the role of the Permian basin in the structure resulting from inversion. It shows that shortening in the cover and basement was of the same order of magnitude at the time of Pyrenean shortening; a low-angle basement short-cut fault and a triangle structure are proposed to account for this geometry. The latter allows the detachment to die out in Triassic evaporites and the two associated thrusts verging northward and southward (back-thrust) respectively form the pop-up. Ba: Hercynian basement; Ci: Early Cretaceous limestone; Cs: Late Cretaceous; Ju: Jurassic; P: Permian clastics; Tr: Triassic evaporites. *Simplified after Roure and Coletta 1996, Fig. 9b, p. 193.*

(Fig. 13.9, Sainte Victoire) that appears to provide the rupture surface allowing thrust displacement. It is possible that the slices of basement micaschists found at the base of the thrust of the Remollon dome occur in this situation (Fig. 13.12) as may the slices of the coal-bearing zone of the Clue de Verdache at the base of the Digne nappe (Geological map of France, 1:50,000, La Javie sheet), though this has not been proven. Although cited widely in the literature of inversion tectonics, footwall short-cut faults have only been identified rarely, and with difficulty, on the edge of extensional faults in the Sub-alpine domain.

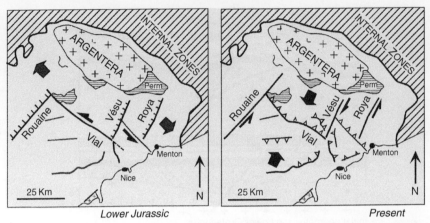

Lower Jurassic   Present

Figure 13.10 *Structural inversion of sub-vertical faults the example of the Maritime Alps.* The Rouaine, Vesubie (Vesu) and Roya faults separate compartments active in strike-slip during Neogene deformation. The Alpine shortening direction was NE–SW and parallel to the strike of faults. These faults were Early and Middle Jurassic rift phase normal faults whose main extension direction was also NW–SE (Fig. 6.21). The emplacement of the Mont Vial thrust was predetermined by an old fault belonging to the NW–SE transfer direction. The main direction of Alpine shortening was perpendicular to the transfer direction of the rift phase. As a result, a complex system of detachments developed in the Neogene leading to the Mont Vial line of thrusts. Location: Fig. 13.5. *Partly adapted from Dardeau et al. 1990.*

5. THE ROLE AND ATTITUDE OF DECOLLEMENT SURFACES DURING INVERSION

The formation of listric decollement faults allowed accommodation of extension during Tethyan rifting. During Alpine compression, many of these surfaces were rejuvenated initially by inversion and in a reverse sense from the beginning of deformation (Fig. 6.2A) to be later mobilized as overthrusts in the external as well as the internal zones (Figs. 13.3 and 13.4).

5.1 The role of decollement surfaces in the External Sub-alpine Chains

The geometry of the decollement surfaces in the sedimentary cover during the shortening phases results from ductility contrasts between beds. Thus, flats detach in ductile beds, i.e. Carboniferous shales, Triassic evaporites,

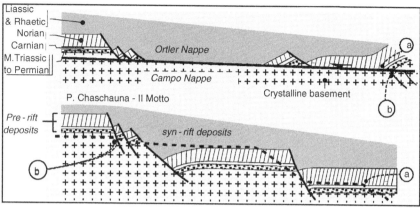

Figure 13.11 *Shear of the basement and tilted blocks: a hypothesis for the genesis of two Austro-alpine nappes in Graubünden (Eastern Alps).* On the Apulia-African margin, the normal fault planes are steep and thus generally not conducive for inversion as reverse faults or thrusts. Here it is supposed that a shear has cut structures inherited from extension to give rise to two superposed nappes. The lower Campo nappe is composed principally of basement while the overlying Ortler nappe consists throughout of the sedimentary cover. As a result, parts of basement blocks have been tectonically transported almost without deformation. Chaschauna and Il Motto: Figs. 10.5 and 10.6, respectively. No scale. Size of sketch: 10–20 km. *After Conti, Manatschal and Pfister 1994 and Lemoine et al. 2000, Fig. 14.3, p. 154.*

black shales of the Late Liassic, Callovian and Oxfordian ('Terres Noires') and Early Cretaceous shales (Fig. 13.2). Ramps cut across brittle beds such as the limestones of the Early Liassic, Tithonian–Berriasian and Early Cretaceous, i.e. the Urgonian platform carbonates as in the Vercors. The location of the thrust ramps may have been guided by rift phase normal faults that in turn determined the locus of rapid variations in the thickness of Mesozoic beds (Fig. 13.6).

On the Vercors transect in the Sub-alpine fold belt, the Alpine decollement surfaces link together at depth to the main thrust surface beneath the External Crystalline Massifs. Towards the top, the thrust envelope rises progressively level by level from the leading thrust edge at the foot of the External Crystalline Massifs as far as the limit of the Molasse basin. However, the decollement surfaces and possibly associated inverted faults appear to die out and are not obvious in the topography of the external domain in front of the main overthrust. The accommodation of the missing shortening is achieved by back-verging folds and by reverse faults of

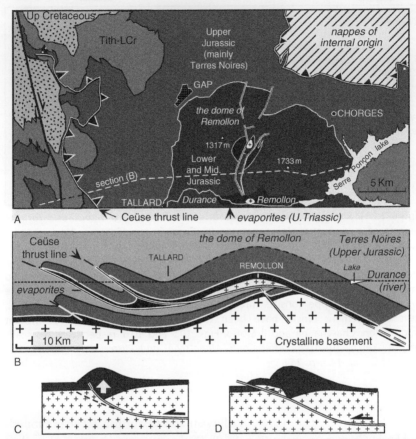

Figure 13.12 *Model for the behaviour of detachment faults at the onset of Alpine shortening: shallow dipping footwall cut off – interpretation of the crystalline basement outcrop of the dome of Remollon (Gap area, Western Alps).* The presence of the unusual small outcrops of Hercynian gneisses at the base of the Mesozoic succession in the core of the dome of Remollon is well known and the subject of a variety of explanations. The interpretation proposed here, but not proven, implies a footwall cut off at the raised edge of a basement tilted block. (A) Geological sketch map. *Simplified from Kerckhove, Gidon and Pairis 1989 and from Gidon 1971.* (B) Structural interpretation. *Simplified from Kerckhove and Gidon 1982 and from Kerckhove, Gidon and Pairis 1989, Fig. 4, p. 8.* (C) and (D) Text-book model for incipient deformation. Reactivation in a reverse sense during compression of an initial listric normal fault allows explanation of the straightening of such faults near the surface. The development of a thrust allows the basement thrusting in the footwall of the extensional fault inducing folds and thrusts in the cover. The appearance of footwall short-cuts can account for the presence of detached slivers of basement along the base of detachments within the cover. Location: Fig. 13.5. *Redrawn from Hayward and Gruham 1989, Fig. 19, p. 35.*

opposite vergence to the main structural assemblage (Fig. 13.1) possibly expressed as pop-ups (e.g. Fig. 13.9, Sainte Victoire) as well as by presence of triangle zones towards the front (Fig. 13.6, Aurel, and Fig. 13.9, Sainte Victoire). This may account for the higher elevation of the Vercors compared to the Valence basin (Fig. 13.1).

5.2 The role of decollement surfaces in the External Crystalline Massif area

With regard to the External Crystalline Massifs, geophysical studies from the ECORS Program have shown that they lie above a common main detachment surface that deepens progressively eastward. This observation recalls the geometric constraint required by the 10–12 km difference in regional elevation between the outcropping basement of the External Crystalline Massifs and that at depth below the adjacent Dauphine basin (Fig. 13.1). As noted earlier, the basement of the External Crystalline Massifs was cut, in common with the whole margin from the time of rifting, by a system of listric faults, all characterized by decreasing dip towards the horizontal with depth. This is the case for the Ornon fault, the master fault that defines the Bourg d'Oisans 'syncline' which is today a sub-aerially exposed decollement surface that cuts the crystalline basement (Figs. 6.2 and 13.3). These surfaces were therefore particularly susceptible to later inversion, especially along their lower sub-horizontal traces, from the moment Alpine compression began.

5.3 Reactivation of decollement surfaces and Tethyan listric faults in internal Alpine units on the Briancon transect

The Brianconnais domain consists of a stack of overthrust nappes each consisting of a Mesozoic succession that includes thick Triassic carbonates resting on Carboniferous strata (Fig. 13.13C). Alpine shortening in this area was propagated along a system of ramps and flats. The flats notably used the Late Werfenian (i.e. Early Triassic) evaporites and probably also the surface-separating unmetamorphosed Carboniferous strata representing remnants of the Hercynian foreland basin, from the underlying metamorphosed pre-Carboniferous basement unknown at outcrop in the Briancon area. The ramps cut across competent strata of Carboniferous and Triassic age. Detailed studies of the stacked nappes

have shown that the ramps nucleated on the traces of Tethyan extensional faults (Fig. 13.13A).

The generality of this system suggests a similar set of ramps and flats in
Brianconnais and Piemontais units (Fig. 13.14). The ramps would have
reused the Tethyan extensional faults that define the main palaeogeographic
units. In depth, they would have coalesced in ductile beds to form flats that
were used as thrust surfaces.

6. CONCLUSION: THE ROLE OF THE EARLIER FAULT FABRIC OF THE PASSIVE MARGIN DURING INVERSION

During the rift phase, the upper crust was subjected to brittle extension and the lower crust was subjected to ductile stretching that is only
rarely observed in the field (Fig. 10.3). Crustal thinning was achieved by a

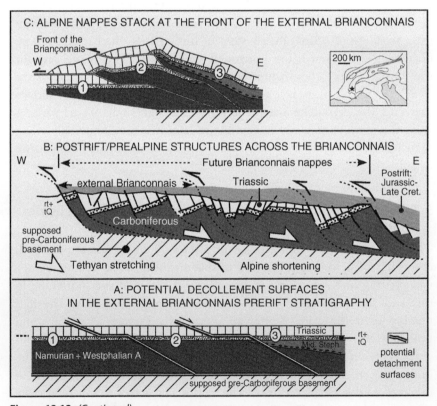

Figure 13.13 (Continued)

network of normal and strike-slip faults defining a system of tilted blocks (Chapter 6). At the start of the orogeny, the lower crust was shortened again in a ductile manner. Thus, while deep decollement surfaces were reused during inversion, those nearest to the upper–lower crust boundary were already preconditioned during rifting. During orogenesis, the new structures resulting from the initiation of shortening were determined by the lithological or mechanical properties of the stratigraphy and also by structures formed in the preceding stretching phase. During inversion, a major role was played by the development of detachment surfaces that appeared in the ductile zones of the cover and accommodated the shortening. Material was transported *en masse* along detachment surfaces whose geometry was influenced by discontinuities introduced by older normal faults.

These relationships explain the well-established observation that main Alpine structural units of the SBR block and Brianconnais are derived from precise palaeogeographic domains and thus carry the same name (Figs. 4.6, 13.4 and 13.14; Chapter 4, Table 4.1). These domains were delimited

---

Figure 13.13 *Inversion of detachment surfaces in the Brianconnais and initiation of thrusts. (A) Pre-rift Brianconnais stratigraphic succession and potential decollement/detachment surfaces. (B) Post-rift structures across the Brianconnais high, with tilted blocks, supposed listric faults and detachment surfaces. (C) The Alpine nappe pile at the front of the external Brianconnais, south of Briancon.* Semi-theoretical sections with no scale. The well-established structure and stratigraphy of the Brianconnais consists of a stack of small overthrust nappes composed of Carboniferous strata overlain by thick Triassic carbonates and in turn a reduced or condensed Jurassic and Cretaceous section. Each succession is specific to individual structural units. On the western Brianconnais front, the nappes exposed in the field often show Mesozoic on Mesozoic contacts only. This observation has given rise in the past to the idea that the whole structure can be explained by a general decollement for most of the Mesozoic succession that roots along favourable surfaces in particular the Scythian evaporites at the top of the Early Triassic: this notion implies that the underlying envelope of Early Triassic and Palaeozoic sandstone behaved as basement that deformed independently of the cover. This view is only partly correct. Detailed mapping has shown that each individual nappe, defined by upper and lower thrust surfaces and characterized by a specific Mesozoic succession, includes also an equally specific Carboniferous succession (a remnant of the Hercynian foreland basin) that is involved in the Alpine deformation. Reconstructions of pre-Alpine structures allow depiction of a system of ramps and flats separating each unit before the advent of thrusting. In schematic figure A, the trace of future Tethyan extensional faults that nucleated the emplacement of ramps in the competent beds has been shown. rt + tQ: clastics of Late Permian to Early Triassic age. *Sections A and C modified from D. Mercier, in J.C. Barfety et al. (1995). Section B completed and modified from Debelmas (1987).*

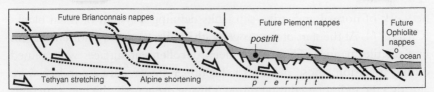

Figure 13.14 *Interpretation of the internal Alps on the Briancon transect in terms of inversion by a system of ramps and flats. Simplified after Debelmas 1987.*

originally by major normal faults or by arrays of major normal faults that separated distinct palaeogeographic domains, themselves defined by characteristic and specific stratigraphic successions. The normal faults in question formed lines of weakness in the brittle upper crust, susceptible to reactivation, whose location would induce inverted faults, overthrusts or nappes from the start of orogenesis.

FURTHER READINGS

Dezes et al. (2004); Dumont et al. (2008); Roure and Coletta (1996).

CHAPTER FOURTEEN

The Birth of the Western and Central Alps: Subduction, Obduction, Collision

Contents

The Western Alps, from Rift to Passive Margin to Orogenic Belt, Volume 14
ISSN 0928-2025, DOI 10.1016/S0928-2025(11)14014-6 © 2011 Elsevier B.V.
All rights reserved.

Summary

The Alpine orogeny was initiated by the northward convergence of Apulia–Africa with respect to Eurasia. The orogenesis followed a succession of characteristic stages, subduction, obduction, collision and hypercollision with back-thrusting that led to the construction of the orogenic wedge.

Only the Western and Central Alps are considered here, the latter are represented by the Matterhorn transect whose outcrop geology and deep structure are well known. The Pre-alpine nappes along this transect have escaped subduction and most of the metamorphism and thus provide precise chronological indicators lost in the internal zones due to their intense deformation.

The onset of subduction is indirectly dated by the Late Cretaceous age of early flysch deposition. However, the outcropping, oceanic (Zermatt, Antrona) and continental (Grand Saint Bernard, Monte Rosa, Sesia) units, were only subducted from the start of the end of the Late Cretaceous (towards 65 Ma for Sesia) and especially by the Eocene.

Flysch, then molasse, basins followed one another in time and space: early flysch of Late Cretaceous to Palaeocene age is linked to Valais subduction (Niesen flysch) and to Liguro-piemontais subduction (Gurnigel and Helminthoid flysch). It was followed in the successive flexural basins by the Brianconnais–Subbrianconnais flysch of Eocene age and the Ultrahelvetic and Helvetic flysch of Late Eocene to Oligocene age. Finally, from the Late Oligocene to the Miocene, molasse was deposited in the subsiding flexural foredeep trough located on the periphery of the fold belt.

The start of convergence at the end of the Late Cretaceous is poorly calibrated chronologically. The Eocene paroxysmal construction of the collisional wedge ended with the succeeding hypercollision in the Late Oligocene.

1. SUBDUCTION, OBDUCTION

The onset of Alpine subduction is supposed, but without consensus, to have resulted from rupture localized at the foot of the Apulia continental margin. The rupture could have been intra-oceanic and induced the onset of ocean–ocean subduction. From its onset, however, subduction has progressively consumed most of the oceanic lithosphere at depth. Slices detached from the upper part of the lithosphere were stacked to form an at-first typically oceanic accretionary prism. The process was not simple scraping at a crustal scale because the metamorphic history of several units shows that they were first entrained at depth in the mantle before subsequent elevation along the subduction zone, and later incorporation in the accretionary prism. Structures in the Queyras and Montviso mountains are inherited from the prism constructed as the Apulia–African margin became an active margin. The oceanic accretionary prism ultimately evolved into a collisional prism during the Tertiary. Above the subduction zone and in front of the prism, a subsiding, asymmetric and elongate, deep-water trench probably no wider than a few tens of kilometres formed the locus of deposition of the Cretaceous flysch, e.g. the Helminthoid flysch.

From the start of continent–continent collision, those parts of the oceanic lithosphere that had escaped subduction were thrust over the continental margin of the lower plate by obduction.

1.1 Lower plate, upper plate

The Tethyan oceanic lithosphere was part of the downgoing, European or lower plate and the upper plate was the Apulia–Africa plate. This relationship is shown by the fact that all the ophiolite nappes are overthrust above continental units situated north of the ophiolitic suture that separates the Monte Rosa (SBR basement) to the NW from the South-alpine massifs of Ivrea and Sesia-Lanzo (Apulia-African basement) to the SE.

1.2 Forced subduction

As shown earlier (Fig. 3.5), the width of the Liguro-piemontais Ocean was no wider than a few hundred kilometres since its spreading rate did not exceed 5–10 mm per year over a duration of less than 60–70 million years. The oceanic lithosphere was thus relatively young, thin and of low density due to the abundance of serpentinites (Chapter 11). In principle, this may imply forced subduction along a gently dipping subduction zone. This idea is compatible with the rarity, if not absence, of calcalkaline volcanism that might be linked to subduction. One exception may be the very small and sparse andesitic lava flows, probably 30–40 Ma in age, which rest on the internal edge of the Sesia massif. The interpretation of this volcanism, located on or near the Peri-adriatic fault array, is debatable and it certainly does not have the signature of subduction because it appears at a very advanced stage of the collision. A proposed explanation is reheating induced by 'slab break off' affecting only the Central Alps. However, the Alps correspond to a collisional fold belt with no remnant known of any island arc that might record subduction.

1.3 Slicing of ophiolite units during subduction and obduction

True Alpine ophiolites are derived mainly from Liguro-piemontais Ocean crust and mantle. Depending on locality, they are composed either of pillow basalts, gabbros or serpentinites with an ophicalcite cover as in the present-day Atlantic (Chapter 11). The thickness of the igneous succession comprising the ophiolite bodies represented in Alpine nappes is variable and determined by the path of the decollement surface along which the ophiolites were detached at the start of subduction or obduction whichever the case.

1.4 Role of slow-spreading ridge segmentation

Units derived from a 'magma–rich' segment (Figs. 11.18 and 11.19) are composed of a foundation of gabbros and relatively thick basalts; in this case, the detachment surface lies close to the base of the gabbros entraining a minor thickness of serpentinites within the slice. In the case of a 'magma-poor' segment extremity composed of rare to absent magmatic rocks (Fig. 11.18), field observations show that the detachment often occurs along the bed of OC2 sedimentary ophicalcites, i.e. at the base of the oceanic sediments (Fig. 11.7A and B). The detachment may also pass below rare and small volcanoes, which lie on OC2-type sedimentary ophicalcites, and have been separated from their original foundation. Elsewhere, the detachment cuts the topographic relief formed by submarine faults, giving rise to distinct serpentinite bodies with or without gabbroic intrusions or basalt cover. The relief of oceanic palaeofaults probably facilitated the break-up and dispersion of ophiolite bodies during the accretion of the oceanic wedge and then Alpine deformation. This model probably stands with few exceptions and accounts for the majority of the observations (Fig. 14.1).

In summary, the slow spreading of the Liguro-piemontais Ocean created oceanic lithosphere whose upper part was composed of serpentinites, 2–3 km in thickness, injected by small gabbros masses and covered locally by basalts (Figs. 11.15 and 11.18). This structure facilitated excision of slices or sheets of thin ophiolites carved from the upper part of the oceanic crust or lithosphere

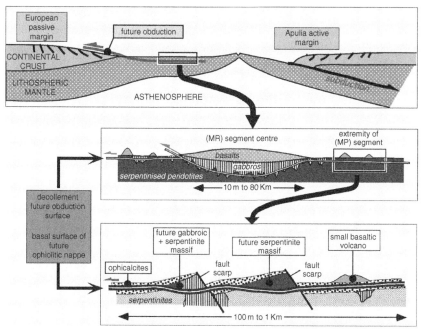

Figure 14.1 *Schematic sketch of the modes of slicing of the upper part of the oceanic lithosphere and shaping of ophiolite nappes. Modified from Lemoine et al. 2000, Fig. 14.4, p. 155.*

(Fig. 14.1). The result was firstly the construction of an oceanic accretionary wedge.

2. THE COLLISION

Subduction first consumed the Tethyan oceanic crust and then the lower part of the continental lithosphere. With the exception of the Piemont units that corresponded to the highly stretched distal part of the continental margin, the lower plate could not sink to great depths within the mantle because of its relatively low density. As a result, subduction was progressively blocked and collision followed subduction due to the continuing convergence between the European and Apulia–Africa plates. The major inverted faults and successive great thrusts were then developed in the continental crust and lithospheric mantle, thereby contributing to the construction of the collisional wedge.

The evolution from the oceanic accretionary wedge into a collisional wedge is supposed to have been gradational and continuous. It occurred during Eocene times. The associated metamorphism was high-pressure–low-temperature (HP–LT). The mantle pressure conditions were typical of eclogites formed by subduction and were followed by less severe conditions that produced blue schists.

2.1 The collisional wedge: a simplified scheme

The present-day Alpine orogenic wedge is largely a collisional wedge that incorporated the original oceanic accretionary wedge mainly constructed at the expense of the subducted lower plate. As in all collisional wedges, the growth of the accretionary oceanic wedge took place by the addition of slices in a basal frontal position corresponding to the same common mechanism of accretion. The nature of the accreted material, oceanic then continental, has in contrast changed with time. Low-angle reverse faults express the main deformation and collectively resulted in horizontal shear of the wedge. The major deformation was thus horizontal shortening associated with vertical thickening.

Different models and graphic representations can be proposed for the structure of the collisional wedge.

The simplified model presented here relies on the interpretation of the deep ECORS-CROP seismic profiles crossing the French–Italian Alps (Figs. 2.12, 2.14 and 2.16). These profiles show that the lithosphere and crust of the lower plate have been successively cut by oblique planes of rupture that have a ramp and flat geometry. Integration of results drawn from surface geology with the interpretation of the deep seismic data has shown that shears affect the lithosphere in the internal part of the fold belt but only the sedimentary cover in the external part.

2.1.1 Ramp and flat geometry

This geometric style results from the difference in rheological properties within the sedimentary succession at all scales from metres to kilometres. This geometry can be observed in its initial state in the structures resulting from inversion (Figs. 13.1 and 13.2). The flats correspond to decollements or detachments which affect ductile beds such as shales interbedded between limestones and sandstones, evaporites deposited between the Hercynian basement and younger cover, or the lower ductile crust. Ramps follow the inverted faults which cut obliquely across competent strata such as thick dolomite, limestone or sandstone beds or the brittle upper part of the continental crust. The slope of the ramps is generally inclined in the direction of the upper plate, a configuration which leads to the thrust rising towards the surface in the direction of the external zones. In consequence, the collisional wedge is successively composed of lithospheric, crustal and then sedimentary slices from the interior to the external part of the fold belt.

2.1.2 Sequencing of thrusts

Thrusts appear in an ordered manner during the construction of an orogenic wedge (Fig. 2.15). The first detachment leads to the transport of unit 1 out of the flexural basin included on the lower plate. The second detachment brings the superposed assemblage 1 plus 2 onto the external domain, and so on. It results in a normal sequence of thrusts in order 1, 2, 3, 4 and 5. When, exceptionally, this sequence is not followed, the thrust is termed *out of sequence*.

2.1.3 Asymmetry of the fold belt

Rocks first affected by thrusting are uplifted because they are thrust above those situated in the most external footwall. In consequence, the wedge reaches its maximum thickness towards the rear, at the border of the upper plate or beneath it from the vertical of the highest part of the fold belt (Figs. 2.15 and 2.16). The Moho deepens slowly from the exterior to the interior of the fold belt from 30 to 55 km (Figs. 2.14 and 2.16).

The architecture of the fold belt is added to by obduction of ophiolite nappes into the orogenic wedge, by overall thrusting of upper plate, Austro-alpine nappes above the wedge and by the deepening of flexural basins located on both sides of the Alpine edifice.

2.2 Flexural basins; flysch and molasse

When collision succeeded subduction, thickening of the lithosphere by stacking of thrust units resulted in loading of the plate edges creating flexural basins (Figs. 14.2–14.4). The foreland flexural basin then replaced the trough superposed on the subduction zone while another flexural basin appeared towards the rear of the arc (i.e. the Po plain in northern Italy: Fig. 14.2). The foreland flexural basin temporarily separated the unde-formed foreland from the mountain fold belt engendered by the collision. Each flexural basin has an asymmetric profile with the axis of maximum depth located at the foot of the deformation front. The flexural basin slopes gently upward towards the foreland, which is undeformed except for the formation of a low relief 'peripheral or flexural bulge', formed as a direct result of the elastic loading of the lithosphere. The early Massif Central may have been initiated as the 'peripheral or flexural bulge' in response to the load of the Alps. During collision, the coupled fold belt–flexural basin

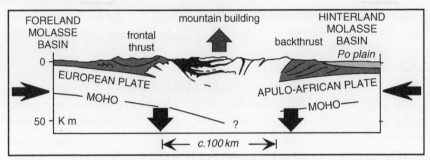

Figure 14.2 *The Central Alps, a doubly verging fold belt.* The double vergence is characterized by thrusts, outward to the NW and SE towards the rear of the fold belt, formed consecutively during hypercollision. The development of two flexural basins (in grey) results from flexural deformation of the two portions of lithosphere adjacent to the fold belt in response to loading caused by tectonic thickening during the collision. *Simplified after Rey et al. 1990, p. 30.*

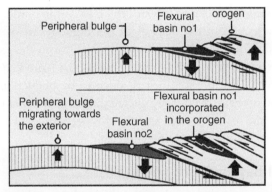

Figure 14.3 *Appearance, deformation and incorporation of successive flexural basins in the collisional wedge.* No vertical scale. A good Alpine example of flexural basin n°2 is given by the Oligocene flysch basin of the Grès d'Annot–Grès du Champsaur–Flysch des Aiguilles d'Arves–Grès de Taveyannaz. Flexural basin n°3 is that of Neogene molasses of the external zones (Fig. 13.5).

migrated from the interior to the exterior of the fold belt. The destiny of successive flexural basins ahead of the fold belt was therefore uplift, folding and incorporation within the collisional wedge while new flexural depression of the lithosphere created a new foreland basin external to its precursor.

Clastic sediments derived from topography undergoing active uplift were captured within the subsiding axes of foredeeps situated adjacent to the

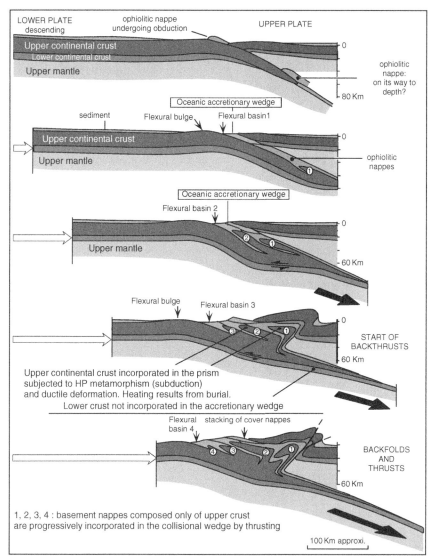

Figure 14.4 *A model for the collisional wedge with ductile deformation of the units of the wedge and detachment of the upper crust.* The scheme owes its inspiration to the sketch of the Matterhorn transect in the Central Alps (Fig. 14.7). It shows how the successive flexural basins formed through time. The model also shows the formation of units composed of upper continental crust in the wedge as well as their detachment by crustal delamination from the lower crust. The ductile deformation of the upper continental crust results from reheating due to burial and the radiogenic heat contribution from this upper crust. The model also takes into account the emplacement of nappes derived from the upper plate. (See colour plate 19) *Modified after Escher and Beaumont 1997 and Lemoine et al. 2000, Fig. 14.7, Table XI.*

mountain front. In the deeper parts of the foredeep, where the subsidence rate exceeded the sedimentation rate, gravity deposits, corresponding to different types of turbidites, were deposited to form a thick and regularly stratified succession. The lateral input of clastics then transported axially for long distances along the trench axis constructed deep sea fans. These turbidites are called flysch, a traditional old and local Swiss German name. In the Western Alps, the *Grès d'Annot* flysch basin is a typical example as is its lateral equivalent to the north, the *Grès du Champsaur* and the *Flysch des Aiguilles d'Arves* of mainly Oligocene age, which has been incorporated into the fold belt. The flexural basin 2 shown in Figs. 14.3 and 14.4 is in this situation.

In those parts where the amplitude of the flexure was lower, subsidence and sedimentation rates were at times equal. There, clastic sediments known as molasse were deposited in tidal, continental, fluvial and lacustrine environments as in the Swiss molasse basin and the Po plain. In the Western Alps, these are exemplified by the Neogene molasse which infills troughs in the Rhone Valley, Lower Dauphine and Valensole basins (Figs. 13.1, 13.2 and 13.8). Flexural basins 3 and 4 shown in Fig. 14.4 occur in this situation.

2.3 Hypercollision, 'Back-thrusts', Fan structures

Hypercollision took place during the very advanced stages of Alpine collision. All the thinned continental crust belonging to the passive margin had been incorporated into the orogenic wedge, which had by then reached its maximum size. From this moment, shears oriented towards the interior of the fold belt were added to thrusts towards the external parts of the fold belt, initiated since the start of the collision. The back-facing overfolds or back-thrusts mainly affected the interior of the fold belt in the direction of the Po plain basin. The high parts of the fold belt thus developed the overall fan structure characteristic of hypercollision. Nonetheless, the profile of the whole fold belt remained asymmetric (Figs. 2.12, 2.15 and 14.4). Contemporaneously, extensional tectonism, commonly nucleated on thrusts and back-thrusts, developed in the axis of the orogenic fan adding its effects to those of superficial erosion to limit and reduce the thickness of the collisional wedge.

2.4 Ophiolite nappes in the orogenic wedge

All the ophiolite nappes of the Alps and Corsica rest today on units linked to the European margin. However, these nappes do not share a common history of deformation or metamorphism. Several are composed of almost undeformed and unmetamorphosed ophiolites and were therefore emplaced early by obduction (Fig. 14.5: CH). Other units were metamorphosed into blue schists or eclogite-grade facies indicative of HP–LT implying fast burial within the oceanic accretionary wedge or subduction down to the mantle (Fig. 14.5). In addition, other units were only subjected to green schist facies metamorphism though some might be retrogressed due to HP–LT metamorphism. This implies relatively modest and/or slower burial during the early subduction stage as well as during the subsequent growth of the collisional wedge in front or below the Austro-alpine nappes, which formed the 'orogenic lid' (see Section 2.6).

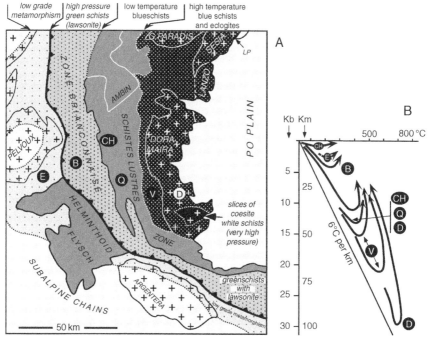

Figure 14.5 *Pressure–temperature paths of several metamorphic units in the Western Alps.* To the left, structural sketch with distribution of metamorphic facies. To the right, some P–T paths. Crosses: crystalline basement. LP: Peri-adriatic fault array; B: Brianconnais; CH: Chenaillet; D: Dora Maira; E: External zone around the Pelvoux; Q: Queyras; V: Viso. In an assemblage of thrust sheets or slices, either of oceanic (CH, V) or continental (B, D) origin, the P–T paths differ from nappe to nappe. Examples are the two structural units of the Chenaillet massif CH (Fig. 11.2) and also within the Montviso, V (Fig. 11.8) compared to the Dora Maira (D). *Modified from Lemoine et al. 2000, Fig. 14.12, p. 168.*

2.4.1 Unmetamorphosed ophiolite nappes emplaced by subduction

These include:

*   The Ybbsitz klippen (between Salzburg and Vienna, on the northern border of the Eastern Alps).
*   The very small ophiolite slices associated with the Cretaceous flysch of the upper Pre-alpine nappes (Col des Gets in the Chablais south of Geneva, Jaun pass, Iberg klippen in Central Switzerland).
*   The upper unit of the Chenaillet (Fig. 11.2; location: Fig. 11.1) and probably also the ophiolites of Balagne, Corsica (location: Fig. 11.1).

Ductile Alpine deformation is almost non-existent as is shown by the well-preserved pillow lavas of the Chenaillet (Fig. 11.4), Ybbsitz (Eastern Alps) and Balagne (Corsica). The low-grade prehnite–pumpellyite metamorphism of the upper unit of the Chenaillet was perhaps first developed in the oceanic crust. It is probable that the nappes obducted on the European margin occupied a more widespread position in the superstructure of the fold belt than today and so the present slices represent only the uneroded remnants.

It should be recalled that the Chenaillet ophiolites also include metamorphic rocks such as flaser gabbros (Chapter 11, Figs. 11.5 and 11.6). Such rocks do not result from Alpine metamorphism but from a phase of much earlier ductile stretching which accompanied Tethyan seafloor spreading.

2.4.2 Subducted ophiolite nappes with HP–LT metamorphism

The subduction-related nappes are far more important at outcrop than the obducted nappes. Good examples are the Glockner nappe (Hohe Tauern window, Eastern Alps), those found at Queyras (Figs. 11.16) and Monviso (Fig. 11.8) in the Western Alps, and Cape Corse (location: Fig. 11.1). Their depth of burial (Fig. 14.5) is estimated (i) at 100–130 km, from coesite-bearing white schists found in the Dora Maira massif, (ii) at 50–80 km, from eclogite-grade metamorphism, both being characteristic of subduction within the mantle and (iii) at 30–50 km from blue schist-grade rocks, implying burial in crustal conditions at the lower part of the accretionary wedge. The well-preserved HP–LT mineral assemblage indicates rapid, i.e. cold, exhumation of these nappes.

Ophiolites metamorphosed to green schist grade were buried to depths estimated at 10–20 km as shown by the Platta-Arosa (Eastern Switzerland) and Tsate (Central Switzerland) nappes (Fig. 2.6). Burial of such units could have been slow and over a long period.

2.5 Post-subduction exhumation of the nappes

The principal mechanism of subduction is burial in the asthenosphere of the continental and oceanic parts of the lower plate representing the downgoing slab. However, several scenarios are possible. For example, the proximal units of the continental margin, whose lithosphere and crust has been stretched least, are too low in density to be subducted to any great depth.

In contrast, units comprising the complete, approximately 100 km thick, oceanic lithosphere are dense and thick. As a result, those units were easily subducted to great depths to disappear within the asthenosphere.

Units exhumed following subduction were either derived from the stretched crust of the distal continental margin or resulted from detachment, or represented the slices of parts of the young, thus relatively less dense, oceanic lithosphere. These slices are relatively thin comprising between 0 and 1000–2000 m of ophiolites without unserpentinized peridotites, overlain by cover sediments and resting on a sole of serpentinites. The ophiolite slices were buried rapidly and deeply enough to be subjected to HP–LT or even UHP–LT (ultra-high-pressure–low-temperature) metamorphism (Fig. 14.5). However, they were also exhumed rapidly enough for their metamorphic paragenesis to be preserved as they were incorporated into the orogenic wedge. The low-temperature metamorphism can be explained by the low thermal conductivity of this rock suite and the relative brief time spent at maximum burial depth. Subsequent uplift was probably favoured by their relatively small size and low density compared to the environment at depth, unserpentinized peridotites having a density of 3.25–3.3 g cc^{-1}. A reconstruction of the P–T path allows assessment of the order of magnitude of the uplift rate of subducted units. The rate of exhumation of specific eclogite units of the Central and Eastern Alps has been estimated at more than 1 mm per year for a small number of tens of millions of years. The examples of Queyras and Monviso show that exhumation was rapid. Such a process uplifted eclogites formed in mantle conditions to crustal conditions with blue schists characteristic of the base of the accretionary wedge. Exhumation from 30 km depth to the surface undergone by units constituting the accretionary wedge was probably very slow, thereby explaining the development of retrograde green schists.

In summary, ophiolite nappes derived by slicing the upper part of the oceanic lithosphere solely comprise outcrops in the fold belt. These ophiolites have been subducted or obducted. Some of those subjected to subduction have been exhumed because their very low density prevented complete consumption. In contrast, the dense and complete parts of the oceanic lithosphere, not susceptible to slicing, were deeply buried and absorbed completely by subduction. In consequence, the ophiolites found in fold belts represent only a modest part of the surface of the oceanic lithosphere from which they were derived.

2.6 The 'orogenic lid' formed by the Austro-alpine nappes

The upper plate Austro-alpine nappes cover the wedge constructed on the foundations of the lower plate from which the ophiolite nappes originated. No less than 5 km thick but of considerable horizontal extent, the Austro-alpine nappes are composed of continental basement covered by sediments mainly of Mesozoic age. The nappes form the 'orogenic lid' of the wedge in the Eastern Alps where they are most widespread (Figs. 2.2 and 2.5). In the Central Alps, erosion has left only the small slice forming the Dent Blanche nappe. It consists essentially of two superposed units of continental basement (Fig. 2.7):

- At the base, the Arolla units are related to the Sesia zone. They comprise only upper crust with local remnants of cover sediments of Triassic to Liassic age.
- Above, the Valpelline units, composed of granulite-grade gneisses, are related to the lower crust of the Ivrea zone.

In the Western Alps, Austro-alpine nappes are in contrast absent at outcrop.

3. CHRONOLOGICAL CONSTRAINTS

3.1 Methods

Biostratigraphic dating: The stages in Alpine orogenesis are well dated from the age of the flysch deposited in superposed subduction-related trenches and in the flexural basins that surrounded the fold belt during collision, and as well as from the ages of the different molasse deposits. Dating of the first appearance of metamorphic minerals in the flysch and molasse provides precise and invaluable constraints on the age of metamorphism. Standard stratigraphic methods allow dating of unconformities resulting from successive episodes of folding and thrust emplacement.

   Radiometric dating: The most recent radiometric dates and, therefore, in principle the most reliable, provide Palaeogene ages, i.e. between

40 and 60 Ma for eclogite-grade metamorphism due to subduction using samples from the Monte Rosa, other ophiolite nappes in the Central Alps and the Monviso (Western Alps). Late Cretaceous ages were proposed earlier but are now considered unreliable.

3.2 The classical "Eo", "Meso" and "Neo-Alpine" events: an artificial, outmoded classification

Various authors have introduced three main periods to classify the stages of Alpine metamorphism identified from radiometric ages.

The Eo-alpine period is supposed to coincide with the Late Cretaceous. Its definition relies wholly on old radiometric dates of metamorphosed eclogites in the Central and Western Alps, allocated to material originating from oceanic crust or the continental margin. More recent and probably more reliable dating made on similar samples provides Eocene ages. The only clearly known signature of Eo-alpine deformation corresponds to the early flysch of Late Cretaceous age, deposited in subduction-related trenches.

However, the problem remains of the age of HP–LT metamorphism in the basement of the Sesia zone, a continental block located between the Penninic and Austro-alpine and called Ultra Penninic by various authors for this reason. Diverse methods have provided a wide range of dates between the Late Cretaceous and the Tertiary (100–60 Ma) for the age of cooling during post-subduction uplift. Recent results give more precise and consistent ages of around 65 Ma, close to the Cretaceous–Tertiary boundary. The newer dates suggest that the Late Cretaceous phase does not exist though more precise and reliable dates are needed to confirm this view.

The Meso-alpine period encompasses the Palaeocene, Eocene and Early Oligocene (60–30 or 35 Ma). It represents the paroxysm of deformation and metamorphism in the Brianconnais and Piemontais zones, and in the Liguro-piemontais ophiolites.

The Neo-alpine period of Late Oligocene and Neogene age is one of tectonism in the external zones and accompanying weak to very weak metamorphism (green schist to prehnite–pumpellyite facies). It represents in principle the period of hypercollision and back-thrusting in the internal domains.

The above division, despite being classical, has the disadvantage of suggesting two great breaks in the development of the Alpine orogenesis when the bulk of the observations indicate a continuum of deformation since the onset of inversion.

3.3 The main results

- In the internal zones, the different biostratigraphic and radiometric ages allow dating of subduction to the start of the Palaeocene (60–40 Ma) and construction of the collisional wedge to the Eocene and Oligocene (approx. 35 Ma; Figs. 14.6 and 14.7). For example, blue schists from the accretionary wedge in the Queyras are dated 60–54 Ma.
- In the external zones of the Central and northern Western Alps, the tectonism dates principally to the Late Oligocene and Neogene. In the middle part of the Sub-alpine chains (Devoluy area and surroundings), the folding began during early Late Cretaceous times (Cenomanian). However, in the southern Sub-alpine chains and Provence, which belong to the Pyrenean domain, the folding began from the end of the Cretaceous (Chapter 8).

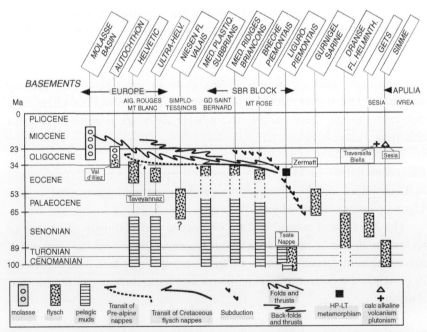

Figure 14.6 *Dating of the main sedimentary, tectonic and metamorphic events on the Matterhorn transect.* Modified from Lemoine et al. 2000, Fig. 14.5, p. 161.

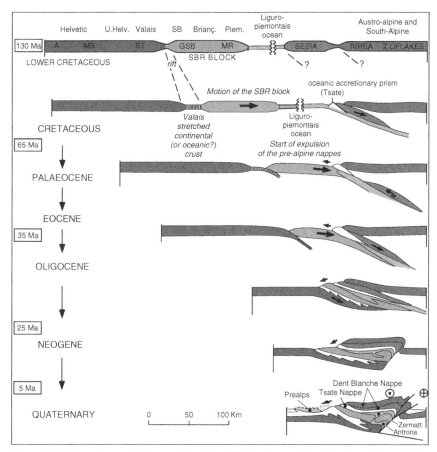

Figure 14.7 *An attempt at reconstruction of the stages of subduction, then Alpine collision along the Matterhorn transect, Central Alps.* This figure was drawn for publication in 1999. From this moment the existence of the Valais Ocean was considered as a reality. In the eventuality where the setting or even the existence of this ocean would be confirmed, the reconstructed sections for the Cretaceous, Palaeocene and Eocene would not be considerably different. However the existence of the Valais Ocean is now in doubt. The key point is that continent–ocean subduction at the Valais would become continent–continent subduction. N to NW to the left; S to SW to the right. Dates shown for the different stages are intentionally imprecise. Compare with Fig. 14.4. A: Aiguilles Rouges; GSB: Grand Saint Bernard (subdivisions; Fig. 2.7); MB: Mont Blanc; MR: Monte Rosa; SBR: Saint Bernard–Monte Rosa block; ST: future Simplon–Ticino nappes, derived from the European margin of the small Valais Ocean. *Modified from Lemoine et al. 2000, Table XI, Fig. 14.7.*

4. MODES OF OROGENESIS IN THE CENTRAL ALPS ALONG THE MATTERHORN (CERVIN) TRANSECT

4.1 Why choose the Matterhorn transect?

The Matterhorn transect is very representative of the structures of the Central Alps. The deepest units of the fold belt outcrop east of the Matterhorn in the Simplon–Ticino area. Their westward extrapolation because of axial plunge allows representation, with a certain degree of plausibility, of structures located at 10 km depth beneath the Matterhorn. In addition, a deep, recently acquired seismic profile provides information on the structure at 20–30 km depth. Lastly, top-quality field studies over the past century following the pioneering work of Emile Argand have provided thorough documentation.

4.2 The orogenic wedge sensu stricto

As noted earlier, the orogenic wedge was constructed on the foundations of the European plate. The wedge consists mainly of a sequence of basement units with their overlying sedimentary cover: the External Crystalline Massifs, Simplon–Ticino nappes, the Great Saint Bernard and Monte Rosa nappes which were both derived from the SBR block (Figs. 2.5–2.7). In addition, the Alpine edifice includes Pre-alpine units which were not incorporated in the wedge and were thrust early towards the exterior of the fold belt.

4.3 The Prealps: cover nappes not incorporated in the orogenic wedge

The Pre-alpine nappes consist uniquely of a sedimentary succession of Triassic to Eocene age. The succession was detached from the original underlying basement before construction of the wedge. These nappes have therefore escaped the metamorphism caused by nappe stacking. Located mainly in the Chablais and the Prealps south of Lake Geneva, they consist of three superposed units (Fig. 2.8). These are, from base to top: the Ultrahelvetic nappes, the three groups of major Pre-alpine nappes and the early flysch nappes (Cretaceous to Early Eocene).

4.3.1 Nappes of early flysch

Their substrate is unknown and they were reputedly expelled from the trenches located above the subduction zones (Fig. 14.7). These are:

- **Valais flysch and Niesen nappe** (Campanian to Palaeocene or Early Eocene?)
- **Liguro-piemontais flysch**. The three nappes (Fig. 2.8), reputedly derived from the trench marking the subduction of the Liguro-piemontais Ocean beneath the Apulia-African continental margin, are Gurnigel–Sarine (Maastrichtian to Early Eocene; 70–50 Ma); Dranses–Helminthoid flysch ('middle' and Late Cretaceous; 90–65 Ma) and Gets (Late Cretaceous).
- **Flysch belonging to the Austro–alpine domain**. This is represented by the conglomerates and flysch of the Simme nappe (start of the Late Cretaceous, 100–80 Ma). It is generally agreed that this nappe was derived from the Canavese zone along the boundary between the basement blocks of Sesia and Ivrea.

4.3.2 Major Pre-alpine nappes

These nappes derive from the 'SBR' block. They are, from N to S, the Prealps Medianes Plastiques nappe (Subbrianconnais zone), the Prealps Medianes Rigides nappe (sedimentary cover of the Saint Bernard basement nappe; Brianconnais zone) and the Nappe de la Brèche (sedimentary cover of the Monte Rosa basement nappe; Piemont zone) (Figs. 2.7, 2.8 and 4.6).

The Eocene flysch belonging to the major nappes is overlain by a 'tectonic cover' composed of early, overthrust flysch represented mainly by the Sarine and Dranses nappes. The thrust transport of flysch deposits into the Tertiary sedimentary basin of the major Pre-alpine nappes interrupted normal sedimentation of Eocene flysch.

4.4 Leapfrog of the early Flysch nappes and the Pre-alpine nappes

The nappes of the Prealps were emplaced in an external position due to the effects of decollement and successive slides in a relatively complex scenario (Fig. 14.8). These nappes were unaffected by metamorphism and as a result are biostratigraphically well dated, especially the flysch. These valuable dates document the deformation of the Great Saint Bernard complex nappes and Monte Rosa nappe from which they originate.

4.5 Ophiolite nappes

The ophiolite nappes on the Matterhorn transect have not all shared a common history as is shown by the different P–T paths followed by individual units. Among the nappes incorporated in the accretionary

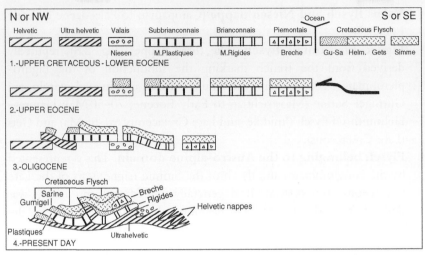

Figure 14.8 *The succession, or 'leapfrog' of thrusting in the Pre-alpine nappes.*
Palaeocene to Early Eocene. Detachment and expulsion of the Cretaceous flysch
nappes led to their emplacement in the Piemontais (future Brèche nappe),
Brianconnais and Subbrianconnais (future Medianes Rigides and Medianes Plastiques
nappes) flysch basins interrupting deposition of the Eocene flysch. The Gurnigel–
Voirons (Gu) element, which belongs to the group of early (Cretaceous) flysch,
followed the same path beyond the Brianconnais until the external domain and was
emplaced during the Oligocene. It now lies on the Ultrahelvetic nappes at the front of
the Prealps. This element is overthrust by the Medianes Plastique nappes. Other
elements of the same 'Gu' nappe which remained behind are found thrust above the
Median Plastique nappes. Late Eocene to Oligocene. The Brèche, Medianes Plastiques
and Medianes Rigides nappes are composed of specific Mesozoic sequences overlain
by Eocene flysch tectonically covered by Cretaceous flysch. These nappes advanced
initially above the Ultrahelvetic domain, which then began its own displacement, and
secondly above the Helvetic domain. The progressive outward displacement of the
Brèche, Medianes Plastiques and Medianes Rigides nappes (including their 'tectonic
cover' of Cretaceous flysch) induced the onset of the thrusting of the Ultrahelvetic and
Helvetic nappes, respectively. All these nappes terminated their displacement on the
external autochthonous Helvetic domain and on the internal part of the newly formed
Molasse basin. Oligocene. Detachment and expulsion of the Niesen nappe after the
passage of the preceding nappes. The Niesen nappe was thrust beyond the
Ultrahelvetic and Helvetic domains to be emplaced behind the Pre-alpine nappes,
which were already in place. Gu: Gurnigel; Sa: Sarine; Helm: Helminthoid flysch.
Modified from Lemoine et al. 2000, Fig. 14.6, p. 162.

wedge, the Zermatt and Antrona units, which envelop the Monte Rosa
nappe (Fig. 2.7), were certainly subducted since they are composed of blue
schist-grade rocks (30–40 km depth) and eclogites (50–70 km depth). Similarly, the Monte Rosa nappe, originating from continental European

basement, was subducted to the 50–70 km depth shown by the eclogites. The Tsate nappe was probably derived from an oceanic accretionary wedge. Parts of the nappe are of blue schist grade but others, to the north, are of green schist grade (10–20 km depth) suggesting subduction to moderate depths. The green schist metamorphism might equally well have been acquired due to the load of the overlying Austro-alpine nappes rather than by subduction.

4.6 The orogenic lid: the Dent Blanche Austro-alpine nappe

The present-day surface area occupied by the Austro-alpine nappes in the Central Alps is modest. Erosion has left only the Dent Blanche nappe (Fig. 2.7), which does not strictly correspond to the continuation of the Austro-alpine nappes of the Eastern Alps (Chapter 2).

4.7 Tentative reconstruction of the succession of events

Knowledge of the Matterhorn transect of the Central Alps has been enriched by many field studies and recent deep seismic profiles have provided new insight into the structure at depth. Nonetheless, the remaining uncertainties are so many that even the best specialists are unable to draw a series of sections for given time intervals that would portray the main events. The main characteristic and well-established data points help illustrate only the general and abstract concepts of these schematic reconstructions.

The reconstruction proposed here (Fig. 14.7) shows the stages in the construction of the fold belt since the start of the Cretaceous. Such a scheme is at best provisional since any new dates, radiometric or biostratigraphic may render the whole concept obsolete.

At the end of the Jurassic or start of the Cretaceous, a new rift phase heralded the opening of the Valais basin. At this time, there is no evidence suggesting subduction had begun in either the Liguro-piemontais Ocean or the Sesia block. During the Early Cretaceous, the continental crust of the Valais zone was subject to extension and subsidence causing a relative displacement of the 'SBR' block towards the continental Apulia–Africa. The age of the flysch suggests that subduction of the Liguro-piemontais Ocean was probably active from the Late Cretaceous. This time also probably marks the initiation of formation of the oceanic accretionary wedge from which the Tsate ophiolite nappe would originate.

During the Eocene, subduction of the SBR block began inducing the expulsion and initial translation of the Pre-alpine nappes (Figs. 14.7 and 14.8).

From the end of the Eocene to the start of the Oligocene, events speeded up: subduction of the SBR block, subduction then rapid exhumation of the ophiolites of Zermatt–Saas-Fee zone and Antrona, progressive formation of the major recumbent fold of Monte Rosa, subduction of the European margin, i.e. successively the Simplon–Ticino nappes (continental basement of the Valais domain) and the future massifs of Mont Blanc and the Aiguilles Rouges. Simultaneously, the Pre-alpine nappes were thrust towards the exterior ending their translation during the Early Oligocene in the flexural molasse basin which had begun to deepen to the north of the fold belt.

From the Late Oligocene to the Neogene, the structuring of the fold belt continued with the emplacement of the Dent Blanche nappe and the progressive formation of back folds and thrusts in the internal zones recording hypercollision. Lastly, major normal faults and sinistral strike-slip faults formed with interpreted displacements of the order of 50–100 km. However, the strike-slip faults cannot be displayed on sections perpendicular to the main Alpine structures.

A simplification has been adopted in which only continental and oceanic crust is shown on this tentative reconstruction. The separation between the upper brittle and lower ductile parts of both the European crust and Sesia block have not been shown. Lower ductile crust is present before deformation. But, at the end of the deformation, all the units derived from the European basement and included in the orogenic wedge consist only of brittle upper crust.

There are several viable hypotheses:

- No oceanic crust is represented here between the Sesia block and the South-alpine continental crust corresponding to the Ivrea zone, for the following reasons. The Ivrea zone comprises lower continental crust only. The Canavese suture line, which is the western part of the Periadriatic line, does not show any ophiolite slices at outcrop. In consequence, the suture corresponding to the Liguro-piemontais Ocean, which runs between the continental blocks of Sesia and Monte Rosa, is believed to be unique on this transect.

- The Tsate nappe consists of a stack of ophiolite slices with associated sediments mainly of Cretaceous age. Remnants of continental crust are also present including Triassic and Liassic strata comparable to those found in the Pre-alpine Breche nappe. The Tsate nappe is thus considered to be derived from part of the oceanic accretionary wedge formed mainly during Cretaceous time.

- The ophiolite nappes of Zermatt–Saas-Fee and Antrona are considered here to have originated from the same oceanic domain (Fig. 3.6B) then to be thrust on the future Monte Rosa nappe. The entire assemblage was then deformed into a large, complex recumbent fold.

The calc-alkaline magmatism mentioned previously near the Sesia zone and located on or near the Peri-adriatic fault array is not well understood. It certainly does not have the signature of subduction because it appears at a very advanced stage of the collision. The explanation proposed is reheating induced by 'slab break off' that affected only the Central Alps.

5. SOME *SINGULARITIES* OF WESTERN ALPS STRUCTURE

5.1 The transition between the Central and Western Alps

One of the singularities of Western Alps structure results from lateral ramp in the Alpine fold belt. This ramp was induced by the movement of the Apulia–Africa block towards the N or NW, matched by lateral escape at the western extremity of the fold belt. One consequence was the absence of a southwestward continuation of the most external flexural basin that extends, as the Molasse basin, from Vienna to Chambery in Savoy. Two other consequences are the disappearance of the Valais domain and the absence at outcrop of Austro-alpine units at the transition between the Central and Western Alps.

5.2 Cross cutting Pyrenean and Alpine folds in the southern Sub-alpine fold belt

Folded Pyrenean structural units continue eastward into Provence after being displaced by the sinistral strike-slip Cevennes fault array (Fig. 7.10). The so-called Pyrenean-provencal folds affected the Provence and Southern Sub-alpine fold belts in the area previously affected by Early Cretaceous rifting (Chapter 7). The Pyrenean–Provence fold axes are oriented E–W (Fig. 13.5) and accompanied by modest thrusting, generally less than 1 or 2 km. The continuation of the North Pyrenean zone today lies beneath the Mediterranean except for the Beausset thrust in the lower Provence basin near the city of Toulon (Fig. 7.10B). Folding began in the Late Cretaceous before the Senonian unconformities (∼80 Ma) in the Devoluy area (Figs. 8.2 and 8.3) and only from the end of the Santonian in Provence (Chapter 8) though the significance of the difference in timing remains to be clearly understood.

By the middle Eocene ($\sim$45 Ma), folding was complete in the Pyrenees
and Provence domains. It was followed by folding in the southern Sub-
alpine fold belt at the same time as folding began in earnest in the northern
Sub-alpine fold belt. Oligocene and Neogene Alpine compression has thus
interfered with the early E–W folds, which have been refolded (Fig. 14.9).

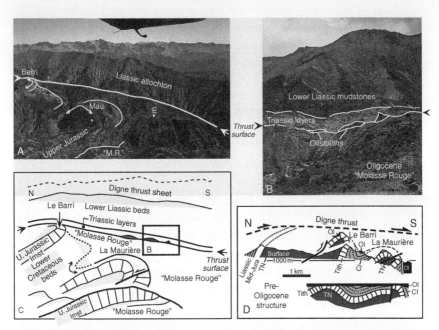

Figure 14.9 *An example of the superposition of Pyrenean and Alpine structures;
left side of the Bès valley, north of Digne.* (A) Aerial view of the left side of the valley.
The so-called Pyrenean–Provence folds are covered discordantly by the Oligocene 'Red
Molasse'. The assemblage, surmounted by the Digne nappe ended its emplacement
during Pliocene time. It is here composed of Triassic and thick Liassic strata. The surface
of the thrust surface is punctuated by a series of limestone blocks of Late Jurassic age.
These blocks have been classically interpreted as tectonic slices. More recently, they
have been considered as olistoliths deposited in the Pliocene depocentre in advance of
the overthrust nappe en route to its present position. E: Piton d'Esclangon from where
photo B has been taken. Mau: La Mauriere; see C, this figure. (B) Close-up view towards
the thrust from the peak near the abandoned village of Esclangon (point E on Photo A).
From base to top: Red Molasse of Late Eocene to Oligocene age – thrust surface
punctuated by limestone blocks of Late Jurassic age – yellow and mauve silts, marls
and dolomites of Late Triassic age – alternating grey limestones and marls of Liassic
age. (C) Geological interpretation of the left part of photo. (D) Interpretation of
geological cross section of the upper left side of the valley and reconstruction of the
section before deposition of the 'Red Molasse' showing E–W fold axes, age not defined
precisely, between the Late Cretaceous and the middle Eocene. Location: Fig. 13.5. (See
colour plate 20) *Modified from Lemoine et al. 2000, Fig. 14.9, Table XII.*

The arc-like shape in map form of the Alpine fold belt has, in consequence, been reinforced by the interference between two generations of differently oriented folds, i.e. first the Pyrenean-provencal folds and later the Alpine folds *sensu stricto*.

5.3 Evolution of the Western and Central Alps from the Cretaceous to the Tertiary

In the external part of the Alpine arc, the rectilinear structural pattern was determined by Jurassic palaeostructure (Fig. 6.16). Structures of Jurassic age parallel either the Cevennes–Belledonne trend of extensional faults or the

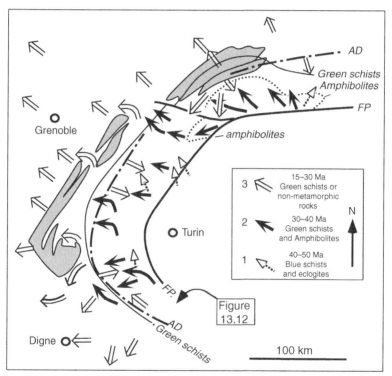

Figure 14.10 *Transport directions of the structural units from the analysis of shear-stretching lineations in the Western Alps.* The stages of successive displacements have been referenced with respect to history of metamorphic events. Initially, the nappes were thrust to the north or NNW. Subsequently, the thrust trajectories were deflected to directions radial to the arc with a dominant NW direction. The distribution of thrusts and back-thrusts became definitely radial in the Oligocene. AD: axis of double over folding; FP: Peri-adriatic fault array. *Modified after Choukroune et al. 1986; Choukroune 1994 and from Lemoine et al. 2000, Fig. 14.10, p. 167.*

sub-perpendicular, Pelvoux–Argentera trend of transfer faults. In terms of present-day structure, the External Crystalline Massifs form two sides of a rectangle parallel to each of the two main Jurassic fault trends. The nappes of internal origin are closely moulded around the rectangle. Palaeomagnetic studies of the Brianconnais nappes and the molasse sediments of the Po plain show that the arcuate shape of the internal units of the south branch of the Alpine arc had already been developed in the Oligocene.

5.3.1 The bend in tectonic trajectories

Study of microtectonic fabrics within thrust sheets has shown that the nappes began their displacement to the N or NNW as in the Central Alps.

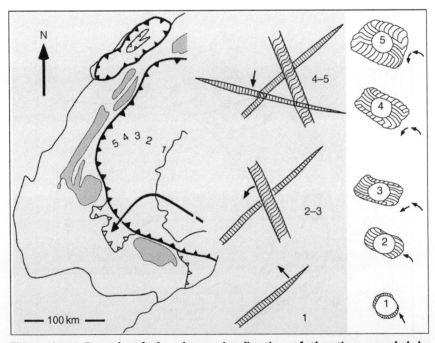

Figure 14.11 *Example of the change in direction of thrusting recorded by microstructures at the base of the Helminthoid flysch nappe known as Parpaillon (Western Alps).* Microstructures have been analyzed in the zone of ductile shear at the base of the Parpaillon nappe, which forms part of the complex Helminthoid flysch nappes. Elongated syntectonic fibres of quartz have crystallized in tension gashes or in pressure shadows against small crystals resistant to deformation, such as pyrite. These quartz fibres indicate the transport direction of the nappe. The orientation of the fibres records two directions of transport, successively towards the NW then towards the SW. The change in direction was probably rapid but continuous. *Modified after Merle and Brun 1984, 6, 711–719, after Merle 1994 and from Lemoine et al. 2000, Fig. 14.11, p. 167.*

The trajectories then curved progressively towards the NW, W or SW leading to a radial disposition with respect to the shape of the arc (Figs. 14.10 and 14.11) as seen from a point on the lateral ramp of the Apulia block.

5.3.2 Comparable tectonic–metamorphic evolution: Western and Central Alps

The tectonic–metamorphic evolution of the internal parts of the Western Alps was comparable to that of the Central Alps. Both are characterized by deep subduction of part of the continental and oceanic units resulting in UHP–LT to HP–LT metamorphism blue schist, eclogite and coesite-bearing white schist grades. Lastly, these units were subjected to retrograde green schist metamorphism during uplift to the surface. The map (Fig. 14.5A) represents the variation in meta-morphic grade from E to W in their present-day location. In addition, the latest displacements of the nappes during hypercollision have led to the radial pattern shown in Fig. 14.10. The units depicted on the latter figure were not arranged in an approximately E–W direction but have an N–S to NNW–SSE alignment as a consequence of the northerly drift of the Apulia-African block.

FURTHER READINGS

Beltrando et al. (2010); Bucher et al. (2004); Bucher et al. (2007); Choukroune et al. (1986); Choukroune (1994); Dal Piaz (2001); Ford et al. (1999); Ford et al. (2006); Nicolas et al. (1990); Pfiffner et al. (2000); Schmid and Kissling (2000); Schmid et al. (1989); Sinclair (1996); Sissingh (2001); Vignaroli et al. (2008).

The Alps – Neotectonics

Contents

Summary

The crustal and lithospheric scale thickening of the Alpine fold belt is a direct result of plate convergence and lithospheric shortening of 300–400 km from the Late Cretaceous to the present.

In the context of active collision, topographic uplift or orogenesis results from two processes: first, the stacking of allochthonous units on thrust ramps at local or regional scales and, second, at the scale of the fold belt, isostatic adjustment of abnormally thick crust, more than 50 km thick beneath the internal (Penninic) zones of the Western Alps.

In such a collisional fold belt, the amount of topographic uplift is principally limited by two processes: (1) denudation by superficial erosion and (2) vertical thinning which accompanies horizontal extension. Tectonic denudation results from displacement on normal faults especially gently dipping normal faults. Although there is a clear continuum of effects of Alpine convergence through to the present, a specific chapter on neotectonics is necessary for two reasons.

First, the Alpine collision continues today but our analytical approach to present-day processes is necessarily different from that used to study past processes, and geophysical methods thus hold a key place at the side of field analysis. New methods, particularly low-temperature thermochronology, have allowed geomorphology to become a quantitative science in a multidisciplinary geophysical–geological approach.

The Western Alps, from Rift to Passive Margin to Orogenic Belt, Volume 14
ISSN 0928-2025, DOI 10.1016/S0928-2025(11)14015-8
© 2011 Elsevier B.V.
All rights reserved.

The physical measurement in real or almost real time of active processes shows that the present fields of displacement, of strain and stress, are incomparably better known than the fossil deformation fields. This leads one to challenge the past history in terms of the validity of changes of scale in space as well as in time. Nonetheless, the extrapolation to geological time scales of the velocities of processes, evaluated in human time scales, i.e. instantaneous in terms of geological time, requires considerable caution.

Second, the Alpine collision is now reaching an advanced stage characterized by a particular tectonic and geomorphic evolution. A remarkable characteristic of this evolution is the importance of two successive extensional processes: initially, syn-orogenic extension limits the growth of the orogenic prism while continued lithospheric convergence is associated with later post-orogenic extension, which reduces the crust to a 'normal' thickness at the moment convergence ends. The two effects are governed by gravity.

In the simplest model, post-orogenic extension affects the axial zone of the fold belt while compressional tectonism expressed by folding and thrusting affects the borders through the effects of tectonic collapse or gravity spreading (sometimes called Camembert or Brie in allusion to the weak behaviour of very mature cheese). 'Late' extension associated with strike-slip faulting parallel to the belt front was first described in the Central and Eastern Alps. The appropriate example is the Peri-adriatic fault array (Fig. 2.2). But the importance of 'late' extension in the Western Alps, particularly in the internal zones has only been properly recognized during the past decade. In the entirety of the Alps, more and more in-depth studies show that from one part of the fold belt to another, different regimes, transtensional or extensional, follow one another with different time scales. As a result it is now possible to begin to integrate the 'late Alpine' history of the past 20 Ma to Recent of evolution with that of the active tectonism of the Plio-Quaternary and present day.

1. SEISMICITY

The distribution of earthquakes on both sides of the Mediterranean defines the diffuse boundary between the Eurasian and African plates with other microplates. Within this large area, the magnitude and frequency of the earthquakes vary considerably. The Aegean area including Greece is the most seismically active while other areas such as northern Tunisia are practically aseismic. In contrast, numerous weak, irregularly distributed earthquakes, at shallow depths, characterize the seismicity of the Alps. An example in the Western Alps is given in Fig. 15.1A.

In the case of the Alps, the coexistence of varied earthquake mechanisms, such as thrusts, strike-slip and normal faults (Fig. 15.2), suggests that indentation of Western Europe by the Apulia-African promontory continues presently in a complex way, which tends to be confirmed by recent continuous GPS measurements.

However, recent reinterpretation of seismicity in the internal arc of the Western and Central Alps has led to the recognition of the prevalence of extensional to transtensional faulting in the upper 10–15 km of the crust. Analysis of the population of seismic faults and their striae as well as focal mechanisms has led to the recognition of vast areas subject to extension both parallel to the fold belt and perpendicular to the fold belt (example Western Alps: Fig. 15.1B). Simpler to analyze, seismicity in the external arc remains interpreted in terms of shortening radial to the arc in response to compression of the same orientation via thrust or strike-slip regimes. Two simple models allow reconciliation of the duality of the tectonic regimes.

(1) Given the continuation of Apulia-African indentation, the nappe pile comprising the internal arc may be subject to surface extension along the convex exterior of the underlying indentor, which pushes at depth the External Crystalline Massifs towards the exterior of the external arc. The weaknesses in this model are the absence of the seismic signature of an indentor at depth as well as the difficulty of explaining extension parallel to the fold belt. It implies the continuation of lithospheric convergence to the limits of the fold belt, which does not seem to be valid from the results of continuous GPS measurements (see below).

(2) The alternative is the model of very soft highly mature Camembert (or Brie!) cheese. In this case, the internal arc collapses under its own weight. The observed extension is then the consequence of thinning of an 'over-thickened' crust. This model agrees with a slowdown of convergence to the limits of the fold belt and with the possibility that the strength of the crust has been diminished by a heating event. One such 'thermal event' has been effectively observed in the Central Alps, where it is attributed to slab break off or detachment of a crustal root. Such a slab break off with uprise of hot asthenosphere at shallow depth below the Western Alps has

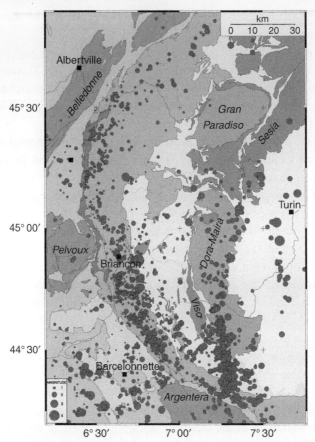

Figure 15.1 *Seismicity of the Western Alps. (A) Map of earthquakes greater than magnitude 1 for the period 1989–1997 (network of IGG and Sisalp seismic stations).* The numerous earthquakes are of moderate magnitude (the large majority are less than 2) concentrated in the upper 10–15 km of the crust and are irregularly distributed. In the internal domains the earthquakes are grouped into two arcuate zones: (1) The 'Brianconnais seismic arc', to the west, just to the rear of the inverted Frontal Brianconnais Thrust. Continuation of the inversion explains the seismicity. (2) The 'Piemontais seismic arc' further east, oblique to surface structures and corresponding to seismicity at greater and greater depths than in the Brianconnais arc; it outlines the corner of Apulia-African mantle detected elsewhere by gravity (Fig. 2.13) and seismic tomography. **(B) Seismotectonic analysis.** Some regions have been defined containing homogeneous focal mechanisms (E1, E2, B1, B2, B3 and P). For each a mean direction of shortening compression (region E1 for the external zone) or extension can be calculated in the horizontal plane. Remarkably, the Brianconnais (B1, B2 and B3) and Piemontais (P) seismic arcs represent a seismotectonic regime in gross extension normal to the fold belt (=radial to the arc). Under the Helminthoid Flysch nappes of the Embrunais-Ubaye, the external zone experiences rather a transtensional regime (E2). The dotted line shows the ECORS-CROP profile (Fig. 2.14). (See colour plate 21) *After Sue et al. 1999, Fig. 2, p. 25–613.*

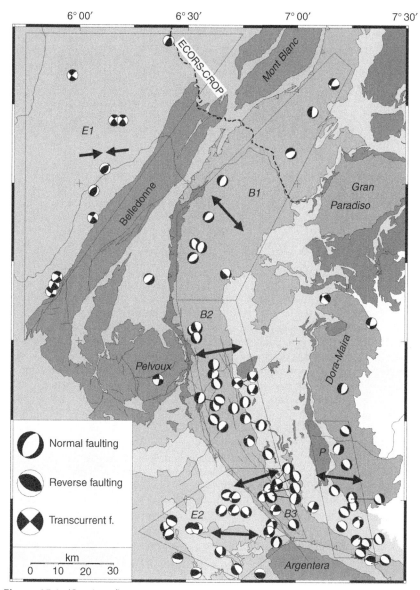

Figure 15.1 *(Continued)*

been recently suggested by seismic tomography. This interpretation remains nevertheless to be confirmed since the thermal effect of such an uprise has never been observed in the Western Alps in contrast to the Central Alps.

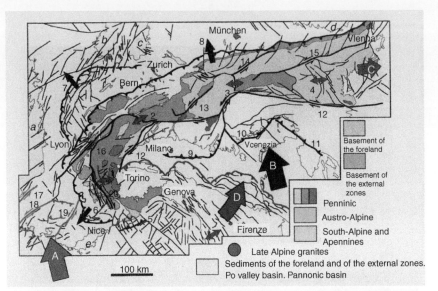

Figure 15.2 Kinematic and dynamic model of the Alps during the last few million years. This model is a synthesis of results of the present-day deformation fields, in particular the focal mechanisms of earthquakes and related results on recent faulting. It mainly documents the ongoing indentation towards the north acting on the fold belt as a whole. Direct and indirect effects of Europe–Africa convergence: (A) Compression transmitted by the basement of the Mediterranean. (B) Indentation of the fold belt by the Apulia-African microplate. (C) Expulsion towards the Pannonian basin. (D) Extension–compression coupled to slab retreat under the northern Apennines. Principal locations in extension: 1: Brianconnais and Queyras. 2: Simplon ductile normal transtensional fault. 3 and 4: Eastern Alps with free eastern boundary. Principal thrust fronts: 5: Inversion of normal faults of the Ligurian margin. 6: Digne nappe and Castellane arc. 7: Jura. 8: North Alpine Front. 9: South Alpine front (Bergamo Alps). 10: South Alpine front (Venetia). 11: Dinaride front. Some major strike-slip faults: 12: Peri-adriatic. 13: Engadine. 14: Inntal. 15: Ennstal. 16: Median fault of the Belledonne External Crystalline Massif. 17: Cevennes fault. 18: Nimes fault. 19: Fault of the middle Durance. 20: Fault of the upper Durance. Backstop massifs of the foreland: a: Massif Central. b: Vosges. c: Black Forest. d: Bohemia. (See colour plate 22) *From Tricart in Lemoine et al. 2000, Table XIII, Fig. 15.5.*

This second model does not require active Alpine convergence to explain the combination of folds and thrusts on the border of the fold belt but it pushes the model to an unrealistic extreme. Continuing discussions and investigations point to complex models combining elements of simple models like the two above. It will be necessary to add the effects of a strong component of sinistral rotation of the Apulia–African indentor which is an independent microplate as suggested from the first results of continuous GPS measurements in the Western Alps (see below).

2. DIRECT MEASUREMENT OF PRESENT-DAY, ACTIVE DEFORMATION OF ROCKS

This technique is based on different methods of varying accuracy such as the contraction of tunnel diameters and the ellipticity of circular bore-holes. Particularly in regions of strong relief, the interpretation of the results is beset by the difficulty of determining the role of gravity slides on mountain flanks. Although the available measurements are irregularly distributed in the fold belt, the shortening directions agree everywhere with those derived from earthquake mechanisms, especially in the external arc of the Western and Central Alps which are in compression radial to the arc. These results confirm the complexity of the present-day Alpine deformation field around the Apulia-adriatic indentor as well as the slow deformation rate.

3. VERTICAL MOVEMENTS

Present-day relative uplift and lowering of ground level has been measured directly over historical time with respect to the fixed reference of sea level. For periods of geological time, the vertical movement of rocks, i.e. their burial or exhumation, is estimated with respect to ground level, itself variable with time. Finally, for geological periods of time, the evolution of the mountainous relief of fold belts like the Alps during their history can be outlined by identifying the areas whose erosion has supplied the sediments infilling the flysch and molasse basins. Methods of tracing the provenance of these sediments have made great progress, e.g. from clastic thermochronology using zircon fission track ages.

3.1 Present-day uplift of the Alps

In contrast to horizontal movements, there is as yet no real-time method of measuring vertical movements when they are only of the order of mm/year. Vertical movements then remain traditionally measured using the geodetic approach of precision levelling around a network of spot heights. Comparison of repeated levelling measurements made over many decades allows measurement of present-day vertical movements. Repeated measurements over long intervals and subsequent statistical analysis are required to detect very slow rates of movement. Implementation of this procedure is much more advanced in the Swiss Alps compared with the Franco-Italian Alps. Using this method, present-day uplift rates of the Alps close to 1.5–2 mm per year have been determined. Comparing the Montgenevre, Gotthard and Grossglockner transects of the Western, Central and Eastern Alps, respectively, it appears that the sectors showing the most rapid uplift do not correspond to comparable structural zones. Moreover, the most rapid zones of uplift do not coincide with the highest elevations, i.e. the External Massifs such as the Ecrins-Pelvoux, Belledonne, Mont Blanc, Aar and Gotthard (Fig. 15.3). In contrast, the internal zones, whose average altitude is somewhat lower, may exhibit faster rates of uplift. The difference is most obviously shown from measurements made over 50 years on the Simplon-Ticino nappes transect. These measurements have been used to construct curve B of Fig. 15.3. In the inner parts of the Western Alps, which are subject to present-day extension, vertical movements of the ground remain to be measured with precision; a better understanding will allow more informed discussion of the validity of the different proposed tectonic models (see below).

3.2 Final exhumation path of Alpine rocks

The final exhumation of rocks, i.e. their approach to the surface until at outcrop, can be deduced from their cooling history on the basis of 'absolute' ages, i.e. the closure age of crystalline systems determined by various methods. With regard to the end of cooling, the method of apatite fission track dating is, for example, particularly useful because it dates cooling below a temperature close to 100°C, i.e. exhumation from less than 3–5 km with the assumption that the geothermal gradient is known. In addition, when possible, track length measurement in apatites allows reconstruction of the thermal history. The cooling of rock samples is largely controlled by the progressive disappearance of the overburden due to erosion, especially in response to uplift, but also, where active, to tectonic thinning by extension, including tectonic denudation along detachment surfaces. In the Alps, these methods provide rates of exhumation indicating that the present fast uplift of the External Crystalline Massifs (of the order of 1 mm per year) began in the Late Miocene (7–8 Ma), with important acceleration during Plio-Quaternary times (the past 3–4 Ma).

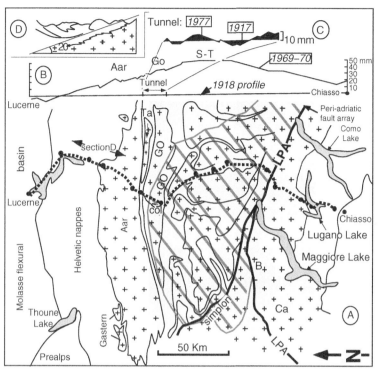

Figure 15.3 *Evolution of the topographic profile across the Central Alps between Lucerne and Chiasso.* The crosses indicate the outcrop of pre-Triassic basement, European in the Central Alps to the north of the Peri-adriatic line (LPA) and Apulia–Africa to the south of this line, in the Southern Alps. The profile crosses from north to south, the external zone (Helvetic nappes), the External Crystalline Massifs, Aar and Gotthard (Go), the Simplo-Ticino zone (S-T) affected by Alpine amphibolite-grade metamorphism (zone A, hatched on the map), and finally, after having cleared the LPA line, the Southern Alps which have escaped Alpine metamorphism. The highest present-day altitudes are found in the External Crystalline Massifs. However, the most rapid uplift in the last 50 years is that of the Simplo-Ticino nappes, which correspond to the deep part of the collisional prism subjected to elevated Alpine metamorphism. A: Location of profile. B: Constant uplift between the geodetic levelling of 1918 and 1960–1970e, i.e. for duration of about 50 years. The Simplo-Ticino zone has been uplifted by 50 mm at a mean rate of 1 mm per year. The Crystalline Massifs have been uplifted by 20–40 mm at a mean rate of 0.5 mm per year for the Gotthard massif. C: Uplift and subsidence in the Gotthard tunnel during the past 60 years. D: Section showing that the basement of the Aar massif, situated below the Helvetic nappes has tilted by about 20°; it is not known when, but perhaps during the emplacement of the nappes in the Neogene. Ta: Tavetsch Massif; Go: Gotthard Massif; Ca: Canavese. *After Schaer et Jeanrichard 1980, Funk et al. (1980), and from Lemoine et al. 2000, Fig. 15.1, p. 171.*

3.3 Morphogenesis of the Alps over the past 30 Ma (Oligocene to Recent)

The major part of the sedimentary products of Alpine erosion is trapped in the Peri-alpine molasse basins. Exemplary studies of the provenance of pebbles in the north and south Alpine basins as well as the Pannonian basin have permitted reconstruction of the topographic evolution of the Central and Eastern Alps since the Oligocene, during the last 30–35 Ma. One classic provenance study has been made on pebbles derived from the Peri-adriatic eruptive suite and the Bergell granite. Another study discriminated pebbles derived from a stack of units with distinct thermal histories using isotopic dating. Similarly, fission track dating of zircons has been applied to the sediments derived by erosion of the nappes covering the Tauern window and the metamorphic dome of the Lepontine Alps. The progressive modification of drainage patterns and the displacement of watersheds can thus be shown to have taken place during a succession of orogenic episodes with relatively rapid growth in relief in direct response to the effects of Alpine convergence (Fig. 15.4).

The major, Miocene orogenic event in the Alps involved northward indentation of Oligocene Peri-adriatic structures by the South Alpine spur, now the Dolomites. N–S shortening was accommodated by major E–W stretching parallel to the Central and Eastern Alpine fold belt accompanied by rapid denudation of the Lepontine dome, Hohe Tauern and Rechnitz windows. Mass balance calculations show that the denudation owes more to the role of extensional detachment surfaces than erosion. During the last 10 million years, extension parallel to the fold belt has decreased in comparison with thrusting in the Venice region, i.e. at the southern margin of the fold belt. In summary, the elongation in the Eastern Alps amounts to about 30% since Oligocene times and the drainage pattern has progressively adapted to stretching throughout this long interval.

In the Eastern Alps, major sub-aerial erosion surfaces were developed by Eocene–Oligocene time, e.g. the undulating Dachstein surface preserved by the continental, Augenstein conglomerates of Oligo-Miocene age. This surface was later uplifted during indentation with the magnitude of the uplift increasing from the east to west. Remnants of the palaeosurface dissected by more recent erosion, found today on the karst plateaus of the Northern Calcareous Alps, suggest an original area of $13,000 \, \text{km}^2$. The existence of this palaeosurface allows assessment of the vigorous rejuvenation of Alpine relief in the Eastern Alps during Miocene indentation. In addition, it provides insight into the scale of the reorganization of the drainage network.

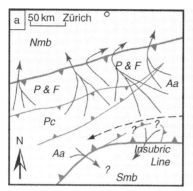
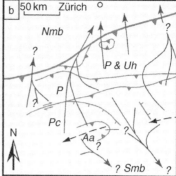

Figure 15.4 *Evolution of relief and drainage pattern during uplift of the Central Alps between the Late Oligocene (30 Ma) and the Miocene (16 Ma).* Palaeorivers are in dark gray. Palaeo-forethrusts, backthrusts and transtensional/extensional faults are shown in light gray. Backfold anticlines in black solid lines. Nmb: Northern molassic basin; P & F: Prealps and Flysch; P: Prealps; P&Uh: Penninic and Ultrahelvetic flysch; Pc: Penninic crystalline core; Aa: Austroalpine; Smb: Southern molassic basin. This type of reconstruction depends largely on tracing the provenance of terrigenous sediments (heavy minerals and geothermochronology). *Simplified after Schlunegger, Slingerland and Matter 1998, Fig. 7, p. 108 and Fig. 9, p. 109.*

There is no equivalent palaeosurface in the Central and Western Alps. Here, the regional effects of erosion have been quantified from estimates of the volume of clastic sediments produced by erosion and deposited in Peri-alpine basins, and more recently in local Alpine lakes and large valleys. These estimates suggest that, since the Eocene (34 Ma), erosion has removed the equivalent of a 5.3 km layer of rock at an average rate of 0.16 mm per year. In comparison, the average rate of denudation by erosion for the whole Quaternary is about 0.44 mm per year. Allowing for the present day, slightly higher rate of 1 mm per year, the rate for the whole post-glacial period (or since 17,000 yr BP) reaches 0.6 mm per year. The maximum, of about 3 mm per year, would have been reached during late glacial periods.

4. HORIZONTAL MOVEMENTS

The spatial variation in horizontal movements can be derived from a comparison of repeated geodetic triangulation networks over several decades. From such data evidence of convergence or divergence between two points can be shown. Allowing for errors, the accuracy is of the order of

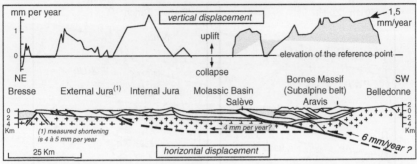

Figure 15.5 *Possible relations between present-day vertical and horizontal movements in the north of the Savoy Alps and the southern Jura.* Vertical movements between successive levelling surveys in the Savoy Alps and southern Jura (NGF network 1886–1907 and IGN69 network 1965–1979) have been calculated for each point with respect to a reference point. These movements, essentially uplift, have been projected along a section showing the main structures in the cover and in the upper part of the basement. This NW–SE-oriented section passes from Oyonnax to 10 km north of Annecy. The uplifted zones coincide with assemblages thrust to the NW. This suggests that these thrusts, which are the most recent Alpine thrusts on this traverse, are still active. The uplift rate is of the order of a millimetre per year and may correspond to horizontal movements of the order of 0.5 mm per year in agreement with the horizontal shortening measured in the Jura. *After Jouanne and Menard 1994 and adapted from Lemoine et al. 2000, Fig. 15.2, p. 173.*

10^{-6} of the measured distance. From the field of displacement of the measured points, the field of deformation can be found easily. For example, a comparison of triangulations made between 1935 and 1980 documents present-day shortening between the Jura and the Belledonne massif (Fig. 15.5) suggesting that the thrust surface running below the Belledonne massif and other parts of the Jura and Sub-alps in Savoy is still active.

Modern GPS studies utilize a network of permanent GPS stations. This technique, first implemented during the 1990s study of the movement along the San Andreas Fault involves continuous measurement of the relative position of the receivers comprising the GPS network. This technique is especially suited to determining the spatial pattern of deformation at slow rates, as in the Alps. By augmenting the signal-to-noise ratio, the time required to detect a significant tectonic signal is greatly decreased. Even so, at least 2 years of continuous measurement are necessary to provide accuracy of the order of 1 mm per year.

A network of permanent GPS stations is currently being installed in the Western Alps. By 2009, 11–13 years of measurements had been made at stations distributed between the main tectonic zones, the two flanks and forelands of the fold belt. The initial, very preliminary results confirm three major results from seismo-tectonic and conventional geodetic analysis: (1) extension roughly oriented transverse to the belt dominates the core (the internal arc) of the Western Alps especially the central part; (2) compression transverse to the belt dominates in the external arc and (3) approximately N–S to NW–SE shortening affects the southern part of the arc as well as the northern margin of the Ligurian basin (Western Mediterranean).

Synthesis of all the available data from the Alps and Mediterranean leads to the hypothesis that the Apulia-adriatic microplate is fragmented. Only its northern part is undergoing sinistral rotation. The location of the pole of sinistral rotation in the north of the Po plain near Milan explains why the Eastern and Southern Alps are subject to a global N–S shortening increasing towards the east. It also explains the contrast with the Western Alps which today escape from the global convergence at their borders. Another important result is that, as in other fold belts, only a modest part of the deformation is reflected by seismic activity; in the studied area, it varies from a few percent to 10–20% at most. The rest of the deformation is aseismic and ductile in nature.

If the disappearance of all important convergence in the Western Alps is confirmed, models of gravity collapse can be given priority in explaining active extension in the internal zones. In consequence, the problem is therefore extrapolation of the instantaneous kinematics to prior periods and to geological time scales.

5. STUDY OF 'LATE' FAULTS IN THE FIELD

'Late' is a convenient, though an imprecise and ambiguous, term to describe the modest brittle deformation which, as seen in the field, succeeds the ductile deformation so important in the Alps.

The network of late normal faults was largely ignored until the past decade. This network formed over a long period of time during the last 10–20 Ma. As a rule, it is impossible to distinguish in the field faults of Miocene age from those formed during Plio-Quaternary time. In contrast, the throw of major faults can be dated indirectly when the displacement has accelerated or slowed down the cooling of the fault blocks (see for example the inverted Brianconnais front to the south of the Pelvoux: Fig. 15.8). In the Alps, particularly in the internal zones, profoundly eroded by Quaternary glaciers, it is rarely possible to characterize unambiguously active faults at outcrop for lack of suitable marker formations and also because of the important masking effects of gravity sliding on mountain flanks. This limits the possibility of dating fault escarpments by cosmogenic isotopes to calibrate neotectonic processes.

During the past decade, the population of 'late ' faults has been analyzed in a quasi-systematic manner in different Alpine zones using conventional methods such as the nature of the displacement and its age. Stress tensors can be deduced from striae on fault planes. The tensors are reduced to the three principal axes of deformation and to the shape ratio of the corresponding strain ellipsoid itself linkable to the type of tectonic regime. These late fields are comparable to those found by other methods such as the *in situ* measurement of surface deformation or earthquake focal mechanism solutions (Fig. 15.1B).

A major result is that continuity of deformation dominates globally from the Neogene to the present in the whole of the fold belt. The network of Miocene faults presently exposed at outcrop in the Queyras and Brianconnais mountains provides a good idea of the presently active brittle structure under the mountains and also to depths up to 10–15 km.

Another result is that different deformation regimes can succeed each other through time following a scenario variable from place to place (see below).

In the Central and Eastern Alps, the northward indentation of the Apulia-adriatic microplate has caused N–S shortening and related orogenic thickening as well as E–W extension. The relative importance of these processes has varied through the past 20 million years. Northward- and southward-verging thrusts on the northern and southern borders of the fold belt demonstrate shortening and thickening. Large normal faults, like the Brenner fault, document extension parallel to the overall strike of the fold belt. The post-Miocene regime of indentation has deformed the corridor delineated, since Oligocene times, by the Peri-adriatic fault array. This corridor resulted initially from both reverse faulting during the Oligocene and dextral strike-slip displacement.

In the Western Alps, the Penninic front separated two very different tectonic regimes during Neogene times.

In front, the external arc (Helvetic-Dauphinois zone) shows the continuation of Oligocene compressive tectonism; orogenesis intensified through thrusts and or strike-slip faults in the basement and cover uplifting the External Crystalline Massifs. Pliocene to current tectonism explains the high altitude of these massifs, which is further accentuated by isostatic rebound linked to the effects of glacial erosion then to Quaternary deglaciation. The main shortening direction is grossly radial to the arc from the Jura to the southern Sub-alpine chains but becomes more northerly in Provence and in the contiguous southerly Sub-alpine chains. This field of compressional/transpressive deformation continues to the present day.

Behind the Penninic front, in the pile of metamorphic nappes, the regime becomes extensional or transtensional and variable from one area to another and through time. Research continues to date the timing that is inevitably diachronous.

(1) In the northern branch of the arc, which is oriented NE–SW, extension parallel to the fold belt dominated first and continued that developed around the Simplon transtensional relay in the Central Alps. This extension, which follows the axis of the fold belt, was controlled by longitudinal strike-slip faults. It expresses lateral extrusion towards the SW, almost symmetric with the extrusion to the east observed in the Eastern Alps, all at the front of the Apulia-African indentor.

Subsequently, and also in the northern branch of the arc, more radial extension developed accompanied by normal faulting often nucleated along ('negative inversion') on thrusts associated with the Penninic front. This change may express the growing role of gravity in governing renewed stretching of the fold belt in the internal part of the arc. This extension, normal to the fold belt, prefigures the present-day extension.

(2) In the central part of the arc, to the E and SE of the Pelvoux, extension may have been established earlier, by the end of the Oligocene. In the Briançonnais nappe pile, early extension was brittle. In the Piemontais Schistes Lustrés and ophiolites of Queyras and Monviso, initial ductile extension was followed by brittle extension during the Late Miocene further to the east. In the two cases, a multidirectional extensional regime was well established. Through time and space, the dominant regime has been extension parallel to the fold belt (Fig. 15.6) or extension perpendicular to the fold belt, i.e. radial to the arc (Fig. 15.7). Regionally, at the scale of all this part of the internal arc, the dynamics resemble radial spreading (in the second sense of the word – in all directions). In the course of time, the regime momentarily became more transtensional with strike-slip rejuvenation of normal faults, exemplified by the dextral, longitudinal Haute Durance–Argentera fault. In the middle part of the arc, rejuvenation in extension of the frontal thrust of the internal

metamorphic arc has been documented and remains the most spectacular (Figs. 15.2 and 15.7). In this part of the arc more than elsewhere, the mechanism called for is gravity collapse of over-thickened crust.

(3) In the less-explored south branch of the arc (NW–SE), the most visible late structures are the longitudinal strike-slip faults dated at least locally as Pliocene: research is in progress to characterize the orogen-parallel extensional structures that preceded the latter.

A large number of detailed studies allows description of the variation in space and time of the 'late Alpine' tectonic regime. Beyond these variations, the major fact is the establishment of an extensional–transtensional regime in the axis of the fold belt of the central and Eastern Alps as well as in the internal arc of the Western Alps for more than 20 Ma. This regime has lasted from the Neogene to the present day.

6. CONCLUSION

The Alps remain an active fold belt due to continuing deformation expressed by uplift and erosion. Though less spectacular than that which caused the mountainous relief, this active deformation, which is distributed throughout the 200–400 km width of the fold belt, characterizes the diffuse boundary between the slowly converging plates. The complexity of the field of stress and deformation within the interior of the fold belt is surprising and has motivated many current studies. The present-day geodynamics of the fold belt revealed by direct measurements of active tectonic processes seems to be a continuation of the recent dynamics shown from field-based analysis of late Alpine faulting. The Alps offer the opportunity of combining both research methods into new interpretative models (Fig. 15.2).

From the longitude of the Alps and Tunisia, the northerly convergence of Europe and Africa, which has lasted for the whole Tertiary, has today an estimated (i.e. not measured directly) velocity of 7–8 mm per year. The southwestern Alps, the oceanic and continental basement of the Western Mediterranean, including the Corsica and Sardinia microcraton, transmitted the compression (A in Fig. 15.2) inducing N–S to NW–SE shortening in the southern Sub-alpine chains and immediate foreland (see strike-slip faults and thrust shown in red in Fig. 15.2).

However, the main effect was N or NW indentation (B in Fig. 15.2) of the Apulia–Adria microplate towards the Central and Eastern Alps. At the western extremity, under the Po plain, a sinistral rotation dominates the

Figure 15.6 *Late Alpine orogen-parallel brittle extension in the Brianconnais zone.* View towards the east of a Brianconnais nappe near Briancon: thick middle Triassic dolomites (mT) covered by thin upper Jurassic (J) limestone beds are covered by Cretaceous (C) calcschists formed from ancient calcareous planktonic oozes. The succession is cut into two blocks tilted towards the south by a normal fault dipping northward (dashes). This fault is planar and not listric suggesting a domino-like motif. Values for the dip of the fault and beds are comparable which suggests locked tilting. The extension is here N–S, i.e. parallel to the fold belt. (See colour plate 23)

Figure 15.7 *Late Alpine orogen-perpendicular brittle extension in the Brianconnais zone.* View towards the NW of a frontal Brianconnais nappe near Briancon: above Early Triassic quartzites (IT) thin Upper Jurassic (J) limestones are covered by ancient Cretaceous marls (C) formed from calcareous planktonic muds. The succession is cut into blocks tilted towards the west by normal faults dipping to the east (dashes). The extension is here E–W and perpendicular to the fold belt. Just beneath these tilted blocks, the major frontal thrust of the Brianconnais zone has been rejuvenated in extension (see Fig. 15.8). Crosscutting relationships indicate that they are contemporaneous forming in response to multidirectional regional extension. (See colour plate 24)

present-day movement (not shown in Fig. 15.2). It may explain the complex deformation in the western Alpine arc that cannot be treated as a simple indentation.

At the scale of the entire Alps, the indentation had two consequences. The first was thrusting to the north and northwest of the Jura and north Alpine front onto the molasse basin and southerly backthrusting of the South Alpine units. The second consequence was strike-slip faulting oriented more or less strike-parallel to the Alps and represented by fault arrays such as the progressively dismembered Peri-adriatic line. These strike-slip faults were induced by the expulsion of structural units towards the extremities of the Alps and were especially active from 20 to 10 Ma. To the east, this lateral extrusion took place towards the free boundary represented by the stretched, subsiding crust of the back-arc Pannonian Basin. To the west, units situated at the heart of the western Alpine arc were expelled to the west, southwest and south in a movement that followed the curvature of the arc.

In the axis of the Alpine fold belt, these movements resulted in syn-orogenic extension which may result from along strike lengthening of the fold belt (3 and 4 in Fig. 15.2) or the bending effects of longitudinal strike-slip faults causing local transtension (2 in Fig. 15.2). In the Western Alps, extension, recently documented in the Brianconnais and Queyras areas (1 in Fig. 15.2), has multiple trends oriented dominantly transverse to the belt that affects most of the internal arc. It is contemporaneous with radial compression in the external arc whose effects are found in the Castellane arc, far to the south of the Jura arc (6 and 7 in Fig. 15.2). Its significance remains a matter for further discussion.

To the south, Alpine and Apennines dynamics acted in interference, likely driven by subduction retreat (roll back). The northerly thrust front and extension behind the chain suggest a general path indicated as D in Fig. 15.2.

Recent and present-day deformation using the example of the Pelvoux to Monviso transect

Whatever the Alpine traverse under consideration, the altitude of the External Crystalline Massifs (example: Mont Blanc, Fig. 16.7) is always higher than the internal zones. Such is the case of the External Pelvoux Massif which can be seen looking west from the Chenaillet, i.e. from the internal zones and French-Italian frontier on this east–west traverse (Fig. 16.8). There, the late normal faults are spectacular in the internal zones (Figs. 16.6–16.8).

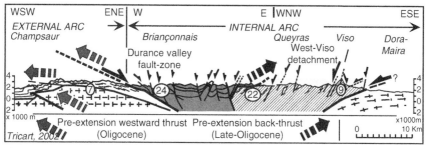

Figure 15.8 *Synthetic section crossing the Western Alps south of the Pelvoux showing late Alpine structures.* From Miocene to the Present, extension has been the dominant mode of deformation in the internal zones of the Western Alps. In contrast, compression continues to propagate westward towards the exterior of the fold belt. The very recent movements of the Digne nappe provide a good example (Fig. 13.5). In the internal zones, the normal faults reoccupy compressional structures, thrusts and backthrusts of Oligocene age. The Brianconnais front is the major thrust plane at the front of the internal arc. It was active during the Oligocene. It became reactivated as an extensional detachment fault from the end of the Oligocene. To the east, possible detachment faults bordering the Monviso range could appear as conjugates to the Penninic Frontal Thrust. The chronology of the deformation is suggested from exhumation rates derived from cooling ages. By way of example, the circled numbers indicate the mean age of passage of rock assemblages below 100°C calculated at four localities from apatite fission track dating. Ages are after Schwarz et al. 2007. *Modified from Tricart et al. 2006, Fig. 8, p. 307*

- The Penninic front or frontal thrust of the internal metamorphic arc corresponds to the emergence at the surface of a major crustal and probably lithospheric scale thrust fault, detaching the Moho at about 80 km as shown by the ECORS-CROP seismic profile (Fig. 2.16). From the Oligocene (c. 30 Ma), this major thrust transported the already-emplaced pile of metamorphic nappes onto the as yet undeformed external domain. By the end of the Oligocene, this same fault reversed its throw to become an extensional detachment fault. Extensional movement continued throughout the Miocene, as shown by the difference in cooling rates (Fig. 15.8). Both local and regional seismicity show that the fault continues to move as an extensional fault at the present day (Fig. 15.1). The study of the Penninic front is thus exemplary because it so clearly links tectogenesis and morphogenesis.
- In the external domain, the Pelvoux Massif has been progressively exhumed by tectonic denudation on its eastern flank while it continued to be thrust westward due to the effect of a late out of sequence thrust. Its uplift has accelerated from the end of the Miocene

until the present, with a very recent, additional acceleration due to Quaternary glaciation effects. Despite undergoing rapid uplift, the Pelvoux block is aseismic (Fig. 15.1A) and is one of the highest massifs of the western Alpine arc (Fig. 15.8) as is Mont Blanc (Fig. 16.6).

- Beyond the Pelvoux, the shortening of the folded external arc that produces the radial pattern is localized in the Sub-alpine chains (Fig. 15.2). Active from during the Miocene to the Present, shortening is expressed at present by seismicity, deformation of Quaternary features and notably by the existence of local uplifts: for example, the uplift of the Barles dome, in the nearby foreland of the Digne thrust sheet.

- The Brianconnais and Piemontais zones which comprise the internal domain on this transect are cut by a relatively dense network of extensional faults. In the Brianconnais, these faults have been active from the beginning of the Miocene (Figs. 15.6 and 15.7). East of the Montgenevre-Chenaillet ophiolite massif, in the high valleys of Queyras and in the Mont Viso range (Figs. 16.7 and 16.8), the extensional deformation was initially ductile but became brittle with extensional faults from Late Miocene time or much later. The present activity of the normal fault network accounts for the regional seismicity at depths from 0 to 10 or 15 km. The seismicity reflects the thinning of the upper crust by extension at a regional scale. This may explain the difference in elevation between the internal and external areas of the Western Alps.

Measurements of vertical and horizontal movements and a better knowledge of active deep structures are required for an integrated and explanatory model of the neotectonism that is the subject of current research.

FURTHER READINGS

Champagnac et al. (2004); Champagnac et al. (2007); Fügenschuh and Schmid (2003); Malusà and Vezzoli (2006); Schaer and Jeanrichard (1974); Sue and Tricart (2003); Tricart et al. (2006); Selverstone (2005).

Summary, Discussion and Conclusion

Contents

© 2011 Elsevier B.V.
All rights reserved.

Summary

The preceding chapters of this book have synthesized classical and modern field- and laboratory-based studies of Western Alpine geology. They have utilized and integrated new foundation studies of mid-ocean ridges, deep seismic sections across present-day passive margins and the Alps as well as studies of sequence stratigraphy in Alpine and extra Alpine basins that coherently integrate tectonics and sedimentation. The Alps, in the context offered here, provide an excellent natural laboratory for examining the structural and stratigraphic history of a passive margin and its subsequent evolution into a fold and thrust belt and associated foreland basins.

In this chapter we recapitulate some key observations and conclude with a brief virtual geological field trip visiting the Alpine peaks of most geological significance in terms of the evolution of the Alpine fold belt from its birth to the present.

Like all ocean basins bounded by passive margins, the history of the Ligurian Tethys can be simply described in terms of the classical succession of pre-, syn- and post-rift phases. The phases of shortening and compression that generated the Alpine fold belt followed later.

Structures acquired in the pre-rift phase, notably Hercynian lineaments and the network of extensional faults that record distant Neotethyan rifting, guided in part the emplacement of more recent structures. Of these, the first formed the network of Jurassic (Tethyan) extensional faults followed by those of the Cretaceous (Atlantic and Valais) rift phase and finally the Alpine overthrusts.

The relative permanence of the main structural lines through time demonstrates the close and fundamental spatial coincidence between Alpine palaeogeographic and tectonic units.

From the end of the Hercynian orogeny, the Pangaea supercontinent, which then included the European craton and its margins, was progressively fragmented, and the ensuing microcontinents reached their maximum dispersal at a time classically dated to the Late Cretaceous. The dispersal process was episodic, giving rise to successive phases of extension followed by subsidence recorded not only in the domain of the future passive margins but also on the adjoining craton where these events contributed notably to the formation of rifted sedimentary basins.

In combination with other processes, successive episodes of extension followed by passive subsidence determined the succession of correlative transgressive–regressive cycles observed on both continental margins as well as in intracratonic basins. The correlation of transgressive–regressive facies cycles at different time scales from one basin to another allows for interpretation of the role of structural control in determining the lithostratigraphic record. This therefore demonstrates the coherence of tectonic process and sedimentary response.

1. THE MAJOR EVENTS THAT CHARACTERIZE TETHYAN HISTORY

The development of the Alpine fold belt cannot be understood without first knowing the history of the Tethys and its margins, the separation of Eurasia from Gondwana in post-Palaeozoic times and the subsequent convergence and collision between Africa and Europe which began at the start of the Late Cretaceous (100 Ma).

The Tethys Ocean has had a long and complex history which can, however, be summarized in terms of three main phases: first, that of the Palaeozoic Palaeotethys, second, that of the Neotethys in the Permo-Triassic, and third, the Ligurian Tethys which formed in the Jurassic. Each of these phases is characterized, without exception, by a similar and typical succession of events.

1. Spreading cutting the Tethyan border of northern Gondwana and separation of microcontinents thus detached from Gondwana.
2. Amalgamation with the southern continental margin of Eurasia of all or part of the new oceanic domain and microcontinents detached from Gondwana.
3. Consecutive creation of folded belts along the southern border of Eurasia.

The Palaeotethys: Opening commenced in the Ordovician in Australia and propagated westward reaching future Western Europe in the Carboniferous. The ensuing fold belts range in age from Devonian to Carboniferous. Microcontinents detached from Gondwana and slices of Palaeotethyan oceanic crust comprise a large part of the Hercynian basement in Europe including that involved in the future Alps.

The Neotethys: Opening began in the Permian in the Middle East and propagated westward during the Triassic (Fig. 3.2). Spreading led to the detachment of the Cimmerian microcontinents and later to the Eo-Cimmerian orogeny in the Middle East (and the Indosinian orogeny further east), as a direct result of subduction of the Cimmerian blocks and Neotethyan oceanic crust beneath the Eurasian plate. The Eo-Cimmerian orogeny had, however, no expression in Western Europe (Section 3.2 in Chapter 14).

Although, spreading did not propagate from the Neotethys into the future western Mediterranean region (Fig. 3.2), a phase of rifting affected much of Western Europe during the Triassic (Fig. 5.1), including the region of the future Southern Alps where it was accompanied by volcanism.

The Jurassic or Liguro-piemontais Tethys: Rifting and spreading led to the separation of Gondwana from the then single Laurussia-North American continent (Fig. 3.4). The line of rupture propagated towards the west from a branch of the Neotethys and eastward from the Central North Atlantic during the course of its opening. The linkage between the two eastern and western spreading centres was accomplished along the complex Newfoundland–Gibraltar transform zone during Jurassic times.

Cretaceous evolution: In the Early Cretaceous, opening of the Bay of Biscay as well as the North Atlantic between Iberia and North America led to the detachment of the Iberian plate which then moved independently from the relative movement between Europe and Africa.

More or less simultaneously during the Early Cretaceous, rifting cut across the northern passive margin of the Jurassic Ligurian Tethys (Fig. 7.10).

Opening of the Mozambique Channel and Somali basin from the Middle Jurassic (Fig. 3.1) and the later Early Cretaceous opening of the South Atlantic Ocean progressively isolated the African plate from the Indian subcontinent and South America. As a result, the motion of Africa with respect to Europe changed to northward during the Cretaceous (115 Ma; Fig. 3.5) leading to the collision that ultimately formed the Alpine fold belt.

2. INHERITANCE: HERCYNIAN, TETHYAN AND ALPINE STRUCTURE

The structural inheritance of the pre-rift phase has profoundly influenced the disposition of structures developed during the later rift phases and Alpine compression.

A characteristic example is provided by the structural role of the Saint Bernard–Monte Rosa block ('SBR' block) and its southward continuation in the Brianconnais and Piemontais.

The 'SBR' block can be considered as originally detached from Gondwana like many others forming the heart of the Hercynian fold belt, and later accreted to southern Europe at an ill-defined date before the Permian. In the Zone Houillere of the Brianconnais, the Carboniferous floras are comparable to those known from Africa or Iberia but not to those of northern Europe and especially the Vosges. Their affinity is Gondwanan and not Laurussian. On the other hand, the Early

Carboniferous (Namurian) of the Brianconnais was not affected by Hercynian metamorphism in contrast to the adjacent European basement which was affected by high-grade Hercynian metamorphism.

Thus, the boundary between the European basement which out-crops in the External Crystalline Massifs and the basement terranes of the 'SBR' block corresponds to a major discontinuity in palaeogeography as well as in Hercynian metamorphic isograds. This important tectonic boundary continued to play a major role throughout the evolution of the Alps.

During Triassic time (Figs. 5.2 and 16.1), this tectonic discontinuity marked the main palaeogeographic boundary between two quite different domains. In the eastern part of the Sub-alpine and Helvetic domains, Triassic sediments are generally a modest 100 m or so in thickness and consist throughout of continental siliciclastics with rare and thin marine interbeds. In marked contrast, the Triassic of the Brianconnais and Piemontais zones consists mainly of subsiding marine platform carbonates or evaporites whose cumulative thickness reaches or exceeds 1000 m (Figs. 5.6 and 16.1).

From the initiation of rifting at the start of the Jurassic, this situation was completely reversed. The external domain began to subside with the accumulation of hundreds and locally 1000+ m of shales, black marls and calcareous mudstones. In contrast, the previously subsident Brianconnais domain remained high becoming sub-aerial from the Middle Liassic through the early Middle Jurassic (Figs. 6.14 and 6.17).

The boundary between the European domain and the 'SBR' block was ruptured during the Early Cretaceous as part of a continuous rift system extending from the Bay of Biscay to the Valais trough. However, the precise kinematics of this rift phase remains poorly understood.

During Tertiary orogenesis, the same boundary between the European domain and the 'SBR' block again played a major role directly linked to the envelope of major thrust nappes affected by Alpine metamorphism and comprising the internal Alpine units. This surface envelope, traditionally called *the Penninic Frontal Thrust* separates internal Alpine units from the external Alpine units which are almost unaffected by Alpine metamorphism and exhibit relatively modest thrusting.

The outer, eastern, boundary of the 'SBR' block is the oceanic crust of the Ligurian Tethys, which began to form in the Middle Jurassic and separated the 'SBR' block from the Apulia-adriatic block, site of the future Southern Alps. This boundary has also consistently played a critical role as it nucleated major thrust surfaces during the Alpine orogeny.

The palaeogeographic limits of Permo–Carboniferous basins coincide spatially with extensional faults linked to rifting and with Alpine thrust surfaces in the whole 'SBR' block and the Brianconnais complex.

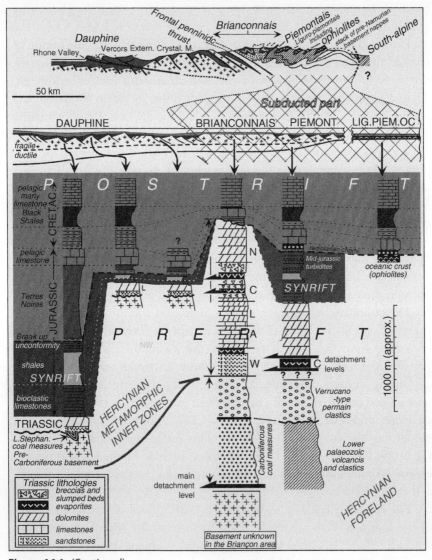

Figure 16.1 *(Continued)*

3. FROM TETHYAN EXTENSION TO ALPINE COMPRESSION

3.1 The Tethyan phase: half-graben, listric faults and detachment surfaces

The Mesozoic extensional regime, which followed basin development by post-orogenic collapse during the Late Palaeozoic, faulted the basement into a series of tilted blocks and half-graben, thereby creating the accommodation space necessary for the accumulation of sediments. The tilted blocks are classically defined by listric faults, whose geometrical characteristics are closely comparable to those observed on deep seismic reflection profiles across present-day passive margins.

Although syn-rift normal faults of Mesozoic age are known throughout the entire Alpine domain, recognition of the listric nature of these faults is often difficult because of Alpine shortening especially in the internal zones. Nonetheless, the Lugano fault provides a rare example where the flattening with depth of the listric fault can be seen at outcrop in the Southern Alps because of modest Alpine deformation and deep erosion (Fig. 10.3). Furthermore, the geometric relationships of listric faults can also be applied to the reconstruction of the Ornon fault which has defined one of the External Crystalline Massifs since Liassic times (Fig. 6.2).

Figure 16.1 *Original location of the main Alpine structural units on the North Tethyan margin and characteristic stratigraphic columns.* Comparison of the stratigraphic columns on the figure illustrates the presence of many marker horizons each characteristic of a stage or group of stages belonging to syn-rift or post-rift deposits from one end of the margin to the other. These are Late Liassic shales, Late Bathonian and Early Oxfordian *Terres Noires* (Late and Middle Jurassic), Tithonian to Berriasian pelagic nannofossil limestones, 'middle' Cretaceous black shales, and the planktonic foraminiferal and nannofossil calcareous marls of the Late Cretaceous. The existence of these markers allows reconstruction of the original stratigraphy of the metasediments where dates are rare or absent. See Figs. 11.21–11.23. The stratigraphy of the pre-rift successions on the proximal parts of the Tethyan margin (Dauphine) is very different from that found on the distal parts of the margin (Brianconnais and Piemontais) for the Triassic, Palaeozoic and basement). The palaeogeographic boundary which separates them is a permanent structural line coincident with the Penninic Frontal Thrust, which also separates almost unmetamorphosed external Alpine units (Dauphine and External Crystalline Massifs) from internal units which have been subjected to subduction (see text). The section is that of Briancon. W: Werfenian (Early Triassic); A: Anisian; L: Ladinian (Middle Triassic); C: Carnian; N: Norian (Late Triassic); SB: Subbrianconnais.

The mechanism of syn–rift extension implies the existence of detachment surfaces which allow accommodation of crustal stretching. In the case of the Lugano fault (Southern Alps, Fig. 10.3), the detachment surface lies as deep as the boundary between the brittle upper and ductile lower crust. By contrast, in the external parts of the Alpine domain along the east border of the Massif Central, the common detachment surface utilizes incompetent Stephanian and Permian beds near the top of the Hercynian basement (Fig. 6.10). A related interpretation of the Brianconnais nappes shows the probable role of late Early Triassic (Late Scythian) evaporites in the formation of flats and the possible role of a common detachment surface separating poly-metamorphosed basement from the overlying Carboniferous cover (Fig. 13.13). However, these detachment surfaces will ultimately root in basement.

3.2 The transition from rifting to oceanic spreading and the formation of the first oceanic crust

In the preliminary phase of rifting, the continental crust was stretched over a large area by normal faulting and then thinned due to the effects of low-angle detachment faulting. The result was to pull apart the continental crust leading to its complete dismemberment and tectonic denudation of the sub-continental lithospheric mantle along the length of a well-defined line of rupture. The normal stratigraphic contact (i.e. not deformed by Alpine tectonism) between serpentinites and pelagic sediments is interpreted as sub-continental mantle tectonic denudation. In effect, the characteristics of a part of Alpine ophiolites are those of the continent–ocean transition zone comparable to the 'oceanic' basement of the continental margins of the North Atlantic between Iberia and Newfoundland.

The ascent of asthenospheric mantle and the separation of the continental lithosphere into two plates resulted in the creation of a very slow-spreading axis.

The oceanic crust of the Liguro-piemontais Ocean is characterized by the absence of dyke complexes and only minor gabbros and basalts with respect to abundant serpentinized peridotites. A part of Alpine ultrabasites results from tectonic denudation of oceanic mantle. Ophicalcites are commonly underlain by serpentinites or gabbros and overlain by pillow basalts or sediments. Moreover, the various ophiolites of the Western Alps including the Chenaillet and Monviso (Figs. 11.8 and 11.16) share common characteristics with slow-spreading mid–ocean ridges such as those of the Central North Atlantic. As a result, the total width of the Tethyan oceanic crust probably did not exceed more than a few hundred kilometres (Fig. 3.5).

3.3 Alpine shortening: the shape and role of detachment surface in the Subalpine zones

In the external zones, the detachments utilize the following main incompetent horizons: Triassic evaporites, Late Liassic marls, the *Terres Noires* of the latest Middle Jurassic and Early Cretaceous marls (Figs. 13.2 and 13.6). Ramps cutting across competent strata link the flats which are commonly located at levels of rapid variation in bed thicknesses (Fig. 13.9) within depocentres controlled by Tethyan syn-rift extensional faults (Figs. 13.6, 13.8 and 13.9). The trace of ramps and flats may also be influenced by the location of syn-rift diapirs composed of Triassic evaporites (Figs. 6.22 and 6.23).

The basement of the External Crystalline Massifs now forms the line of present-day peaks, even though they represent deep structural units of the Alpine edifice. Their present exposure is a direct result of the large sub-horizontal displacement along the deep thrust decollement surface which cuts the Hercynian basement to root at the brittle/ductile boundary in the crust, as shown from the results of the deep seismic 'ECORS' programme.

The existence of detachment surfaces cutting the basement at depth has only been recently been demonstrated following publication of classical industry seismic sections crossing the Sub-alpine domain (Fig. 13.2) and the scientific 'ECORS' deep seismic profiles. While the existence of these detachments was suspected for geometric reasons, they were rarely pictured in a systematic manner. Exceptionally, the sections of Staub (1926), drawn and integrated solely from field studies, represent basement shortening by simple inversion faults (Fig. 16.2). On these classic sections, folds in the cover accommodate shortening in the basement, demonstrating that the

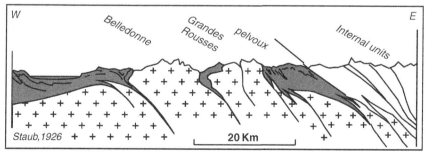

Figure 16.2 *Historical structural cross section across the External Crystalline Massifs based on outcrop geology from the latitude of Mont Blanc to the Pelvoux.* Simplified after Staub 1926.

notion of disharmonic deformation between cover and basement was already well known from this early date.

3.4 The shape and role of detachment surfaces in the Helvetic domain and the internal zones: half-grabens to nappes

Structural and stratigraphic analysis shows that the formation of nappes results in the first instance from the expulsion of the Mesozoic infill of half-graben formed during Tethyan rifting at the time of Alpine wedging. This interpretation allows a clear understanding of why each Alpine nappe is characterized by a specific stratigraphic succession, reflecting the nature of the pre-rift succession as well as the proximal or distal position of half-graben on the passive margin. This observation has been known since the start of classic Alpine field geology. For example, the Morcles nappe is interpreted as resulting from the almost complete expulsion of the Mesozoic infill of a half-graben separating the Aiguilles Rouges and Mont Blanc massifs (Fig. 13.4).

In the internal zones, the main levels of detachment are the boundaries between poly-metamorphic pre-Carboniferous basement and cover, and evaporites of the Scythian (Early Triassic) and Carnian (Late Triassic) evaporites (Figs. 13.13, 13.14 and 16.1). The existence of such detachments explains why the basement of the nappes constituting the Zone Houillere of the Brianconnais in France and Switzerland is unknown at outcrop. Likewise the Peyre Haute nappe of the Brianconnais nappes, as well as the Piemontais nappes, is composed only of Late Triassic (Norian Hauptdolomit) and a condensed or attenuated Jurassic and Cretaceous cover. This is because the detachment is rooted in the Carnian evaporites that lie just below and are locally preserved under the main thrust mass.

It can be considered that most of the listric Tethyan extensional faults would be simply and systematically rejuvenated in an inverse sense along the same plane to accommodate basement shortening during Alpine deformation. While this may well be the case for their sub-horizontal planes, there are no direct observations to demonstrate this.

On the other hand, the steeply dipping parts of normal faults located close to the land surface are commonly obliquely cut by new shear planes allowing detachment. One example, from the Austro-alpine nappes of the Grisons shows a thrust plane separating the Ortler nappe above from the Campo nappe below (Fig. 13.11). The flats on this surface exploit ductility contrasts between beds but the ramps intersect sub-vertical extensional faults themselves detaching in basement slices. This figure can be compared to the interpretation of the gneiss slice situated beneath the Remollon dome (Fig. 13.12) and similarly the top thrusts depicted on the east flank of the Bourg d'Oisans 'syncline' (Fig. 13.3).

These observations are summarized in Fig. 13.13 which represents the transformation of Tethyan half-grabens into Alpine nappes. The insight provided by this concept is very powerful and valuable and facilitates understanding of structural as well as stratigraphic relationships. However, the detachment surfaces also have multiple origins with new shears ensuring linkage between pre-existing structures.

4. DEVELOPMENT OF THE LIGURIAN TETHYS AND TRANSGRESSIVE–REGRESSIVE CYCLES AT THE EUROPEAN SCALE

The development of the Ligurian Tethys took place during part of the long period of dispersal of the continents derived from the break-up of Pangaea after the end of the Hercynian cycle (Fig. 3.1). As shown above, the Tethyan rifting consisted of a succession of phases closely correlated in time along the European margin (Fig. 6.19) as well as on the conjugate Apulia margin in the Grisons (Fig. 10.7).

This period also corresponds to a very long-term Mesozoic transgressive phase whose acme was reached during middle Cretaceous times (Vail et al. 1977) prior to the onset of closure of the Tethyan Ocean due to Alpine orogenesis (Fig. 3.7). The Mesozoic transgression was modulated by a series of transgressive–regressive cycles whose duration was the order of a system or sub-system (25–40 Ma depending on specific cases). While these cycles were initially defined in the intracratonic basins of Western Europe (Fig. 16.3) they are broadly linked to regional patterns of extension and passive subsidence.

The history of the Triassic transgressive-regressive facies cycle is linked to that of the earlier Neotethys. In the principle, this cycle formed part of the pre-rift phase of the Ligurian Tethys (Chapter 5 and Fig. 5.3A).

4.1 The Early and Middle Jurassic Transgressive–Regressive Facies cycle and the Ligurian Tethys rifting phases (+35 Ma)

This cycle records the major part of the rifting events of the Ligurian Tethys. Each episode of rifting typically comprised a phase of extension followed by a phase of rapid subsidence of the entire margin then

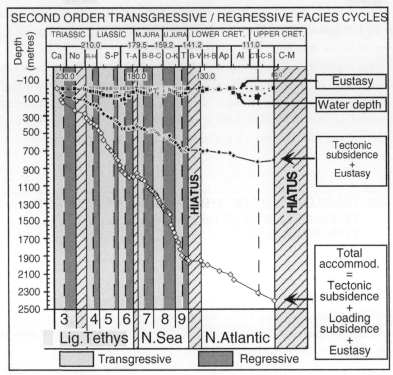

Figure 16.3 *Succession of long- to medium-term transgressive–regressive facies cycles on the west European craton and timetable of main events in the adjacent Tethyan and Atlantic domains.* Rifting episodes J1, J2, J3 refer to Fig. 6.19. *Adapted from Jacquin and Graciansky 1988, Fig. 32, p. 42.*

quiescence before the next phase (Fig. 6.20). As a consequence, the newly created accommodation space was passively infilled by sediments during the period of tectonic quiescence that followed the faulting. The syn-rift sequence thus consists of a series of tectonically controlled transgressive–regressive facies cycles (Fig. 16.3). In the Tethyan realm, these are characterized by an overall transition from shallow marine environments during the initial latest Triassic to earliest Liassic rift phase (Figs. 6.7 and 6.9) to widespread, open, deep marine environments in the axes of half-grabens and the rift prior to the onset of spreading in Middle Jurassic time. The wider effects of extension prior to spreading of the Ligurian Tethys are also recorded in the adjacent, intracratonic Aquitaine, London–Paris and Saxony basins of Western Europe (Fig. 16.5). The peak of the long-term transgressive half cycle is dated as Early Toarcian (*Falciferum* ammonite zone; start of the Late Liassic). It corresponds to the organic-rich beds called Paper shales or *Schistes Carton* that are widely distributed and of constant thickness throughout the Early Jurassic of European sedimentary basins including the European Tethyan margin.

The end of the long-term cycle was characterized by widespread regression with emergence and erosion of the European platforms as well as the Central North Sea. Correlative accumulation of several hundred metres of well-bedded marls took place in half-grabens formed by rifting as shown by the Late Toarcian to Aalenian infill of the Dauphinois Basin (Figs. 6.16 and 6.17). The rapid Bajocian transgression (Middle Jurassic) across the European-wide so-called mid-Cimmerian discontinuities ended the cycle.

4.2 The Late Middle Jurassic to lowest Cretaceous Transgressive–Regressive Facies cycle and the Ligurian Tethys spreading phase

4.2.1 Unconformities at the base of the Late Bathonian and the break-up unconformity of the Ligurian Tethys

One set of main unconformities records a major event dated to the boundary between the Middle and Late Bathonian or a little later to the Late Bathonian depending on the locality. In the Alpine region, it marks the last of the extensional episodes, which just precede the onset of spreading in the Ligurian Tethys (Fig. 6.19) on both the European and Apulia margins (Fig. 10.7) and the reorganization of depocentres on the adjacent margins. It is therefore the Tethyan break-up unconformity, which is recorded on the conjugate margins by a system of surfaces with hiatuses, erosion and commonly conglomeratic horizons. This marks the lower boundary of the corresponding major transgressive–regressive facies cycle. The unconformity is particularly well developed on the Brianconnais platform where Late Bathonian platform carbonates rest on karstified Triassic dolomites and carbonates (Figs. 6.14 and 16.1). The unconformity is also well shown on the eastern border of the Massif Central and on the Provencal platform where the Bathonian also rests directly on Triassic beds.

4.2.2 The deposition of the Terres Noires: a characteristic of the long-term transgressive phase

The Jurassic of the Dauphinois domain is characterized by deposition of the Late Bathonian to Oxfordian *Terres Noires* whose erosion causes the 'badlands' topography typical of the Sub-alpine foothills. These black shales are 1000–2000 m in thickness on those parts of the margin where relative post-break-up subsidence was maintained by continued activity along extensional faults inherited from the Tethyan phase (Fig. 6.16). Fault activity may possibly reflect distant rifting in the future Bay of Biscay and in the future North Sea at the same time as the entire Tethyan margin was undergoing thermal subsidence.

The *Terres Noires* can be correlated in part with other shales of the same facies, for example, the Argiles de Woevre of the Paris Basin, the Gold-schenkton of the French and German Jura, the Oxford Clay of England and the Heather Formation of the North Sea. The period of *Terres Noires* deposition is marked by two transgressions of very great geographic extent dated first to the Early Callovian (Gracilis ammonite zone, top of Middle Jurassic) and second to the Early Oxfordian (Mariae ammonite zone; start of the Late Jurassic). These two transgressive episodes resulted in the temporary drowning of the carbonate platforms and allowed argillaceous material originating from erosion of emergent areas such as the Massif Central, Armorican massifs, the Ardennes and Corsica–Sardinia block to be transported and deposited in the main subsiding areas.

The period of *Terres Noires* deposition represents what can be called a siliceous shale signal in the history of Jurassic sedimentation in Europe.

4.2.3 End Jurassic and basal Cretaceous (Tithonian–Berriasian) limestones: a characteristic of the long-term regressive phase

On the west European craton, the end of the Jurassic and start of the Cretaceous was marked by a regression of possible eustatic origin whose maximum was reached in the Ryazanian in the Boreal domain or during the Berriasian in the Tethyan domain. It closes with deposition of continental sediments such as the Purbeckian in England, the Paris Basin and the Jura. It is coincident with rifting in the Wessex basin, Bay of Biscay, Aquitaine basin, the Valais and the end of rifting in the North Sea.

In the proximal parts of the Tethyan margin, the same long-term regression is recorded by the maximum progradation of platform carbonates. In the pelagic environment on the distal parts of the margin, carbonate production was caused by biotic events. Blooms of calcareous nannoplankton are marked by deposition of the nannofossil and ammonite-rich limestones of Tithonian and Berriasian age that typify the Tethyan domain.

Deposition of these Tithonian–Berriasian limestones represents what may be called a 'limestone signal' in the history of Tethyan sedimentation.

4.3 The Early Cretaceous Transgressive–Regressive facies cycle and the North Atlantic rift phases (Late Berriasian to Early Aptian; 20 Ma approx)

Although the Tethyan margin had continued to passively subside as a whole after the Middle Jurassic, one part at least of the fault system was still active during the Early Cretaceous times. Much as in the early to Middle Jurassic, a network of Early Cretaceous extensional faults cut the entire European craton forming an associated system of tilted blocks (Chapter 7). To the west, the faults record closely the opening of the North Atlantic west of Iberia and the Bay of Biscay. On the Atlantic margin, contemporaneous fault systems are very well developed on the margins of Iberia as well as in the onshore Pyrenean and Aquitaine domains. In the Alpine area, these fault arrays determine the boundaries of the Vocontian Gulf and surrounding Urgonian carbonate platforms (Fig. 7.10) and probably the margins of the Valais basin (see discussion, Chapter 7).

The long-term transgressive–regressive cycle of the Early Cretaceous can be subdivided into several transgressive–regressive cycles of medium duration, i.e. less than 10 Ma. Within the Tethyan realm, the medium-term transgressive intervals were characterized, when the environment was suitable, by development of pelagic and hemipelagic marls with ammonites. The regressive intervals are characterized by construction of basinward prograding carbonate platforms on rift-related bathymetric highs. The better exposed and better studied of these is the thick Urgonian platform of which the uppermost surface defines the maximum progradation towards the marine domain and, in turn, the maximum of the Cretaceous regression all around on the Tethyan margins. Time equivalents of these Urgonian deposits in the Boreal domain are regressive prograding sandstones. Deposition of the Urgonian limestones represents what may be called another limestone signal in the history of Tethyan sedimentation.

4.4 The 'middle' to Late Cretaceous transgressive–regressive facies cycle: onset of North Atlantic opening

The start of the long-term transgressive–regressive facies cycle of the Cretaceous, also a global event, was marked by the drowning of platform carbonates from the tropics to the Arctic, from Australia to Mexico to the Tethyan domain and to Spitsbergen. The age is Early Aptian (Deshayesi ammonite zone). This period dates the opening of the Bay of Biscay and the ocean west of Iberia. The transgressive phase was characterized by widespread deposition of black shales in the North and South Atlantic Ocean as well as in the subsiding parts of the Tethys Ocean (Fig. 16.1). At the same time, condensed sections, mainly of Albian age, were deposited on the Tethyan platforms. The peak of the transgression, considered to mark the highest sea level during in the Mesozoic, was reached at around the Cenomanian–Turonian boundary.

The regressive phase corresponds approximately with the onset of Alpine deformation. Calcareous sedimentation resumed and was sustained by a new biotic evolutionary event. This was the voluminous bloom of calcareous planktonic foraminifera, which resulted in deposition of the widespread foram-nanno oozes of the Late Cretaceous of the Atlantic and Tethys. It was followed by a widespread regression that ended the Mesozoic.

4.5 Coherence between the stratigraphic record and geodynamic events: application to the interpretation of metamorphic terrains

Establishing the stratigraphy of sedimentary formations in the internal zones of fold belts is a major global challenge. A well-defined stratigraphic frame-work is essential for structural interpretation and for the reconstruction of their history. However, this is a difficult problem in those belts formed by inversion of a passive margin such as the Alps. This is because the sediments comprising the internal zones, such as the *Schistes Lustrés* originate from the distal parts of the passive margin or indeed the oceanic basin *sensu stricto*. These sediments are commonly poor in original fossil content compared to the adjacent platform and adjoining slope of the proximal parts of the margin. There is a further negative complication. The internal zones of mountain belts commonly comprise terranes subject to poly-phase defor-mation and metamorphism. Any original fossils have disappeared as a result and indisputable dates are rare. Obviously, if there are no stratigraphic markers, reconstruction and the structural framework are very difficult to establish and thus remain uncertain. Until relatively recently, this was the case for the *Schistes Lustrés* of the Alps which represent the major part of the sedimentary volume in the internal zones. However, since the 1980s, recognition of the widespread limestones deposited in response to short-lived biotic evolutionary events in the Jurassic and Cretaceous has provided a useful key to resolve the vexed problem of stratigraphic correlation in the 'Schistes Lustrés' aided by rare but indisputable dates. These limestones owe their existence to two phenomena whose effects combine and cause wide-spread and largely synchronous deposition of the Tithonian marine lime-stones: (i) the bloom of calcareous nannofossils and (ii) the progradation of platform carbonates towards the open marine domain in conjunction with the general regression (Figs. 16.4 and 16.5) at least on the west European

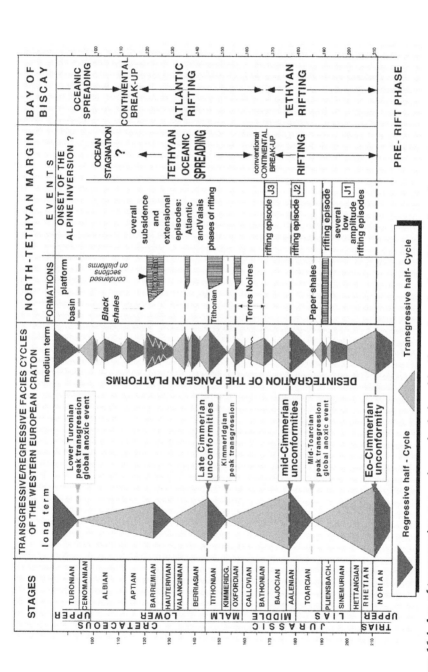

Figure 16.4 **Age of transgressive–regressive cycles induced by the main episodes of Jurassic Tethyan rifting.** This diagram is valid for the basins on the west European craton as well as in the Alpine domain. See discussion on the break-up definition in Chapter 12 and Fig. 12.3. (See colour plate 25)

Adapted and completed from Jacquin and Graciansky 1988, Fig. 2, p. 16. Ages from Gradstein et al. 1994.

craton dated to the Jurassic–Cretaceous boundary. It appears therefore legitimate to consider that the limestones (which form a sandwich between other formations which record the background sedimentary signal) can be assigned to the end of the Late Jurassic. Therefore, the interpretation shown in Figs. 11.21–11.23 is dramatic and has, in our firm view, a serious foundation that provides the basis for long-range stratigraphic correlation in the 'Schistes Lustrés' also applicable in principle to the internal zones of other Tethyan folded belts of the same age.

It should be noted that there are additional biotic event markers that further aid correlation in the internal zones. These include the siliceous signals marking environments and periods favourable for blooms of radiolaria, for example, in the Early Oxfordian that correlate in time with a major transgression. Another example is the calcareous signal of the late Cretaceous marking the bloom of planktonic foraminifera and calcareous nannofossils that precedes the latest Cretaceous long-term regression.

5. CONCLUSION: NEOTECTONICS AND ALPINE PEAKS: A COMMON HISTORY

5.1 Europe: the southwest side of Mont Blanc

Mont Blanc, the highest point, almost 5000 m, of the Alps and Europe belongs to the European continent and is composed of Hercynian basement rocks (Figs. 13.5 and 16.6). During Liassic rifting, it was a submarine topographic high as shown by the condensed sedimentary section on its eastern flank. During collision, the Mont Blanc block was thrust to the NW, onto the sedimentary 'syncline' of Chamonix and towards the neighbouring Aiguilles Rouges block. As a consequence the half-graben syn-rift fill has been expelled onto the Aiguilles Rouges block to form the Morcles nappe (Fig. 13.4).

From being a basement high during Jurassic rifting, the basement of Mont Blanc today lies at a very high altitude, notably with respect to the internal zones which lie at its eastern foot. At altitudes not more than 2000 m, the dark rocks correspond to the internal zones, mainly the Brianconnais zone originally carried over the Mont Blanc massif along the Pennine Frontal Thrust. Their relatively low altitude results from inversion of the throw on the thrust which has been recently rejuvenated as a normal fault. The black line outlines the internal thrust zones.

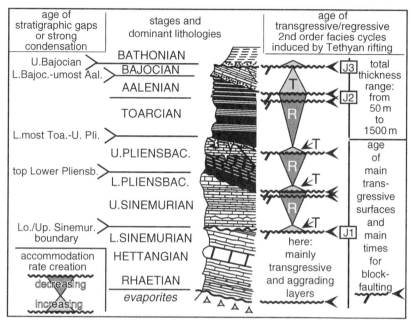

Figure 16.5 *Variation in subsidence and accommodation space and transgressive–regressive facies cycles in the Alpine foreland.* Each phase of extension is preceded by a decrease in accommodation space with a matching correlative regression. In contrast, each extension phase is followed by an increase in accommodation space with a matched, correlative transgression. The variations in accommodation space driven by episodic extension modulate the long-term evolution of subsidence. The example is from a well in the Central Paris Basin, an intracratonic basin located in the Alpine foreland. Early and Middle Jurassic, Late Jurassic and then Early Cretaceous episodes record at a distance the effect of pre-Tethyan, North Sea and Bay of Biscay rift phases. The major Early/Middle Jurassic event is recorded on the European craton by a widespread hiatus or condensed section. The Campanian-Maastrichtian hiatus corresponds to the onset of Alpine folding. Ca: Carnian; No: Norian; R-H: Rhaetic-Hettangian; S-P: Sinemurian-Pliensbachian; T-A: Toarcian-Aalenian; B-B: Bajocian-Bathonian; O-K: Oxfordian-Kimmeridgian; T: Tithonian; B-V: Berriasian-Valanginian; H-B: Hauterivian-Barremian; Ap: Aptian; C: Cenomanian; T-C-S: Turonian-Coniacian-Santonian; C-M: Campanian-Maastrichtian. *Adapted and completed from Jacquin and Graciansky 1988, Fig. 17, p. 600. Ages from Gradstein et al. (1994).*

The External Crystalline Massifs are more elevated with respect to the internal zones despite their probably higher rates of present uplift as shown on the Aar transect (Fig. 15.3). The difference in average altitude is due to young, Plio–Quaternary extension on the Penninic frontal thrust (Fig. 15.8).

In brief, the highest summit of the Alps carries the imprint of: (1) rifting, (2) inversion during the Alpine collision and (3) later extension (or negative

Figure 16.6 ***Massif of Mont Blanc, view from the south.*** The black line is the Frontal Penninic Thrust. It separates the Mont Blanc crystalline core in the snowy background from the Brianconnais Zone Houillere in the foreground. (See colour plate 26) *After Lemoine et al. 2000, Table XIV, Fig. 16.1.*

inversion), during the tectonic collapse of the internal part and heart of the Alpine fold belt as its relief became too high and unstable.

5.2 The Ocean: the northwest side of Monviso

The summit of Monviso (Fig. 16.7) consists of oceanic basalts metamorphosed to eclogite facies grade then retrogressed. These basalts, with visible pillows and pillow breccias, exhibit the traces of a tectonic history linked to spreading followed by subduction of the Liguro-piemontais oceanic lithosphere, and in turn collision and uplift of the fold belt.

The basalts were formed by ocean ridge submarine volcanism in depths of approximately 2500–3500 m. Subsequent, thermal subsidence of the ridge flanks created water depths of 4000 m though possibly less given the modest width of the ocean and sedimentation.

Spreading was replaced by plate convergence in the Late Cretaceous and the initiation of subduction. The basalts were subducted to a depth of more than 50 km demonstrated by eclogite facies metamorphism and they now form part of an assemblage of nappes which are very thin due to detachment of the most superficial part of the oceanic lithosphere (Figs. 11.2 and 14.5). These nappes were later thrust onto the European continental margin during the Europe–Africa collision.

Figure 16.7 *Monviso, looking west, view from the peak of Chateau Renard (Western Alps).* (See colour plate 27) *After Lemoine et al. 2000, Table XV, Fig. 16.2.*

How this thrusting took place in time with respect to the initial sub-duction of Liguro-piemontais lithosphere cannot be defined precisely. Later uplift took place probably at a rate of about a millimetre per year. This was mainly compensated by horizontal spreading or neotectonic collapse at a regional scale demonstrated by sets of late Alpine, steeply dipping conjugate normal faults. These multiple faults control the pyramid-like shape of Monviso.

In brief, the Monviso shows traces of: (1) seafloor spreading, (2) sub-duction, thrusting, then post-collision uplift, and finally (3) normal faulting during late orogenic collapse.

5.3 The Ocean and European continent: a view to the Ecrins–Pelvoux External Crystalline Massif from the Chenaillet

The photograph (Fig. 16.8), which looks westward, shows the oceanic crust in foreground, composed of the undeformed pillow basalts of the Chenaillet. The continental crust is in the background where the dark triangle is the Barre des Ecrins and the Pelvoux is slightly to the left. Between the Chenaillet and Pelvoux, the topographic profile of nappes of the Brianconnais zone is outlined by a black line. These nappes form the external part of the internal zones derived from the European continental margin and are found at a much lower altitude than the

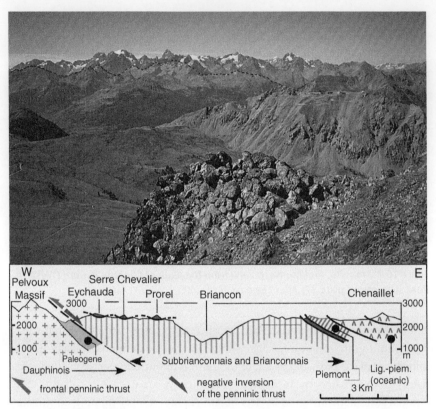

Figure 16.8 *Two crusts, oceanic and continental, seen from the summit of the Chenaillet looking west.* (See colour plate 28) *After Lemoine et al. 2000, Table XVI, Fig. 16.3.*

crystalline massif of the external zone on which it has been thrust. The change in relief is probably due to recent displacement along a normal fault rejuvenating the Pennine frontal thrust as shown on a section oriented E–W (Fig. 15.8).

The Chenaillet ophiolites have experienced almost no Alpine deformation or metamorphism (Fig. 11.2, left and Fig. 14.5) except for slight prehnite–pumpellyite-grade metamorphism which may be of oceanic origin and related to the spreading process. These ophiolites have quite clearly not been subjected to the high-pressure–low-temperature metamorphism characteristic of subduction. In contrast to the Monviso ophiolites, their emplacement on the European margin by thrusting could only have occurred by obduction and probably from the start of the collision.

5.4 Africa emplaced on Europe: The Matterhorn

The Matterhorn pyramid (Fig. 16.9) consists of a variety of gneisses derived from Apulia basement. In detail, two superposed units can be recognized. The higher Valpelline series (3 in Fig. 16.9) consists of granulite facies gneisses of the lower continental crust (3b) resting on the less highly metamorphosed Arolla gneiss of the upper crust (3a). At the base of the pyramid, partly hidden by ice and snow (2 in Fig. 16.9), Liguro-piemontais ophiolites and their sedimentary cover *(Schistes Lustrés)* are present. The glacier to the left on the slopes of the Monte Rosa covers European basement (1 in Fig. 16.9). In the Matterhorn area, the oceanic basement with its sedimentary cover is sandwiched between the Apulia and European continental margins.

Figure 16.9 *Matterhorn: a fragment of Africa resting on Europe.* (See colour plate 29) *After Lemoine et al. 2000, Table XVI, Fig. 16.5.*

REFERENCES

Agard, P., Lemoine, M., 2003. Visage des Alpes: structure et evolution geodynamique *Commission de la Carte Geologique du Monde*, Paris, *48p*.

Airy, G.B., 1855. On the compilation of the effect of the attraction of mountain ranges as disturbing the apparent astronomical latitude of stations in geodetic surveys. Proc. Roy. Soc. Lond. 7, 51–75, More than 70 years.

Alhilali, K.A., Damuth, J.E., 1987. Slide block (?) of Jurassic sandstone and submarine channels in the basal upper Cretaceous of the Viking Graben: Norwegian North Sea. Mar. Petrol. Geol. London, 4, 35–49.

Amaudric Du Chaffaut, S., 1982. Les unites alpines à la marge orientale du massif cristallin Corse. *Thesis, Ecole Normale Superieure Paris*, 133p.

Amaudic Du Chauffaut, S., Fudral, S., 1986. De la marge oceanique à la chaîne de collision dans les Alpes du Dauphine. Bull. Soc. Geol France, Paris, 8(11), 197–231.

Antoine, P., 1968. Sur la position structurale de la 'zone du Versoyen' (nappe des Brèches de Tarentaise sur les confins Franco-italiens). Trav. Lab. Geol. Univ. Grenoble, 44, 5–26.

Antoine, P., 1971. La zone des Brèches de Tarentaise entre Bourg Saint Maurice (Vallée de l'Isère) et la frontière Italo-suisse. Trav. Lab. Geol. Univ Grenoble, 9, 367p.

Antoine, P., Loubat, H., Vatin-Pérignon, N., 1973. Hypothèses nouvelles sur l'origine et la signification des 'ophiolites' du domaine pennique externe (Savoie, Valais). Géol. Alpine 49, 21–39.

Antoine, P., Barféty, J.C., Vivier, G., Gros, Y., Fudral, S., Landry, P., et al., 1992. Carte geologique de la France à 1/50 000, feuille 727: Bourg Saint Maurice, second ed. Brgm Ed, Orléans, 120p.

Argand, E., 1916. Sur l'arc des Alpes occidentales. Eclogae Geol. Helv. 14, 145–191, 70 years, Basel.

Argand, E., 1934. La zone pennique. Guide Geol. Suisse, *Wepf*, Basel, 3, 149–189.

Arnaud, H., 1974. Carte geol. France (1/50 000), feuille Mens (XXXII-37–844). *BRGM Ed*, Orleans.

Autran, A., Carreras, J., Durand-Delga, M., Laumonier, B., 1995. In: Barnolas, A., Chiron J.C. (Eds.), Synthèse geologique et Geophysique des Pyrenees. *BRGM* and ITGE Printers, Orléans, France. pp. 679–693.

Avedik, F., Camus, A.L., Ginsburg, A., Montadert, L., Roberts, D.G., Whitmarsh, R.B., 1982. A seismic refraction and reflection study of the continent-ocean transition beneath the North Biscay margin. Phil. Trans. Roy. Soc. London, A 305 (1489), 5–35.

Bailey, E.B., 1935. Tectonic Essays, mainly Alpine, Oxford Univ. Press, Oxford., 200 p, 49 fig.

Bally, A.W., Gordy, P.L., Stewart, G.A., 1965. Structure, seismic data and orogenic evolution of the Southern Canadian Rocky mountains. Bull. Canad. Petr. Geol. 14 (3), 337–381.

Bally, A.W., Snelson, S., 1981. Realms of subsidence. Can. Soc. Pet. Geol. Mem 6, 9–75.

Bally, A.W., Tari, G.C., 2004. Interpretation of seismic data in a regional context: Developing Frontier Exploration Opportunities. *AAPG Winter Education Conference*.

Barfety, J.C., Lemoine, M., de Graciansky, P.C., Tricart, P., Mercier, D., 1995. Notice explicative, carte geol. France (1/50 000), feuille Briançon (823). *BRGM Ed.*, Orleans.

Baud, A., Septfontaine, M., 1980. Presentation d'un profil palinspastique de la nappe des Prealpes medianes en Suisse occidentale. Eclogae. Geol. Helv. 73 (2), 651–660.

Baud, A., Marcoux, J., Guiraud, R., Ricou, L.E., Gaetani, M., 1993. Late murghabian map. In: Dercourt, J., Ricou, L.E., Vrielinck, B. Eds., *Atlas Paleoenvironmental Maps. BEICIP-FRANLAB*, Rueil Malmaison.

Baudrimont, A.F., Dubois, P., 1977. Un bassin mesogeen du domaine peri-alpin: le Sud-Est de la France. Bull. Cent. Rech. Explor. Prod. Elf-Aquitaine, Pau 1 (1), 109–320.

Bailey, E.B., 1935. Tectonic Essays, mainly Alpine, Oxford University Press, Oxford, 200 p, 49 figs.

Beck, R.H., Lehner, P., 1974. Oceans, New Frontier in Exploration. Am. Ass. Petrol. Geol. Bull. 58 (3), 376–395.

Beltrando, M., Rubato, D., Compagnni, R., Lister, G., 2007. Was the Valaisan basin floored by oceanic crust? Evidence of Permian magmatism in the Versoyen Unit (Valaisan domain, NW Alps. Ophioliti 32 (2), 85–99.

Beltrando, M., Rubato, D., Manatschal, G. (2010). From passive margins to orogens: the link between ocean-continent transition zones and (ultra)high-pressure metamorphism. *Geological Society of*, America, 38, 6, 559–562.

Beltrando, M., Compagnoni, R., Lombardo, B. (2010). (Ultra-) Highpressure metamorphism and orogenesis: An Alpine perspective. *Gondwana Res.* doi: 10.1016. 20 p., in press.

Berger, W.H., Winterer, E.L., 1974. Plate stratigraphy and the fluctuating carbonate line. Spec. Publ. Intern. Assoc. Sediment. I, 11–48.

Bernoulli, D., 1964. Zur Geologie des Monte Generoso. Beitr. Geol. Karte Schweiz, N.F. 118, 134 p.

Bernoulli, D., Jenkyns, H.C., 1974. Alpine, Mediterranean and Central Atlantic Mesozoic facies in relation to the early evolution of the Tethys. *In*: Dott Jr, R.H., and Shaver, R.H. (Eds.), Modern and Ancient Geosynclinal Sedimentation, *Spec. Public. Soc. Econ. Paleont. Mineral.*, 19, 129–160.

Bernoulli, D., Lemoine, M., 1980. Birth and evolution of the Tethys: the overall situation. *26e Congrès géol. intern., Paris, Coll.* C5, 168–179.

Bernoulli, D., Bertotti, G., Froitzheim, N., 1990. Mesozoic faults and associated sediments in the Austro-alpine - south Alpine passive continental margin. Mem. Soc. Geol. Ital 45, 25–68.

Bernoulli, D., Manatschal, G., Desmurs, L., Müntener, O., 2003. Where did Gustav Steinmann see the Trinity, Back to the roots of an Alpine ophiolite concept. *In*: Dilek, Y., Newcombe, S. (Eds.), Ophiolite Concepy and the Evolution of Geological Thought, *Geol. Soc. Am., Spec. Public.*, 373, 93–110.

Bernoulli, D., Jenkyns, H.C., 2009. Ancient oceans and passive margins of the Alpine-Mediterranean Tethys; deciphering clues from Mesozoic pelagic sediments and ophiolites. Sedimentology 56, 149–190.

Bertotti, G., Picotti, V., Bernoulli, D., Castellarin, A., 1993. From rifting to drifting: Tectonic evolution of the south-Alpine upper crust from the Triassic to the Early Cretaceous. Sediment. Geol. 86, 53–76.

Bertrand, M., 1884. Rapports de structure des Alpes de Glaris et du Bassin Houiller du Nord *Bull. Soc. Geol. France*, Paris, 3, (XII), 318–330.

Bigi, G., Cosentino, D., Parotto, M., Sartori, R., 1990. *Structural model of Italy, map scale 1:500 000, sheet 1.*

Bilotte, M., 2007. Permanence, au Cretace superieur, de la position de la limite plateforme/bassin dans la zone sous-pyrenenne orientale (Aude, France) Implications geodynamiques *Geologie de la France*, BRGM Ed., Orléans, 1, 33–53, 13 fig, 4 pl.

Blackett, P.M.S., Clegg, J.A., Stubbs, P.H.S., 1960. An analysis of rock magnetic data. Proc. Roy. Soc. London, A 256 (1286), 291–322.

Bogdanoff, S., Schott, J.J., 1977. Etude paleomagnetique et analyse tectonique dans les schistes rouges permiens. du sud de l'Argentera. *Bull. Soc. Geol. France*, Paris, 7a (19), 909–916.

Boillot, G., Grimaud, S., Mauffret, A., Mougenot, D., Kornprobst, J., Mergoil_Daniel, J., et al., 1980. Ocean-continent boundary off the Iberian margin: a serpentinite diapir west of the Galicia Bank. Earth and Planet. Sci. Lett. 48, 23–34.

Boillot, G., Lemoine, M., Montadert, L., Biju-Duval, B., 1984. Les marges continentales actuelles et fossiles autour de la France. Masson Editions, Paris, 342 pages.

Boillot, G., Recq, M., Winterer, E.W., Meyer, A.W., Applegate, J., 1987. Tectonic denudation of the upper mantle along passive margins: a model based: a model based on drilling results (ODP leg 103, western Galicia margin, Spain. Tectonophysics 132, 335–342.

Bonijoly, D., Perrin, J., Roure, F., Bergerat, F., Courel, L., Elmi, S., et al., 1996. The Ardeche palaeomargin of the South-East Basin of France: Mesozoïc evolution of a part of the Tethyan continental margin (Geologie Profonde de la France) *Mar. Petrol Geol.*, 13, (6), 607–623.

Bonnatti, E., Honnorez, J., 1976. Sections of the Earth's crust in the Equatorial Atlantic. J. Geophys. Res. 81, 4087–4103.

Bosellini, A., Winterer, E., 1975. Pelagic limestones and radiolarites of the Tethyan Mesozoic; A genetic model. Geology 3, 279–282.

Bousquet, R., Oberhänsli, R., Goffe, B., Wiederkehr, M., Koller, F., Schmid, S., et al., 2008. Metamorphism of metasediments at the scale of an orogen: a key to the Tertiary geodynamic evolution of the Alps. *In*: Siegesmund S., Fügenschuh B., Froitzheim N. (Eds.) Tectonic Aspects of the Alpine-Dinaride-Carpathian System. *Geol. Soc. Special Publications*, 298, 393–411.

Brongniart, A., 1813. Essai de classification mineralogique des roches mélanges. Journal des Mines v. XXXIV, 190–199.

Bucher, S., Bousquet, R., 2007. Metamorphic evolution of the Brianconnais units along the ECORS-CROP profile (Western Alps): new data on metasedimentary rocks. Swiss J. Geosci. 100, 227–242.

Bucher, S., Ulardic, C., Bousquet, R., Ceriani, S., Fugenschuh, B., Gouffon, Y., et al., 2004. Tectonic evolution of the Briançonnais units along a transect (ECORS-CROP) through the Italian Western Alps. Eclogae Geol. Helv. 97, 321–345.

Bullard, E.C., Everett, J.E., Smith, A.G., 1965. The fit of the continents around the Atlantic. Phil. Trans. Roy. Soc. London, A 258 (1088), 291–322.

Butler, R.H., 2008. Faulted foredeep and outer rise, Calabrian trench. www.seismicatlas.org.

Caby, R., Michard, A., Tricart, P., 1971. Decouverte d'une brèche polygenique à elements granitoïdes dans les ophiolites metamorphiques piemontaises (Schistes Lustrés du Queyras, Alpes françaises). C.R. Acad. Sc. Paris, 273, 999–1002.

Cannat, M., 1993. Emplacement of mantle rocks in the seafloor at mid-ocean ridges. J. Geophys. Res. 98 (B3), 4163–4172.

Cannat, M., Lagabrielle, Y., Bougault, H., Casey, J., De Coutures, N., Dmitriev, L., et al., 1997. Ultramafic and gabbroic exposures at the Mid-Atlantic Ridge: geological mapping in the 15°N region. Tectonophysics 279 (1), 193–213.

Cannat, M., Sauter, D., Mendel, V., Ruellan, E., Okino, K., Escartin, J., et al., 2006. Modes of seafloor generation at a melt-poor ultraslow-spreading ridge. Geology 34, 605–608.

Carey, S.W., 1958. Continental Drift- a symposium. University of Hobart, 375p.

Caron, C., 1972. La nappe superieure des Prealpes; subdivisions et principaux caractères du sommet de l'edifice prealpin. Eclogae Geol. Helv. 65, 57–73.

Caron, J.M., Bonnin, B., 1980. La Corse, France: Introduction à la géologie du Sud-Est. *26th Intern. Geol. Congress, guide book*, Paris, 80–90, and 149–168.

Champagnac, J.D., Sue, C., Delacou, B., Burkhard, M., 2004. Brittle deformation in the inner NW Alps: from early orogen-parallel extrusion to late orogen-perpendicular collapse. Terra Nova 16, 232–242.

Champagnac, J.D., Molnar, P., Anderson, R.S., Sue, C., Delacou, B., 2007. Quaternary erosion-induced isostatic rebound in the western Alps. Geology 35, 195–198.

Chenet, P., Montadert, L., Gairaud, H., Roberts, D.G., 1982. Extension ratios on the Galicia, Portugal, and Northern Biscay Continental Margins: Implications for evolutionary models of Passive Continental Margins,. *In*: Watkins, J., Drake, C.L. (Eds.), Studies in Continental Margin Geology, AAPG Memoir 34, 703–716.

Chenet, P.Y., Francheteau, J., 1980. Bathymetric Reconstruction Method: Application to the Central Atlantic Basin between 10°N and 40°N. *Initial Reports of the Deep Sea Drilling Project, US Government Printing Office* Washington DC, vol. LI, LII, LIII, 1501–1513.

Chevalier, F., 2002, Vitesse et cyclicite de fonctionnement des failles normales de rift, implication sur le remplissage stratigraphique des bassins et sur les modalites d'extension d'une marge passive fossile. Application au demi-graben de Bourg d'Oisans (Alpes Occidentales, France). *Thesis, Université de Bourgogne,* Dijon (France), *313p. Unpublished.*

Chevalier, F., Guiraud, M., Garcia, J.P., Dommergues, J.L., Quesne, D., Allemand, P., et al., 2003. Calculating the long-term displacement rates of a normal fault from the high resolution stratigraphic record (early Tethyan rifting, French Alps). Terra Nova 15, 410–416.

Choukroune, P., Ballèvre, M., Cobbolod, P., Gautier, Y., Merle, O., Vuichard, J.P., 1986. Deformation and motion in the Western Alpine arc. Tectonics 5 (2), 215–226.

Choukroune, P., 1994. Deformations et deplacements dans la croute terrestre. *Masson Editions,* Paris, 226 p.

Claudel, M.E., Dumont, T., Tricart, P., 1997. Une preuve d'extension contemporaine de l'expansion oceanique de la Tethys ligure en Brianconnais: les failles du Vallon Laugier. C.R. Acad. Sc Paris, 325, 273–270.

Conti, P., Manatschal, G., Pfister, M., 1994. Syn-rift sedimentation, Jurassic and Alpine tectonics in the central Örtler nappe (Eastern Alps, Italy). Eclogae Geol. Helv. 87, 63–90.

Cordey, F., Bailly, A., 2007. Alpine ocean seafloor spreading and onset of pelagic sedimentation: new radiolarian data from the Chenaillet - Montgenèvre ophiolite (French - Italian Alps. Geodynamica Acta 20 (3), 131–138.

Courel, L., coordinateur 1984. Trias. *In:* Debrand-Passard, S. (Ed.), Synthese Geologique du Sud-Est de la France. *Mem. Bureau Rech. Geol et Min. (BRGM),* Orleans, France, n° 125, 61–118.

Courel, L., Poli, E., Vannier, F., Le Strat, P., Baud, A., Jacquin, T., 1998. Sequence stratigraphy along a Triassic transect on the Western Peri-tethyan Margin in Ardeche (SE France): Correlations with Subalpine and Germanic Realms. *In:* de Graciansky, P.C. et al. (Eds.), Mesozoic and Cenozoic Sequence Stratigraphy of European Basins, *SEPM Special Publication, 60,* Tulsa, OK, 691–700.

Dal Piaz, G.V., 2001. History of tectonic interpretations of the Alps. J. Geodyn. 32, 99–114.

Dana, J.D., 1873. On some results of the Earth's contraction from cooling. Am. J. Sci. 5 (3), 423–443.

Dardeau, G., de Graciansky, P.C., 1990. Halocinèse et jeu de blocs dans les Baronnies: Diapirs de Propiac, Montaulieu, Condorcet (Departement de la Drôme, France) *Bull. Centres Rech. Explor. Prod. Elf-Aquitaine,* Pau, 14, 1, 111–151 and 14, (2), 443–464.

Debelmas, J., 1955. Les zones Subbrianconnais et Briançonnais occidentale entre Vallouise et Guillestre (Hautes Alpes) *Memoire pour servir a l'explication de la carte geologique detaillee de la France,* Paris, 171pp.

Debelmas, J., 1987. Les Alpes: un exemple de chaîne de collision. - *Bulletin Pedagogique trimestriel n°2bis. Special Geologie, Association des Professeurs de Biologie et Geologie de l'Enseignement Public,* Paris, p. 115.

Debrand-Passard, S., Autran, A., 1984. Grandes lignes et principales etapes de l'evolution geodynamique du Sud-Est de la France. *In:* Debrand-Passard, S. (Ed.), Synthese Geologique du Sud-Est de la France. *Mem. Bureau Rech. Geol. et Min. (BRGM),* Orleans, n° 125, p. 585.

Decandia, F.A., Elter, P., 1972. La zona ofiolitifera del Bracco nel settore compresso fra Levanto e Val Gaveglia (Appenino Ligure). Memorie della Società Geologica Italiana Roma, 11, 503–530.

Dercourt, J., Ricou, L.E., Vrielink, B. (Eds.) 1993. Atlas Tethys Paleoenvironmental Maps, *BEICIP-FRANLAB, Gauthier-Villars Publishers*, Rueil-Malmaison, France 307p, 14 maps, 1 pl.

Dercourt, J., Gaetani, M., Vrielink, B., Barrier, E., Biju-Duval, B., Brunet, M.F., Cadet, J.P., Casquin, S., Sandulescu, M. (Eds.), 2000. Atlas Peri-Tethys, Paleogeographical maps *CCGM. CGMW*, Paris, 24 maps and explanatory notes, 269p.

Desmurs, L., Müntener, O., Manatschal, G., 2002. Onset of magmatic accretion with magma-poor passive margins: a case study from the Err-Platta ocean-continent transition? Eastern Switzerland. Contrib. Mineral. Petrol. 144, 365–382.

de Saussure, H.B., 1790. Voyages dans les Alpes. Vol. IV, Neuchâtel.

Dewey, J.F., Helman, M.L., Knott, S.D., Turco, E., Hutton, D.H.W., 1989. Kinematics of the Western Mediterranean. *In:* Coward, M., Dietrich, D. (Eds.), *Geol. Soc. London Sp. Publ.*, 45, 265–283.

Dezes, P., Schmid, S.M., Ziegler, P.A., 2004. Evolution of the European Cenozoic Rift System: interaction of the Alpine and Pyrenean orogens with their foreland lithosphere. Tectonophysics 389, 1–33.

Dietz, R.S., 1961. Continent and ocean basin evolution by spreading of the sea floor. Nature 190, 854–857.

Dietz, R.S., 1963. Collapsing continental rises: an actualistic concept of geosynclines and mountain building. J. Geol. 71 (3), 314–333.

Doglioni, C., 1987. Tectonics of the Dolomites (Southern Alps, Northern Italy). J. Struct. Geol. 9 (2), 181–193.

Drake, C.L., Ewing, M., Sutton, G.H., 1959. Continental margins and geosynclines, the east coast of North America north of Cape Hatteras. *Phys. Chem. Earth*, Pergamon publishers, New York, 3, 110–198.

Dromart, G., Allemand, P., Quiquerez, A., 1998. Calculating rates of syn-depositional normal faulting in the western margin of the Mesozoic Subalpine Basin (south-east France). Basin Res. 10, 235–260.

Dromart, G., Ader, M., Allemand, P., Curial, F., Guillocheau, F., Vidal, G., 1996. Delineation of hybrid and carbonate reservoirs through genetic stratigraphy in the Lower Mesozoic of southeastern France: procedures and benefits. Mar. Petrol. Geol. 14 (6), 653–670.

Dumont, T., 1988. Late Triassic – Early Jurassic evolution of the Western Alps and their European foreland; initiation of Tethyan rifting. Bull. Soc. Geol. France Paris, 8 (IV), 601–611.

Dumont, T., Champagnac, J.D., Crouzet, C., Rochat, P., 2008. Multistage Alpine shortening in the Central Dauphiné (French Western Alps): implications for pre-Alpine restoration *Swiss J. Geosci.*, doi: 10.1007/s00015-008-1280-2.

Durand, C., Gente, P., Dauteuil, O., 1995. Caracteristiques morphologiques des segments axiaux de la dorsale medio-Atlantique (20°N-24°N). C.R.Acad.Sci. Paris 320, 411–419.

Du Toit, A., 1937. Our Wandering Continents, Oliver and Boyd publishers, London and Edinburgh, 366pp.

Duval, B., Cramez, C., Jackson, M.P.A., 1992. Raft tectonics in the Kwanza Basin, Angola. Mar. Petrol. Geol. London 9, 389–404.

Eberli, G.P., 1988. The evolution of the southern continental margin of the Jurassic Tethyan Ocean as recorded in the Allgäu Formation of the Austro-alpine Nappes of Graubünden (Switzerland). Eclogae Geol. Helv. 81 (1), 175–214.

Elmi, S., 1984. Tectonique et Sedimentation Jurassique. *In:* Debrand-Passard, S., Ed., Synthese Geologique du Sud-Est de la France. *Mem. Bureau Rech. Geol et Min. (BRGM)*, Orleans, France, n° 125, 166–175.

Escher, A., 1866. Sur la geologie du canton de Glaris. *Actes Societe Helvetique Sciences Naturelles*, 71–75.

Escher, A., Masson, H., Steck, A., 1987. Coupes geologiques des Alpes occidentales suisses. Rapport Geologie n° 2 du Service Hydrologique et Geologique National, Bern.

Escher, A., Beaumont, C., 1997. Formation, burial and exhumation of basement nappes at crustal scale: a geometric model based on the Western Swiss-Italian Alps. J. Struct. Geol. 19 (7), 955–974.

Floquet, M., Hennuy, J., 2003. Evolutionary gravity flow deposits in the Middle Turonian - Early Coniacian Southern Provence basin (SE France): origins and depositional processes. In: Locat, J., Mienert, J. (Eds.), Submarine mass movements and their consequences. Kluwer Academic Publishers, Dordrecht, the Netherlands, Adv. Nat. Technol. Hazards Res. 19, 417–424, 3 fig.

Floquet, M., Gari, J., Hennuy, J., Leonide, P., Philip, J., 2005. Sedimentations gravitaires carbonatees et siliciclastiques dans un bassin en transtension, series d'age Cenomanien a Coniacien moyen du Bassin Sud-Provencal. 10 ème Congres Français de Sedimentologie. Livret d'excursion Geologique, Association des Sedimentologistes Française, Paris, Special volume 52, 80 p.

Ford, M., Lickorish, W.H., Kusznir, N.J., 1999. Tertiary foreland sedimentation in the Southern Subalpine Chains, SE France: A geodynamic appraisal. Basin Res. 11, 315–336.

Ford, M., Duchene, S., Gasquet, D., Vanderhaeghe, O., 2006. Two-phase orogenic convergence in the external and internal SW Alps. J. Geol. Soc. London 163, 815–826.

Fraser, S.I., Fraser, A.J., Lentini, M.R., Gawthorpe, R.L., 2007. Return to rifts – the next wave: fresh insights into the petroleum geology of global rift basins. Petrol Geol 13, 99–104.

Frey, M., Desmond, J., Neubauer, F., 1999. Metamorphic maps of the Alps 1:1000000 Schweiz. Mineral. Petrogr. Mitt., Zürich 79/1.

Froitzheim, N., Eberli, G.P., 1990. Extensional detachment faulting in the evolution of a Tethys passive continental margin, Eastern Alps, Switzerland. Geol. Soc. Am. Bull 102, 1297–1308.

Fügenschuh, B., Schmid, S.M., 2003. Late stages of deformation and exhumation of an orogen, constrained by fission-track data: a case study in the Western Alps. Geol. Soc. Am. Bull. 115, 1425–1440.

Fugiwara, T., Lin, J., Masumoto, T., Kelemen, P.B., Tucholke, B.E., Casey, J.F., 2003. Crustal Evolution of the Mid-Atlantic Ridge near the Fifteen-Twenty Fracture Zone in the last 5 Ma. Geochem. Geophys. Geosyst., 4, 3, (1024), 1–25, doi: 10.1029/2002GC000364, ISSN: 1525–2027.

Funk, H., Gubler, E., Rybach, L., Lambert, A., 1980. Hoehenaenderungen der Fixpunkte im Gotthard-Bahntunnel zwischen 1917 und 1977 und ihre Beziehung zur Geologie. Eclogae Geol. Helv. 73, 583–592.

Gaetani, M., Gnaccolini, M., Jadoul, F., Garzanti, E., 1998. Multiorder Sequence Stratigraphy in the Triassic System of the Western Southern Alps In: de Graciansky, P.C. et al. (Eds.), SEPM Special publication, 60, Mesozoic and Cenozoic Sequence Stratigraphy of European Basins, Tulsa, OK, 704–717.

Gaskell, T.F., Hill, M.N., Swallow, J.C., 1959. Seismic measurements made by H.M.S. Challenger in the Atlantic, Pacific and Indian Oceans, and in the Mediterranean Sea, 1950–1953. Phil. Trans. Roy. Soc. London, A 251, 23–85.

Gauthier, A., Rehault, J.P., 1988. Dérive de la plaque corso-sarde Publ. Centre Régional de Documentation Pédagogique de Corse (Ajaccio), 2 slides.

Gente, P., Pockalny, R.A., Durand, C., Deplus, C., Maiaa, M., Ceuleneer, G., et al., 1995. Characteristics and evolution of the segmentation of the Mid-Atlantic Ridge between 20°N and 24°N during the last 10 million years Earth and Planet. Sci. Lett., 129, (1–4), 55–71.

Gidon, M., 1971. Notice explicative, carte geol. France (1/50 000), feuille Gap (N°XXXIII-38), BRGM, Orleans. France.

Gignoux, M., 1950. Geologie Stratigraphique, 4eme edition. Masson Ed., Paris, 737p.

Gillchrist, R., Coward, M., Mugnier, J.L., 1987. Structural inversion in the external French Alps. Geodynamica Acta, Paris 1, 5–34.

Ginsburg, L., Montenat, C., 1966. Carte geol. France (1/50 000), feuille Roquesteron (XXXVI-42-972). BRGM Ed, Orleans.

Goffé, B., Schwartz, S., Lardeaux, J.M., Bousquet, R., 2004. Exploratory notes to the map: metamorphic structure of the Alps, Western and Ligurian Alps. Mitteilungen der Osterereichischen Mineralogischen Gesellschaft 149, 125–144.

Graciansky P.C. de, Busnardo, R., Doublet, R., Martinod, J., 1987. Tectonique distensive d'âge cretace Inferieur aux confins des Baronnies (Chaînes subalpines meridionales); liaison avec le rifting atlantique; consequence sur la tectonique alpine. Bull. Soc. Geol. France, Paris, 8, III, 6, 1211–1214.

Graciansky P.C. de, Lemoine, M., 1988. Early Cretaceous extensional tectonics in the southwestern French Alps: A consequence of North-Atlantic rifting during Tethyan spreading Bull. Soc. Geol. France, Paris, 8, (IV), 733–737.

Graciansky P.C. de, Dardeau, G., 1990. Halocinèse et jeu de blocs pendant l'evolution de la marge europeenne de la Tethys - Les diapirs des Baronnies et des Alpes maritimes Bull. Centres Rech. Explor. Prod. Elf-Aquitaine, Pau, 14, (2), 109–110.

Graciansky P.C. de, Dardeau, G., Dommergues, J.L., Durlet, C., Marchand, D., Dumont, T., et al. and others 1998. Ammonite biostratigraphic correlation and Early Jurassic Sequence Stratigraphy in France – Comparisons with some UK sections In: de Graciansky, P.C. et al. (Eds.), SEPM Special publication, 60, Mesozoic and Cenozoic Sequence Stratigraphy of European Basins, 583–622, Tulsa, OK.

Gradstein, F.M., Agterberg, F.P., Ogg, J.G., Hardenbol, J., Van Veen, P., Thierry, J., et al., 1994. A Mesozoic time scale. J. Geophys. Res. 99, 24051–24074.

Groupe Galice, 1979. The continental margin of Galicia and Portugal: Acoustic stratigraphy, dredge stratigraphy and structural evolution. In: Sibuet, J.C., Ryan, W.B.F. (Eds.), Initial Reports of the Deep Sea Drilling Project, Vol. 47B, U.S. Gov. Print. Off, Washington, D.C, 663–662M.

Hall, J., 1859. Paleontology Geol. Sur. New York, Albany, 3, (1), 66–95.

Haug, E., 1900. Les geosynclinaux et les aires continentales. Contribution a l'etude des regressions et transgressions marines Bull. Soc. Geol. France, Paris, 28, (3), 617–711.

Hayward, A., Graham, R., 1989. Some geometrical characteristics of inversion In: Williams, G., Powell, C.M., Cooper, M.A. (Eds.), Inversion Tectonics, Geol. Soc. London, Sp. Publ., 44, Fig. 19, 17–40.

Heezen, B.C., Tharp, M., Ewing, M., 1959. The Floor of the Oceans.1: The North Atlantic. Geol. Soc.Am. Spec. Paper 65, 122p.

Heim, A., 1891. Untersuchungen über den Mechanismus der Gebirsbildung im Anschluss an die geologische Monographie der Tödi-Windgällen Gruppe, Schwabe, Basel., Bd1, (346p.), Bd 2 (246 p.), 1878.

Helmert, F.R., 1909. Die Tiefe Der Ausgleichflache Bei Des Prattsche Hypothese Fur das Gleicagewicht De Erdkruste und De Verlauf Der Schewerestorung von Innern Der Kontinente Und Ozeane Nach Den Kusten: Sitzber Deut. Kgl. Preusz. Akad. Wiss., 18, 1192–1198.

Hess, H.H., 1962. History of the Ocean Basins. In Petrologic Studies – a volume in honour of A.F. Buddington, Mem. Geol. Soc. Am., 599–620,

Holmes, A., 1928. Radioactivity and earth movements. Trans. geol. soc. Glasg. 18 (3), 559–608.

Hughes, B.D., Baxter, K., Clark, R.A., Snyder, D.B., 1996. Detailed processing of seismic reflection data from the frontal part of the Timor Trough accretionary wedge, eastern Indonesia In: Hall, R., Blundell, D. (Eds.) Geol. Soc. London. Sp. Pub. 106, Tectonics of South east Asia 75–83.

Jacquin, T., Arnaud-Vanneau, A., Arnaud, H., Ravenne, C., Vail, P., 1991. Carbonate production and stratigraphic architecture of a shelf-margin. Mar Petrol Geol., London 8, 122–139.

Jacquin, T., de Graciansky, P.C., 1988. Major Transgressive - Regressive Facies Cycles: Stratigraphic Signature of European Basin Development. In: de Graciansky, P.C. et al. (Eds.), SEPM Special publication, 60, Mesozoic and Cenozoic Sequence Stratigraphy of European Basins, 15–29, Tulsa, OK.

Jacquin, T., de Graciansky, P.C., 1988. Transgressive - Regressive (Second Order) Facies Cycles: the effects of tectono-eustasy, In: de Graciansky, P.C. et al. (Eds.) SEPM Special publication, 60, Mesozoic and Cenozoic Sequence Stratigraphy of European Basins, Tulsa, OK, 31–42,.

Jarvis, G.T., Mckenzie, D.P., 1980. The development of sedimentary basin with finite extension rates. Earth Planet. Sci. Lett 48, 42–52.

Jouanne, F., Menard, G., 1994. Quantification des mouvements verticaux actuels du Sud du Jura et des Alpes nord - occidentales par comparaison de nivellements: première analyse. C.R. Acad. Sci II, Paris, 319, 691–679.

Kaczmarek, M.A., Müntener, O., Rubato, D., 2008. Trace element chemistry and U-Pb dating of zircons from oceanic gabbros and their relationship with whole rock composition (Lanzo, Italian Alps). Contrib. Mineral. Petrol. 155, 295–312.

Kay, M., 1951. North American geosynclines. Mem. Geol. Soc. Am 48, 143pp.

Kay, M., 1967. On geosynclinal nomenclature. Geol. Mag. 104 (4), 311–316.

Keppie, J.D., 1994. The Pre-mesozoic Terranes in France and Related Areas, Springer Verlag, Heidelberg.

Kerckhove, C., Gidon, M., 1982. Le Dôme de Remollon Programme Geologie Profonde de la France. BRGM Ed, Orleans, 381–385.

Kerckhove, C., Gidon, M., Paris, J.L., 1989. Notice explicative, carte geol. France (1/50 000), feuille Chorges, N° 870. BRGM Ed, Orleans. France.

Kornprobst, J., 1994. Les roches metamorphiques et leur signification geodynamique. Masson editions, Paris, 224p.

Kuenen, P., 1950. Marine Geology. Wiley, New York, p568pp.

Lagabrielle, Y., Cannat, M., 1990. Alpine Jurassic ophiolites resemble the modern central Atlantic basement. Geology 18, 319–322.

Lagabrielle, Y., Lemoine, M., 1997. Ophiolites des Alpes, de Corse et de l'Apennin: le modèle des dorsales lentes. C. R. Acad. Sci. Paris, 325 (XX), 909–920.

Lagabrielle, Y., 2009. Mantle exhumation and lithospheric spreading: An historical perspective from investigations in the Oceans and in the Alps-Apennines ophiolites. Ital. J. Geosci. (Bull. Soc. Geol. It.) 128 (2), 279–293.

Lardeaux, J.M., Schwartz, S., Tricart, P., Paul, A., Guillot, S., Bethoux, N., et al., 2006. A crustal-scale cross-section of the south-western Alps combining geophysical and geological imagery. Terra Nova 18, 412–422.

Lavier, L., Manatschal, G., 2006. A mechanism to thin the continental lithosphere at magma-poor margins. Nature 04608, Letters, 440/16, 324–328.

Leleu, S., Ghienne, J.F., Manatschal, G., 2005. Upper Cretaceous to Paleocene alluvial-fans documenting interaction between tectonic and environmental processes (Provence, SE France). Alluvial fans. In: Harvey, A.M., Mather, A.E., Stockes, M. (Eds.), Geomorphology, Sedimentology, Dynamics, Geol. Soc. London, Sp. Publ., 251, 217–239.

Lemoine, M., 1972. Rhythmes et modalites des plissements superposes dans les chaînes subalpines meridionales des Alpes Occidentales françaises. Geol. Rundsch 61 (3), 975–1010.

Lemoine, M., et al., 1975. Mesozoic Sedimentation and Tectonic Evolution of the Brianconnais Zone in the Western Alps - Possible Evidence for an Atlantic-type

Margin between the European Craton and the Tethys. *International Sedimentology Congress, Nice, Theme 4, Tectonics and Sedimentation,* Vol. 2, p. 211.

Lemoine, M., Trumpy, R., 1987. Pre-oceanic rifting in the Alps. Tectonophysics 133, 305–320.

Lemoine, M., 1988. Des nappes embryonnaires aux blocs bascules: evolution des idée's et des modèles sur l'histoire Mesozoique des Alpes Occidentales. *Bull. Soc. Geol. France,* Paris, 8, IV, S, 787–797.

Lemoine, M., Dardeau, G., Delpech, P.Y., Dumont, T., de Graciansky, P.C., Graham, R., et al., 1989. Extension Jurassique et failles transformantes jurassiques dans les Alpes Occidentales. *C. R. Acad. Sc.,* Paris, II, 309, 1711–1716.

Lemoine, M., de Graciansky, P.C., Tricart, P., 2000. De l'Ocean a la Chaine de Montagnes: Tectonique des plaques dans les Alpes. Gordon and Breach Publishers, Paris, 207p.

Le Pichon, X., 1968. Sea floor spreading and continental drift. J. Geophys. Res 73 (12), 3661–3697.

Liati, A., Gebauer, D., Fanning, C.M., 2003. The youngest basic oceanic magmatism in the Alps (Late Cretaceous; Chiavenna Unit, Central Alps): geochronological constraints and geodynamic significance. Contrib. Mineral. Petrol 146, 144–158.

Liati, A., Froitzheim, N., Fanning, C.M., 2005. Jurassic ophiolites within the Valais domain of the Western and Central Alps: geochronological evidence for re-rifting of the oceanic crust. Contrib. Mineral. Petrol 149, 446–461.

Lippitsch, R., Kissling, E., Ansorge, J., 2003. Upper mantle structure beneath the Alpine orogen from high-resolution teleseismic tomography - art. n°2376. J. Geophys. Res. Solid Earth 108, 48–62.

Lister, G.S., Etheridge, M.A., Simon, P.A., 1991. Detachment models for the formation of passive continental margins. Tectonics 10, 1038–1064.

Lombardo, B., Rubatto, D., Castelli, D., 2002. Ion microprobe U-Pb dating of zircon from a Monviso metaplagiogranite: implications for the evolution of the Piedmont-Liguria Tethys in the Western Alps. Ophioliti 27 (2), 109–117.

Lugeon, M., 1902. Les grandes nappes de recouvrement des Alpes du Chablais et de la Suisse. Bull. Soc. Geol. France 4 (1), 723–825, Paris.

Mckenzie, D.P., Parker, R.L., 1967. The North Pacific: an example of tectonics on a sphere. Nature 216, 1276–1280.

Mckenzie, D.P., 1978. Some remarks on the development of sedimentary basins. Earth Planet. Sci. Lett. 40, p. 25–32.

Malusa, M.G., Vezzoli, G., 2006. Interplay between erosion and tectonics in the Western Alps. Terra Nova 18, 104–108.

Manatschal, G., 1995. Jurassic rifting and formation of a passive continental margin (Platta and Err nappes, Eastern Switzerland): geometry, kinematics and geochemistry of fault rocks and a comparison with the Galicia margin. - Geol. Inst. ETH and Univ. Zurich, *unpubl. Ph.D. thesis,* N° 11188.

Manatschal, G., Nievergelt, P., 1997. A continent – Ocean transition record in the Err and Platta nappes (Eastern Switzerland). Eclogae Geol. Helv. 90 p. 8, 3–27.fig. 4.

Manatschal, G., Bernoulli, D., 1998. Rifting and early evolution. Mar. Geophys. Res. 20, 371–381.

Manatschal, G., Bernoulli, D., 1999. Architecture and tectonic evolution of non-volcanic margins Present day Galicia and ancient Adria. Tectonics 18, p. 1105, 1099–1119. Fig. 4b.

Manatschal, G., 2004. New models for evolution of magma-poor rifted margins. Int. J. Earth Sci. (Geol. Rundsch.) 93, 432–466.

Manatschal, G., Engström, A., Desmurs, L., Schaltegger, U., Cosca, M., Müntener, O., et al., 2006. What is the tectono-metamorphic of continental break-up: The example of the Tasna Ocean - Continent Transition. J. Struct. Geol. 28, 1849–1869.

Manatschal, G., Müntener, O., Lavier, L.L., Minschul, T.A., Peron-Pindivic, G., 2007. Observations from the Alpine Tethys and Iberia – Newfoundland margins pertinent to the interpretation of continental breakup. *In*: Karner, G.D., Manatschal, G., Pinheiro, L.M. (Eds.), *Imaging, mapping and modeling Continental Lithosphere Extension and Breakup*, Special Publications, 282, 291–324. Geological Society, London, p. 300, Fig. 6, doi: 10.1144/SP282.14 0305-8719/07.

Manatschal, G., Müntener, O., 2009. A type sequence across an an ancient magma-poor ocean-continent transition: the example of the western Alpine Tethys ophiolites. Tectonophysics 473, 4–19, doi: 10.1016/l.tecto.2008.07.021.

Marchant, R., 1993. The Underground of the Western Alps. Mem. Geol. Lausanne, n° 15, 137p.

Marcoux, J., Baud, A., Ricou, L.E., Bellion, Y., Besse, J., Gaetani, M., et al., 1993. Late Anisian and Late Norian maps. In: Dercourt, J., Ricou, L.E., Vrielinck, B.(Eds.), Atlas Paleoenvironmental Maps, BEICIP - FRANLAB, Rueil Malmaison.

Marthaler, M., 2001. Le Cervin est-il africain? Lep publisher, Lausanne, 96p.

Mascle, A., Vially, R., Deville, E., Biju-Duval, B., Roy, J.P., 1996. The petroleum evaluation of a tectonically complex area: the Western margin of the Southeast Basin (France). Mar. Petrol. Geol. 13 (8), 941–961.

Masson, F., Verdun, J., Bayer, R, Debeglia, N., 1999. Une nouvelle carte gravimetrique des Alpes occidentales et ses consequences structurales et tectoniques. C.R. Acad. Sci. Paris, Sciences de la Terre et des Planètes, 329, 865–871.

Masson, H., 2002. Ophiolites and other (ultra)basic rocks from the West-central Alps: new data for a puzzle. Bull. Geol. Lausanne, 356 and Bull. Soc. vaudoise Sci. Nat., 88, 263–276.

Masson, H., Bussy, F., Eichenberger, M., Giroud, N., Meilhac, C., Presniakov, S., 2008. Early Carboniferous age of the Versoyen ophiolites and consequences: non-existence of a 'Valais ocean' (Lower Penninic, western Alps). Bull. Soc. Geol. France Paris, 179 (1), 337–355.

Matte, P., 1986. La chaîne varisque parmi les chaînes paleozoïques periatlantiques, madele d'evolution et position des grands blocs continentaux au Permo-Carbonifère. Bull. Soc. Geol. France. Paris, 8 (II), 9–24.

Matte, P., 1991. Accretionary history and crustal evolution of the Variscan belt in Europe. Tectonophysics 196, 309–337.

Matte, P., 1998. Continental subduction and exhumation of HP rocks in Paleozoic orogenic belts: Uralides and Variscides. GFF 120, 209–222.

Mayne, H., 1962. Common reflection point horizontal stacking techniques. Geophysics 27 (6), 927–938.

Megard-Calli, J., Faure, J.L., 1988. Tectonique distensive et sedimentation au Ladinien superieur - Carnien dans la zone briançonnais. Bull. Soc. Geol. France. Paris, 8 (IV), 705–715.

Menot, R.-P., Von Raumer, J.P., Bogdanoff, S., Vivier, G., 1994. Variscan Basement of the Western Alps: the External Crystalline Massifs. In Keppie, J.D. (Ed.), The Pre-Mesozoïc Terranes in France and Related Areas, Springer Verlag, Heidelberg, 458–466. Fig. 1 p 459.

Merle, O., Brun, J.P., 1984. The curved translation path of the Parpaillon nappe (French Alps). J. Struct. Geol. 6, 711–719.

Merle, O., 1994. Nappes et Chevauchements. Masson Ed., Paris, 137 p.

Montadert, L., Roberts, D.G., de Charpal, O., Guennoc, P., 1979. Rifting and subsidence on the northern continental margin of the Bay of Biscay. In *Initial Reports of the Deep Sea Drilling Project*, vol. 48. Washington D.C. (U.S Government Printing office), 1025–1061.

Montadert, L., Roberts, D.G., de Charpal, O., Guennoc, P., Sibuet, J.C., 1979. Northeast Atlantic passive continental margins: Rifting and subsidence. *In*: Talwani, M., Hay, W., Ryan, W.B.F. (Eds.), *Deep Drilling Results in the Atlantic Ocean: Continental Margins and Paleoenvironments, Maurice Ewing Ser.*, vol. 3, 154–186, AGU, Washington, D.C.

Müntener, O., Hermann, J., Trommsdorff, V., 2000. Cooling history and exhumation of of lower crustal granulite and upper-mantle (Malenco, eastern central Alps. J. Petrol. 41, 175–200.

Müntener, O., Desmur, L., Pettke, T., Meier, M., Schalteger, U., 2002. Melting and malt/ rock reaction in extending mantle lithosphere: trace elements and isotopic constraints from passive margin peridotites. Geochim. Cosmochim. Ac. 66, A536.

Müntener, O., Pettke, T., Desmur, L., Meier, M., Schalteger, U., 2004. Refertilization of mantle peridotite in embryonic ocean basins: trace element and Nb isotopic evidence and implications for cust – mantle relationship. Earth Planet Sci. Lett. 221, 293–308.

Murray, J., Renard, A.F., 1891. Report on deep sea deposits based on specimens collected during the voyage of H.M.S. Challenger in the years 1873–1876, *Challenger Reports*, MSO, Edinburgh, 525.

Neubauer, F., von Raumer, J.F., 1993. The alpine basement- linkage between variscides and east-mediterranean mountain belts. In: von Raumer, J.F., Neubauer, F. (Eds.), Pre-Mesozoic Geology in the Alps. Springer-Verlag, Heidelberg, 641–663, Fig. 14.

Nicolas, A., Polino, R., Hirn, A., Nicolich, R., and Ecors-Crop Working Group 1990. ECORS-CROP traverse and deep structure of the western alps: asynthesis alpine survey *In:* Roure, F., Heitzman, P., Polino, R. (Eds.), Deep Structure of the Alps, *Mem. Soc. Geol. France,* Paris, *156, 99–104.*

Nicolas, A., 1996. General conclusions of the ECORS-CROP alpine survey. *In:* Roure, F., Bergerat, F., Damotte, B., Mugnier, J.L., Polino, R. (Eds.), The ECORS – CROP Alpine Seismic Traverse, *Mem. Soc. Geol. France,* Paris, *170, 99–104.*

Nicolas, A., 1999. Les Montagnes sous la mer. *BRGM Ed.,* Orleans, France, 187 p.

Olivet, J.L., 1996. La cinematique de la plaque iberique. Bull. Centres Rech. Explor. - Prod. ELF Aquitaine 20 (1), 131–195.

Patriat, C., et al., 1982. Les mouvements relatifs de l'Inde, de l'Afrique et de l'Eurasie. Bull. Soc. Geol. France Paris, 7 (24), 363–373.

Paul, A., Cattaneo, M., Thouvenot, F., Spallarossa, D., Bethoux, N., Frechet, J., 2001. A three-dimensional crustal velocity model of the southwestern Alps from local earthquake tomography. J. Geophys. Res. Solid Earth 106, 19367–19389.

Peddy, C., Pinet, B., Masson, D.G., Scrutton, R.A., Sibuet, J.C., Warner, M.R., et al., 1989. Continental structure of Goban Spur from deep seismic reflection profiling. J. Geol. Soc. Lond. 146, 427–437.

Penrose field conference on ophiolites 1972. Geotimes, 17, 24–25.

Peron-Pindivic, G., Manatschal, G., 2009. The final rifting evolution at deep magma-poor passive margins from Iberia. – Newfoundland: a new point of view. Int. J. Earth Sci. 98 (7), 1581–1597, DOI 10.1007/s00531-008-0337-9.Fig. 6.

Pfeifer, H.R., Biino, G., Menot, R.P., Stille, P., 1993. Ultramafic rocks in the pre-mesozoic basement of the central and external western alps. In: von Raumer, J.F., Neubauer, F. (Eds.), Pre-Mesozoic Geology in the Alps. Springer-Verlag, Heidelberg, 119–143.

Pfiffner, O.A., Lehner, P., Heitzmann, P., Mueller, S., Steck, A., 1997. Results of NPR-20. Deep structure of the Alps. Birkhauser Verlag, Basel, 380 p.

Pfiffner, O.A., Ellis, S., Beaumont, C., 2000. Collision tectonics in the Swiss Alps: Insight from geodynamic modeling. Tectonics 19, 1065–1094.

Pickup, S.L.B., Whitmarsh, R.B., Fowler, C.M.R., Reston, T.J., 1996. Insight into the nature of the ocean-continent transition off West Iberia from a deep multichannel seismic reflection profile. Geology 24 (12), 1079–1082.

Ravnas, R., Steel, R.J., 1997. Contrasting style of Late Jurassic syn-rift turbidites sedimentation: a comparative study of the Magnus and Oseberg areas, northern North Sea. Mar. Petrol Geol. London, 14, 417–449.

Razin, P., Bonijoly, D., Le Strat, P., Courel, L., Poli, E., Dromart, G., et al., 1996. Stratigraphic record of the structural evolution of the western extensional margin of the Subalpine Basin during the Triassic and Jurassic, Ardèche, France. Mar. Petrol Geol. 13 (6), 625–652.

Reston, T.J., Pennell, J., Stubenrauch, A., Walker, M., Perez-Gussinye, M., 2001. Detachment faulting, mantle serpentinisation and serpentinite volcanism beneath the Porcupine basin, southwest of Ireland. Geology 27, 587–590.

Rey, D., Quarta, T., Mouge, P., Miletto, M., Lanza, R., Galdeano, A., et al., 1990. Gravity and aeromagnetic maps on the western Alps: contribution to the knowledge of the deep structures along the ECORS - CROP seismic profile. Mem. Soc. Geol. France, Paris, 156, Mem. Soc. geol Suisse, Zurich, 1, Vol. Spec. Soc. geol. It., Roma, 1, p. 30.

Richardson, A.N., Blundell, D.J., 1996. Continental collision in the banda arc In: Hall, R., Blundell, D.J. (Eds.), Geol. Soc. London Sp. Publ., 106, 47–60.

Ring, U., Richter, C., 1994. The Variscan structure and metamorphic evolution of the eastern Southalpine basement. J. Geol. Soc. Lond. 151 (4), 755–766.

Rosenbaum, G., Lister, G.S., Duboz, C., 2002. Relative motions of Africa, Iberia and Europe during Alpine orogeny. Tectonophysics 359, 117–129.

Roure, F., Brun, J.P., Coletta, B., Van Den Driessche, J., 1992. Geometry and kinematics of extensional structures in the Alpine Foreland Basin of southeastern France. J. Struct. Geol. 14 (5), 503–519.

Roure, F., Brun, J.P., Colletta, B., Vially, R., 1994. Multiphased extensional structures in the alpine foreland of southeastern france. In: Mascle, A. (Ed.), Exploration and petroleum geology of France, European Assoc. Petrol. Geol., Spec. Pub. N°4. Springer Verlag Publisher, 237–260.

Roure, F., Coletta, B., 1996. Cenozoic inversion structures in the foreland of the Pyrenees and the Alps. Mem. Museum National Hist. Nat. Paris, n°170, 173–209.

Roure, F., Bergerat, F., Damotte, B. J.L. Mugnier, Polino, R. Eds 1996. The ECORS – CROP Alpine Seismic Transverse Mem. Soc. Geol. France, Paris, n° 170.

Runcorn, S.K., 1956. Paleomagnetism, polar wandering and continental drift. Geol. Mijnbouw. Amsterdam, 8, 253–256.

Sassi, F.P., Spiess, R., 1993. The south-alpine metamorphic basement in the eastern alps, In von Raumer J.F., Neubauer F. (Eds.) Pre-Mesozoic Geology in the Alps, Springer-Verlag, Heidelberg, 599–607.

Schaer, J.P., Jeanrichard, F., 1974. Mouvements verticaux anciens et actuels dans les Alpes suisses. Eclogae. Geol. Helv. 67, 101–119.

Schardt, H., 1893. Sur l'origine des Prealpes Romandes Ach. Sciences phys. Nat. Genève, 3, 30, 570–583 and Eclogae. Geol. Helv., 4, 129–142.

Schlee, J.S., Klitgord, K.D., 1988. The structure of Georges Bank US Atlantic margin. In: Sheridan, R.E., Grow, J.A. (Eds.), The Geology of North America, Geol. Soc. Am. I 2, 243–269.

Schlunegger, F., Slingerland, R.L., Matter, A., 1998. Crustal thickening and crustal extension and crustal extension as controls on the evolution of the drainage network of the Central Alps between 30 Ma and the present: constraints from the stratigraphy of the North Alpine foreland basin and structural evolution of the Alps. Basin Res. 10, 197–212.

Schmid, S.M., Aebli, H.R., Heller, F., Zingg, A., 1989. The role of the Peri-adriatic Line in the tectonic evolution of the Alps. In: Coward, M., Dietrich, D., Park, R.G. (Eds.), Alpine Tectonics, Geol. Soc. London Sp. Publ, 45, 153–171.

Schmid, S.M., Kissling, E., 2000. The arc of the western Alps in the light of geophysical data on deep crustal structure. Tectonics 19, 62–85.

Schmid, S.M., Fügenschuh, B., Kissling, E., Schuster, R., 2004. Tectonic map and overall architecture of the Alpine orogen. Eclogae. Geol. Helv. 97, 93–117.

Schwarz, S., Lardeaux, J.M., Tricart, P., Guillot, S., Labrin, E., 2007. Diachronous exhumation of HP-LT metamorphic rocks from south-western Alps: Evidence from fission-track analysis. Terra Nova 19, 133–140.

Sclater, J.G., Hellinger, S., Tapscott, C., 1977. The paleo-bathymetry of the Atlantic Ocean from Jurassic to present. J. Geol. 85, 509–552.

Schmid, S.M., Ruck, P., Schreurs, G., 1990. The significance of the Schams nappe for the reconstruction of paleo-tectonic and orogenic evolution of the Penninic zone along the NFP-20 traverse (Grisons, Eastern Switzerland). Mem. Geol. Soc. Fr, N.S, 156; Mem. Geol. Soc. Suisse 1;, Mem. Soc. Geol. Italia 1, 263–287.

Selverstone, J., 2005. Are the Alps collapsing?. Annu. Rev. Earth Planet Sci. 33, 113–132.

Septfontaine, M., 1995. Large scale progressive unconformities in Jurassic strata of the Prealps S of Lake Geneva: interpretation as synsedimentary inversion structures; paleo-tectonic implications. Eclogae Geol. Helv. 88, 553–576.

Sibuet, J.C., Srivastava, S., Manatschal, G., 2007. Exhumed mantle forming transitional crust in the Newfoundland - Iberia rift and associated magnetic anomalies. J. Geophys. Res. 12 (003856, B06105.doi:10.1029/2005JB.

Sinclair, H.D., 1996. Plan-view curvature of foreland basins and its implications for the paleo-strength of the lithosphere underlying the western Alps. Basin Res. 8, 173–182.

Sissingh, W., 2001. Tectonostratigraphy of the West Alpine Foreland: correlation of Tertiary sedimentary sequences, changes in eustatic sea-level and stress regimes. Tectonophysics 333, 361–400.

Skelton, A.D.L., Valley, J.W., 2000. The relative timing of serpentinzation and mantle exhumation at the ocean – continent transition, Iberia: constraints from oxygen isotopes. Earth Planet Sci. Lett. 178, 327–338.

Snyder, D.B., Milsom, J., Prasetyo, A.N., 1996. Geophysical evidence for local indentor tectonics in the Banda arc east of Timor. In: Hall, R., Blundell, D. (Eds.), Tectonic Evolution of Southeast Asia. Geol. Soc. London Sp. Publ, 106, 61–73.

Stampfli, G.M., Borel, G.D., 2002. A plate tectonic model for the Paleozoic and Mesozoic constrained by dynamic plate boundaries and restored synthetic oceanic isochrons. Earth Planet. Sci. Lett. 196, 17–33.

Staub, R., 1926. Beitrage zur geologischen Karte der Schweiz, NiF, Leifg, 52.

Stampfli, G.M., 1993. Le Brianconnais, terrain exotique dans les Alpes?. Eclogae. Geol. Helv. 86 (1), 1–45.

Stampfli, G.M. (Ed.), 2001. Geology of the Western Swiss Alps, a guide book, Mem. Geol., Lausanne, N° 36, 195p.

Stampfli, G.M., Borel, G.D., 2002. A plate tectonic model for the Paleozoic and Mesozoic constrained by dynamic plate boundaries and restored synthetic oceanic isochrons. Earth Planet. Sci. Lett. 196, 17–33.

Staub, R., 1926. Beitrage zur geologischen Karte der Schweiz, NiF, Leifg. 52.

Steinmann, G., 1927. Die ophiolotischen Zonen in den mediterranen Kettengebirgen. Comptes Rendus XIVème Congrès Geologique International, Madrid, Graficas Reunidas, 2, 637–667.

Steinman, M., 1994. Ein Beckenmodell für das Nordpenninikum der Ostschweiz Jahrbuch der Geologischen Bundesanstalt, Wien, 137, 675–621.

Steiman, G., 1905. Geologische Beobachtungen in den Alpen Schardtsche Ueberfaltungstheorie und die geologische Bedeutung der Tiefseeabsätze und der ophiolitischen Massengesteine. Berichte der Naturforchenden Gesellschaft zu Feiburg im Brisgau 16, 18–67.

Stockmal, G.S., Beaumont, C., Boutilier, R., 1986. Geodynamic models of convergent margin tectonics: Transition from rifted margin to overthrust belt and consequences for Foreland-basin development. Am. Assoc. Petrol. Geol. Bull 70 (2), 181–190.

Sue, C., Thouvenot, F., Frechet, J., Tricart, P., 1999. Widespread extension in the core of the western Alps revealed by earthquake analysis. J. Geophys. Res. 104, n°B11, 25611–25622.

Sue, C., Tricart, P., 2003. Neogene to ongoing normal faulting in the inner western Alps. Tectonics 22, doi:10.1029/2002TC001426.

Suess, E., 1885. The face of the Earth, Engl. Ed.1904, 5V, Oxford University Press, New York.

Schwartz, S., Lardeaux, J.M., Tricart, P., Guillot, S., Labrin, E., 2007. Diachronous exhumation of HP-LT metamorphic rocks from south-western Alps: Evidence from fission-track analysis. Terra Nova 19, 133–140.

Tankard, A.J., Welsink, H.J., 1989. Mesozoic Extension and styles of basin formation in Atlantic Canada. *AAPG Memoir*, 46, 175–196.

Taylor, F.B., 1910. Bearing of the Tertiary mountain belt on the origin of the earth's plan. Geol. Soc. Am. Bull 21, 179–226.

Termier, P., 1903. Les nappes des Alpes orientales. Bull. Soc. Geol. France. Paris, 4 (III), 712.

Thelin, P., Sartori, M., Burri, M., Gouffon, Y., Chessex, R., 1993. The Pre-Alpine Basement of the Brianconnais (Wallis, Switzerland). In von Raumer, J.F., Neubauer, F. (Eds.), Pre-Mesozoic Geology in the Alps. Springer-Verlag, Heidelberg, 297–315.

Thöni, M., 2006. Dating eclogite facies metamorphism in the Eastern Alps - approaches, results, interpretations: a review. Mineral Petrol. 88, 123–148.

Thouvenot, F., Paul, A., Frechet, J., Bethoux, N., Jenatton, L., Guiguet, R., 2007. Are there really superposed Mohos in the southwestern Alps? Tau Beta New seismic data from fan-profiling reflections. Geophys. J. Int. 170, 1180–1194.

TRANSALP Working Group, 2002. First deep seismic reflection images of the Eastern Alps reveal giant crustal wedges and transcrustal ramps. Geophys. Res. Lett. 29, 92.1–92.4, 10, GL014911.

Tricart, P., Gout, C., Lemoine, M., 1985a. Mosaïque de blocs failles et injection de serpentinite dans la croûte oceanique tethysienne: l'exemple des ophiolites de Chabrières (Haute Ubaye, zone piemontaise des Alpes occidentales). C.R. Acad. Sc. Paris, 300, serie II, 817–820.

Tricart, P., Gout, C., Lemoine, M., 1985. Tectonique synsedimentaire saccadee d'âge cretace inferieur dans l'ocean tethysien ligure: un exemple dans les Schistes Lustrés à ophiolites de Chabrière (Haute Ubaye, Alpes occidentales françaises). C.R. Acad. Sc. Paris 300, serie II, 879–884.

Tricart, P., Lemoine, M., 1983. Serpentinitic ocean bottom in South Queyras ophiolites (French Western Alps): record of the incipient oceanic opening of the Mesozoic Ligurian Tethys. Eclogae. Geol. Helv. 76, 611–629.

Tricart, P., 2003. Notice explicative de la feuille Aiguilles – Col Saint Martin. *Service de la Carte géologique de France*, B.R.G.M. Ed. Orléans, 150 p.

Tricart, P., Lardeaux, J.M., Schwartz, S., Sue, C., 2006. The late extension in the inner western Alps: a synthesis along the south-Pelvoux transect. *Bull. Soc. Geol. France*, Paris, 177, 299–310.

Trümpy, R., 1972. Zur Geologie des Unterengadins. Ergebnisse der wissenschaftlichen. Untersuchungen im Schweizerischen Nationalpark 12, 71–87.

Trumpy, R., 1980. Field trips guide, Switzerland, 26th Int. Geol. Congr.

Trümpy, R., 2001. Why plate tectonics was not invented in the Alps. Int. J. Earth Sci. (Geol. Rundsch.) 90, 477–183.

Tucholke, B., Lin, J., 1994. A geological model for the structure of ridge segments in slow spreading ocean crust. J. Geophys. Res 99 (B6), 11937–11958.

Tucholke, B.E., Sibuet, J.C., 2007a. Leg 210 synthesis: tectonic, magmatic, and sedimentary evolution of the Newfoundland-Iberia rift: a synthesis based on ocean drilling through ODP Leg 210. *In* Tucholke, B.E., Sibuet, J.C., and Klaus, A. (Eds.), Proc. ODP, Sci. Results, 210: College Station, TX (Ocean Drilling Program), 1–56, doi:10.2973/odp. proc.sr.210.101.2007, Figure 1.

Tucholke, B., Sawyer, D.S., Sibuet, J.C., 2007b. Breakup of the newfoundland – iberia rift. In: Karner, G.D., Manatschal, G., Pinheiro, L.M. (Eds.), Imaging, Mapping and Modelling Continental Lithosphere Extension and Breakup, Geological Society, London, Special Publications, 282, 9–46.

Vail, P.R., Mitchum, R.M., Todd, R.G., Widmier, J.M., Thompson, S., Sangree, J.B., et al., 1977. Seismic stratigraphy and global changes in sea level. In: Payton, C.E. (Ed.), Seismic stratigraphy - applications to hydrocarbon exploration. AAPG Memoir. 26, 49–212.

Vening Meinesz, F.A., 1941. Gravity over the continental edges. Koninkl. Ned. Akad. Wetenschap. Proc.V. 44.

Vignaroli, G., Faccenna, C., Jolivet, L., Piromallo, C., Rossetti, F., 2008. Subduction polarity reversal at the junction between the Western Alps and the Northern Apennines, Italy. Tectonophysics 450, 34–50.

Vine, F.J., Matthews, D.H., 1963. Magnetic anomalies over mid- ocean ridges. Nature 199, 947–949.

von Raumer, J.F., Neubauer, F., 1993. Pre-Mesozoic Geology in the Alps. Springer Verlag publishers, Berlin, Heibelberg, 677 p.

Warburton, J., Burnhill, T.J., Graham, R.H., Isaac, K.P., 1990. Evolution of the Oman mountain foreland basin. In Robertson, A.H.F. et al. (Eds.) The geology and tectonics of the Oman region. Geol Soc Spec Publ. 49, 419–427.

Wegener, A., 1924. The origin of continents and oceans. Methuen Press, London.

Wiesinger, M., Neubauer, M., Handler, R., 2006. Exhumation of the Saualpe eclogite unit, Eastern Alps: constraints from 40Ar/39Ar ages and structural investigations. Mineral Petrol. 88, 149–180.

Weissert, H., Bernoulli, D., 1985. A transform margin in the Mesozoic Tethys: evidence from the Swiss Alps. Geol. Rundsch 74, 665–679.

Welsink, H.J., Srivastava, S.P., Tankard, A.J., 1989. Basin architecture of the Newfoundland continental margins and its relationship to ocean crust fabric during extension In: Tankard, A.J., Balckwill, HR. (Eds.), Extensional Tectonics and Stratigraphy of the North Atlantic Margins, AAPG Memoir 46, 197–213.

Wernicke, B., 1985. Uniform sense simple shear of the continental lithosphere. Can. J. Earth Sci. 22, 676–681.

Wilson, J.T., 1966. Did the Atlantic close and then reopen?. Nature 211, 676–681.

Wilson, R.C.L., Hiscott, R.N., Willis, M.G., Gradstein, F.M., 1989. Lusitanian basin of west-central portugal: mesozoic and tertiary tectonic, stratigraphy and subsidence history. In: Tankard, A.J., Balckwill, H.R. (Eds.), Extensional Tectonics and Stratigraphy of the North Atlantic Margins, AAPG Memoir 46, 197–213.

Wilson, R.C.L., Manatschal, G., Wise, S., 2001. Rifting along non-volcanic passive margins: stratigraphic and seismic evidence from the Mesozoic successions of the Alps and western Iberia. In: Wilson, R.C.L., Whitmarsh, R.B., Taylor, B., Froitzem, N. (Eds.) Non-volcanic Rifting and Continental Margins: Evidence From Land and Sea. Geological Society, Special Publication, London, 187, 429–452.

Witmarsh, R.B., Wallace, P.J., 2001. The rift-to-drift development of the west Iberia non-volcanic continental margin: A summary and review of the contribution of Ocean Drilling Program Leg 173. In: Beslier, M.O., Witmarsh, R.B., Wallace, P.J., Girardeau, J. (Eds.), Proceedings of the ocean Drilling Program, Scientific Results, 173, College Station, TX, 1–36.

Whitmarsh, R.B., Manatschal, G., Minshull, T.A., 2001. Evolution of magma-poor continental margins from rifting to seafloor spreading. Nature 413, 150–154.

Winkler, W., Bernoulli, D., 1986. Detrital high-pressure/low-temperature minerals in a late Turonian flysch sequence of the Eastern Alps (western Austria); implications for early Alpine tectonics. Geology 14, 598–601.

Winterer, E.J., Bosellini, A., 1981. Subsidence and sedimentation on Jurassic passive continental margin, Southern Alps, Italy. Am. Ass. Petrol. Geol. Bull. 65, 394–421.

Renaut, R.,, Ashley, G.M. (Eds.), Continental Rift Basin Sedimentology, *SEPM Special Publication* No. 73, 57–81.

Withjack, M.O., Schlische, R.W., Olsen, P.E., 2008. Development of the passive margin of eastern north america: mesozoic rifting, igneous activity, and breakup. In: Bally, A.W. (Ed.), Phanerozoic Regional Geology of the World, Vol. 1. Elsevier, Amsterdam, in press.

Worrall, D.M., Snelson, S., 1989. Evolution of the northern Gulf of Mexico, with emphasis on Cenozoic growth faulting and the role of salt. *In*: Bally, A.W., Palmer, A. (Eds.), The Geology of North America, *Geol. Soc. Am. A*, 97–139.

Wortmann, U.G., Weissert, H., Funk, H., Hauck, J., 2001. Alpine plate kinematics revisited: the Adria problem. Tectonics 20, 134–147.

Ziegler, P.A., Roure, F., 1996. Architecture and petroleum systems of the Alpine orogen and associated basins. *In*: Ziegler, P.A., Horvath, F. (Eds.), Peri-Tethys Memoir 2, *Mem. Mus. Hist. Nat*, 170, 15–45.

GEOGRAPHICAL INDEX

Note: Page numbers with '*b*' and '*t*' in the index denote boxes and tables respectively and page numbers in boldface refer to figures

SUBJECT INDEX

A

Airy, G.B., 9

Alpine shortening, 269–270, 273–275, 345–346
  detachment fault behaviour, 284, 287–288
  near perpendicular to strike extensional faults, 276–279
  oblique to strike extensional faults, 279–281

Alps
  deep structure from geophysical studies, 45–53
  evolution of, 74–75
  geographical definition, 30
  structural units, 30–31, 32, 33
  See also Central Alps; Eastern Alps; Southern Alps; Western Alps

Antoine, P., 152, 160

Antrona nappe, 41

Apennines, 32, 235

Apulia-Africa margin, 189–190, 203–204, 254
  displacements, 71–72
  extension, 193, 195–196, 198–199, 200–203
  Grisons transect, 196–203
  Southern Alps transect, 190–196

Aroley limestone, 153, 155–156, 159

Arpenaz waterfall, France, 4, 5

Austro-alpine, 41
  compression, 171
  Palaeozoic development of basement, 87–88
  thrust, 35

B

Bally, A.W., 15, 18

Balma unit, 160–161

BANDA seismic sections, 181

Basalts, 225, 239
  temporal succession, 226–230

Basin-forming mechanisms, 13–15

Bausset thrust, 165

Bay of Biscay, 18, 19, 263

Belledonne massif, 83, 85, 86

Belledonne trend, 131, 132–133, 263

Belluno Trough, 191

Beltrando, M., 159–162

Berger, W.H., 23–24

Bertrand, M., 6

Besson Lakes, 125

Biellese zone, 193

Blackett, P.M.S., 11

Bourg d'Oisans, 272–273, 275
  half-graben, 111, 120–125, 275, 279
    crustal extension modes, 124–125
    development of, 119
    geometry in syn-rift phase, 122–124

Breakup unconformity, 19, 255–258, 350

Breccias, 177–179
  ophicalcitic, 221–222, 224, 225

Breche nappe, 163, 166, 179

Brenner fault, 330

Brianconnais, 108–111, 128, 144, 336
  decollement surface reactivation, 285–286
  extension, 333
  inversion of detachment surfaces, 287
  Mesozoic facies succession, 26
  pelagic sediments, 177
  syn-rift unconformities, 130
  thrust, 35

C

Calcite compensation depth (CCD), 24–26

Canavese zone, 193

Caporalino units, 186

Carnian crisis, 103–104

Central Alps, 31, 33, 34–41
  Austro-alpine, 41
  compression, 170
  Late Cretaceous deformation, 179
  linkage with Western Alps, 37
  morphogenic evolution, 326–327
  pelagic sediments, 175–179
  South-alpine, 41–42

Printed and bound by CPI Group (UK) Ltd, Croydon, CR0 4YY

08/05/2025

01864815-0003